AF358737

Thermochemical Conversion Processes for Solid Fuels and Renewable Energies

Thermochemical Conversion Processes for Solid Fuels and Renewable Energies

Editors

Falah Alobaid
Jochen Ströhle

MDPI • Basel • Beijing • Wuhan • Barcelona • Belgrade • Manchester • Tokyo • Cluj • Tianjin

Editors
Falah Alobaid
Technical University of Darmstadt,
Institute for Energy Systems and
Technology,
Germany

Jochen Ströhle
Technical University of Darmstadt,
Institute for Energy Systems
and Technology,
Germany

Editorial Office
MDPI
St. Alban-Anlage 66
4052 Basel, Switzerland

This is a reprint of articles from the Special Issue published online in the open access journal *Applied Sciences* (ISSN 2076-3417) (available at: https://www.mdpi.com/journal/applsci/special_issues/thermochemical_conversion_solid_fuels_renewable_energies).

For citation purposes, cite each article independently as indicated on the article page online and as indicated below:

LastName, A.A.; LastName, B.B.; LastName, C.C. Article Title. *Journal Name* **Year**, *Volume Number*, Page Range.

ISBN 978-3-0365-1096-5 (Hbk)
ISBN 978-3-0365-1097-2 (PDF)

Cover image courtesy of Falah Alobaid and Jochen Ströhle.

Contents

About the Editors

Falah Alobaid received his Ph.D. degree with the highest distinction from the Technical University of Darmstadt, Germany. His dissertation was awarded the energy special prize of the Technical University of Darmstadt in 2014. He has since been the research group leader at the Institute of Energy Systems and Technology at the Technical University of Darmstadt. In 2018, he received his habilitation degree in energy systems (habilitation lecture: Energy Storage Technologies) and then followed by the academic title "Privatdozent" in 2019. He has vast experience and a record of accomplishment in the field of the numerical simulation of energy systems and the associated engineering processes using one-dimensional dynamic process simulation and three-dimensional computational fluid dynamics.

Jochen Ströhle took his Ph.D. degree in Mechanical Engineering from the University of Stuttgart, Germany. He has large experience in the field of power plant technology, particularly in solid fuel combustion/gasification processes and the related CO_2 capture technologies, with key expertise in the modelling and simulation of multiphase flows and power plant processes, as well as experimental investigations in pilot-scale test facilities.

Editorial

Special Issue "Thermochemical Conversion Processes for Solid Fuels and Renewable Energies"

Falah Alobaid * and Jochen Ströhle

Institute for Energy Systems and Technology, Technical University of Darmstadt, Otto-Berndt-Straße 2, 64287 Darmstadt, Germany; jochen.stroehle@est.tu-darmstadt.de
* Correspondence: falah.alobaid@est.tu-darmstadt.de; Tel.: +49-(0)-6151-16-23004

Abstract: The world society ratifies international measures to reach a flexible and low-carbon energy economy, attenuating climate change and its devastating environmental consequences. The main contribution of this Special Issue is related to thermochemical conversion technologies of solid fuels (e.g., biomass, refuse-derived fuel, and sewage sludge), in particular via combustion and gasification. Here, the recent activities on operational flexibility of co-combustion of biomass and lignite, carbon capture methods, solar-driven air-conditioning systems, integrated solar combined cycle power plants, and advanced gasification systems, such as the sorption-enhanced gasification and the chemical looping gasification, are shown.

Keywords: thermochemical conversion technologies; combustion; carbon capture and storage/utilization; gasification; solar-driven air-conditioning; integrated solar combined cycle; energy and exergy analyses; thermodynamic modeling; dynamic process simulation

Citation: Alobaid, F.; Ströhle, J. Special Issue "Thermochemical Conversion Processes for Solid Fuels and Renewable Energies". *Appl. Sci.* **2021**, *11*, 1907. https://doi.org/10.3390/app11041907

Academic Editor: Frede Blaabjerg

Received: 10 February 2021
Accepted: 17 February 2021
Published: 22 February 2021

Publisher's Note: MDPI stays neutral with regard to jurisdictional claims in published maps and institutional affiliations.

1. Introduction

Human beings find themselves at the beginning of the 21st century in a contradictory situation in which, on the one hand, significant growth in global demand for energy is expected while, on the other hand, human activities have posed a dangerous rise in the global average temperature by approximately 1.0 ± 0.2 °C above pre-industrial levels. Global warming is likely to reach 1.5 °C in the period between 2030 and 2050 if the consumption of fossil fuels continues to increase at the current rate [1]. It is generally accepted that a great share of greenhouse gas emissions is anthropogenic and originated from utilizing fossil fuels, with contributions coming from manufactured materials (e.g., concrete), deforestation, and agriculture (including livestock). Societies around the world actively support measures towards a flexible and low-carbon energy economy to attenuate climate change and its devastating environmental consequences. These measures include process improvement, new thermochemical conversion technologies, such as gasification or combustion of alternative energy sources, such as biomass [2,3], implementation of carbon capture and storage/utilization technologies [4,5], and promotion of renewable energy sources for power generation and district heating or cooling [6,7], as briefly described below:

- Process improvement of thermal power plants, cement, and metallurgical industries represents an effective method to reduce greenhouse gas emissions. A variety of measures could be considered here, such as an increase in process efficiency and flexibility, and enhancement of operation mode concerning the load change times and the rate of shutdown/start-up procedure [8], as well as process retrofitting with modern flue gas cleaning devices for particulate matter, nitrogen oxides (NO_x), sulfur oxides (SO_x) and carbon dioxide (CO_2).
- The carbon capture and storage/utilization (CCS/U) technologies may offer a rapid response to the global challenge by significantly reducing CO_2 emission from major emitters (e.g., power and cement plants). Depending on the oxidation of fossil fuels and the manner of CO_2 capture, it is distinguished between three CO_2 capture

1

methods, namely, oxy-fuel, pre-combustion, and post-combustion [9]. In the oxy-fuel process, fossil fuel is combusted using pure oxygen with circulated flue gas to obtain lower adiabatic combustion temperature. The generated flue gas consists of carbon dioxide, where the steam can be easily separated by a condensation process. The main drawback is separating oxygen from air using an air separation unit that is energy-intensive [10]. The chemical-looping process is considered an energy-efficient oxy-fuel method [11,12]. Solid particles of metal oxide are applied as oxygen carriers and these particles circulate between two coupled fluidized beds, namely, air, and fuel reactor. In the pre-combustion method, the solid fuel is gasified using steam and oxygen as a gasification agent (usually at higher-pressure levels in a fluidized bed system or an entrained-flow gasifier). The produced gas consists essentially of hydrogen, carbon monoxide, carbon dioxide, and trace gases. Using a gas-cleaning unit, the carbon dioxide and the trace gases can be separated and the producer gas can be converted into value-added chemicals or combusted in a combined-cycle power plant (integrated gasification combined cycle (IGCC)) [13]. The post-combustion approach has the advantage that existing processes can be retrofitted with CO_2 capture. Two technologies can be used, namely, the chemical scrubbing of flue gas or the carbonate-looping process. The latter uses limestone as a solid sorbent, circulating between interconnected fluidized bed reactors (carbonator and calciner) [14].

- The increased use of renewable energy sources (e.g., biomass, wind power, and photovoltaics) contributes to a decrease in CO_2 emissions in the power generation sector. Through the substitution of fossil fuels by using alternative energy sources such as refuse-derived fuel (RDF), solid recovered fuel (SRF), tire-derived fuel (TDF), and sewage sludge, a considerable reduction in emissions can be further achieved [15]. The electrification of heating and transport sectors offers also a great opportunity for achieving zero emissions. However, variable renewable energy sources can lead to a seemingly paradox situation of negative electricity prices at times of high renewable electricity output and/or low demand, as well as peak electricity prices at times of low renewable electricity output and/or high demand. To maintain the security of supply, there are several potential solutions such as the expansion of high-voltage transmission infrastructure, the use of flexible power plants with CCS/U technologies, and the implementation of large-scale energy storage [16]. The solutions differ in their potential impact, technological maturity, and economic viability so that according to the opinion of authors, the future electricity system will contain all of these concepts to varying degrees with the possible integration of value-adding processes beyond electricity such as the power-to-fuel technology. The carbon-neutral fuels (e.g., hydrogen, methane, gasoline, diesel fuel, or ammonia) can be generated from renewable energy sources by the electrolysis of water to make hydrogen that hydrogenates carbon dioxide or nitrogen captured from thermal power plants or air.

According to the above background and in support of the development of thermochemical conversion processes for solid fuels and renewable energies, this Special Issue contains nominated contributions to:

- Gasification and combustion of alternative fuels (e.g., biomass, refuse-derived fuel, solid recovered fuel, tire-derived fuel, sewage sludge, and low-rank coal);
- Technological combinations of conversion processes based on renewable sources (power-to-fuel);
- Carbon capture and storage/utilization CCS/U technologies (carbon capture-to-fuel);
- Renewable energy for heating and cooling purposes to reduce peak demand, including energy storage systems to mitigate grid imbalances;
- Thermodynamic studies, computational fluid dynamics (CFD), and process simulation of the above-mentioned issues.

The Editors are pleased to bring the best and recent advancements in this field of research to the scientific community in this compact, peer-reviewed Special Issue. Manuscripts that included the latest research progress in terms of development and optimization of conversion processes and concepts, especially for intermittent renewable energy sources, with thermodynamic analysis, CFD and process simulation of these systems were submitted and reviewed by recognized and expert reviewers. In the Special Issue, manuscripts of high quality and that made an explicit contribution to the technical and scientific knowledge were accepted, highlighting the main developments and the new findings. Accordingly, 10 papers were accepted and published in this Special Issue. All articles can be accessed freely online.

2. Special Issue Findings

In the following, a summary of the accepted papers with their most relevant contributions is illustrated.

- The first paper, accepted in this Special Issue, authored by Gallucci, K.; Taglieri, L.; Papa, A.A.; Di Lauro, F.; Ahmad, Z.; Gallifuoco A. from the University of L'Aquila, Italy. In this study, the authors investigated the CO_2 sorption capacity of hydrochar for the upgrading of biogas to bio-methane [17]. The hydrochar was prepared based on a waste product (silver fir sawdust) available in Central Europe and Abies species available worldwide. Experiments were performed using a 316-stainless steel batch reactor at different temperatures and residence times. The hydrochar, obtained hydrothermal carbonization, was activated with potassium hydroxide impregnation and subsequent thermal treatment. The morphology and porosity of the hydrochar, characterized through Brunauer–Emmett–Teller, Barrett–Joyner–Halenda (BET–BJH), and scanning electron microscopy (SEM) analyses, were first evaluated and the sorbent capacity was then compared with traditional sorbents. The authors claimed that the developed hydrochar conceivably offers a new, feasible, and promising option for CO_2 capture using low cost and environmentally friendly materials.
- The authors of the second paper (Heinze, C.; Langner, E.; May, J.; Epple, B.) from the Technical University of Darmstadt, Germany, introduced a new char gasification model that represents all conditions in a fluidized bed gasifier [18]. For abundantly available low-rank coal, the conversion in fluidized bed gasifiers is a feasible technology to produce valuable chemicals or electricity while also offering the option of carbon capture. In this study, the non-isothermal thermogravimetric method was applied to gasify the char of Rhenish lignite at atmospheric pressure by using steam and carbon dioxide as a gasification medium. Two reaction models, namely, Arrhenius and Langmuir–Hinshelwood, as well as four conversion models (volumetric model, grain model, random pore model, and Johnson model), were fitted and evaluated with the measurement data. For both steam and carbon dioxide gasification, the authors stated that the Langmuir–Hinshelwood reaction model together with the Johnson conversion model is the most suitable method to describe the char conversion of the used Rhenish lignite, showing a coefficient of determination 98% and 95%, respectively.
- The third paper, authored by Almoslh, A.; Alobaid, F.; Heinze, C.; Epple, B. from the Technical University of Darmstadt, Germany, compared two mathematical models, namely, the rate-based model and the equilibrium-stage model, when both are applied to simulate the tar absorption process from syngas using soybean oil as a solvent in a research lab-scale test rig [19]. Experimental data at different operation points, published by Bhoi [20], were used to validate the developed models. The authors claimed that the rate-based model has higher accuracy than the equilibrium model. However, a minor deviation between the rate-based model and the experimental data was reported, which increases by increasing the bed height. An analysis study of the tar absorption process was also performed, revealing the influence of height-packed bed, temperature, and flow rate of the soybean oil on tar removal efficiency.
- The fourth paper, accepted in this Special Issue, authored by Savuto, E.; May, J.; Di Carlo, A.; Gallucci, K.; Di Giuliano, A.; Rapagnà, S. from University of Teramo, Italy. In this study, steam gasification experiments for lignite in a bench-scale fluidized-bed gasifier were carried out to evaluate the quality of the gas produced at different oper-

ating conditions [21]. Olivine was used as bed material and the steam/fuel ratio was maintained at approximately 0.65. The influence of temperature and air injections in the freeboard was evaluated in terms of the conversion efficiencies, gas composition, and tar produced. Furthermore, the obtained ashes during the gasification tests were analyzed with X-ray Diffraction (XRD) and Scanning Electron Microscope/Energy-dispersive X-ray Spectroscopy (SEM/EDS) analysis, and an affinity between calcium and sulfur was reported. The authors stated that the increase in the operating temperature leads to an improvement of the gas quality and a lower amount of tar produced. The experiments with air injections in the freeboard did not result in the desired effect on tar reduction. Compared to other tests performed with biomass at similar operating conditions, the amount of tar produced was, however, lower.

- The main contribution of the fifth paper is related to a solar-driven air-conditioning system utilizing absorption technology. In this study, the authors Al-Falahi, A.; Alobaid, F.; Epple, B. from the Technical University of Darmstadt, Germany, proposed a solar driven-absorption cooling system as an alternative technology to the conventional air conditioning of a house under hot and dry climate in Baghdad, Iraq [22]. The effect of different parameters on the solar cooling performance was evaluated. The results show that the weather conditions have a crucial influence on the performance of the solar absorption air-conditioning system, with the peak loads during the summer months. The highest performance was achieved in August with an average coefficient of performance (COP) of 0.52 and a solar fraction of 59.4%. The authors claimed that this study provides a roadmap for engineers, showing that all of the operating and design variables should be considered when developing a solar-driven air-conditioning system under the Iraq climate.

- The sixth paper included in this Special Issue dealt with an important topic that is now under research investigation as an effective gasification technology. By avoiding the use of the costly air separation unit, chemical looping gasification (CLG, see Figure 1) is a novel gasification method, allowing for the production of a nitrogen-free high calorific synthesis gas from solid hydrocarbon feedstocks (e.g., biomass and refuse-derived fuel). An equilibrium process model for an autothermal chemical looping gasification process of biomass was developed by Dieringer, P.; Marx, F.; Alobaid, F.; Ströhle, J.; Epple, B. at the Technical University of Darmstadt, Germany [23]. The results show that pursuing continuous CLG operation leads to challenges in terms of the oxygen carrier (OC) circulation, which is responsible for both, oxygen and heat transport between the air and fuel reactor. According to the authors, the CLG faces an essential dilemma. Here, higher OC circulation rates are necessary to fulfill the process heat balance (i.e., retain constant temperatures in the fuel reactor), whereas significantly lower circulation rates are required in terms of the necessary oxygen transport. Therefore, two strategies to achieve the autothermal CLG behavior through a de-coupling of oxygen and heat transport were suggested and evaluated. The findings of this study encourage deeper numerical modeling of the chemical looping gasification of biomass, as only through the deployment of elaborate models considering hydrodynamics and reaction kinetics can in-depth inferences regarding the process efficiency be offered.

- The authors of the seventh paper, published by Almoslh, A.; Alobaid, F.; Heinze, C.; Epple, B., presented a combined experimental/numerical study on CO_2 absorption [24]. Here, the effect of pressure on the gas/liquid interfacial area was investigated experimentally in the pressure range of 2 to 3 bar using an absorber tray column test rig, erected at the author's institute. Furthermore, a rate-based model was generated based on the design data of the real test rig. A simulated waste gas, consisting of 30% carbon dioxide and 70% air, and distilled water as an absorbent were used in this work. Two gas flow rates were applied. The results predicted by the rate-based model agrees very well with the experimental data. At a higher inlet gas flow rate, the gas/liquid interfacial area was significantly decreased. A pressure increase leads to a decrease in the gas/liquid interfacial area and thus decreases the absorption rate of carbon dioxide.

- The eighth paper resulted from the collaboration of two universities (Technical University of Darmstadt, Germany) and (Military Technical College, Egypt). The paper, authored by Temraz, A.; Rashad, A.; Elweteedy, A.; Alobaid, F.; Epple, B. investigated the performance of an existing 135 MW integrated solar combined cycle (ISCC) power plant in Kureimat, Egypt [25]. The existing ISCC power plant that consists of a solar field and a solar steam generator integrated into a combined cycle power plant (CCPP) was thermodynamically studied under Kureimat climatic conditions using the concept of energy and exergy analyses. The overall thermal efficiency, the exergetic efficiency, and the exergy destruction of each component in the power plant were calculated at different ambient temperatures (5, 20, and 35 °C) and different solar heat inputs (0, 50, 75 MW). The results show that the solar field has the lowest exergetic efficiency, followed by the condenser. Furthermore, it was found that the thermal efficiency and the exergetic efficiency of the ISCC and the CCPP (when no solar field heat input is supplied) decrease with increasing the ambient temperature.

- The authors (Peters, J.; Alobaid, F.; Epple, B.) from the Technical University of Darmstadt, Germany presented a combined experimental/numerical study on circulating fluidized bed boilers (CFBs) [26]. The ninth paper of this Special Issue contributes to close the knowledge gap for the operational flexibility of CFB. Corresponding to industrial standards, a long-term campaign on Polish lignite combustion during transient operation has been performed at a 1 MW_{th} scale (see Figure 2). A load following sequence for fluctuating electricity generation/demand was reproduced experimentally by four load changes from 60% to 100% load and vice versa. Based on the design data obtained from the test facility, a core-annulus dynamic process simulation model was developed. The core-annulus model was tuned with experimental data of a steady-state test point and validated with the load cycling tests. The simulation results reproduce the key characteristics of CFB combustion with good accuracy. Further numerical results can also be found in [27]. Detailed measurement data were provided during the load change for the most important parameter in the system, such as the pressure and temperature profiles along the riser, the flue gas concentrations, and the solid compositions at different locations of the test facility.

- The last paper of this Special Issue was published by Beirow, M.; Parvez, A.M.; Schmid, M.; Scheffknecht, G. A., from the University of Stuttgart, Germany. In this work, a novel sorption enhanced gasification (SEG) in a dual fluidized bed gasification system was presented [28]. The SEG system is considered a promising and flexible method for the tailored syngas production to be used in chemical manufacturing or power generation (see Figure 3). A simulation model was developed, describing the hydrodynamics in a bubbling fluidized bed gasifier and the kinetics of gasification reactions and CO_2 capture (defined by the number of carbonation/calcination cycles and the make-up of fresh limestone). Experimental data of a 200 kW pilot plant were applied to model validation. The authors claimed that the developed model can successfully predict the performance of the pilot plant at different operation conditions. With the help of the validated model, different operational parameters such as gasification temperature, steam-to-carbon ratio, solid inventory, and fuel mass flow were investigated. The parametric study shows a larger dependence on the limestone make-up, especially for gasification temperatures below 650 °C. The obtained results were summarized in a reactor performance diagram, showing the syngas power depending on the fuel feeding rate and the gasification temperature.

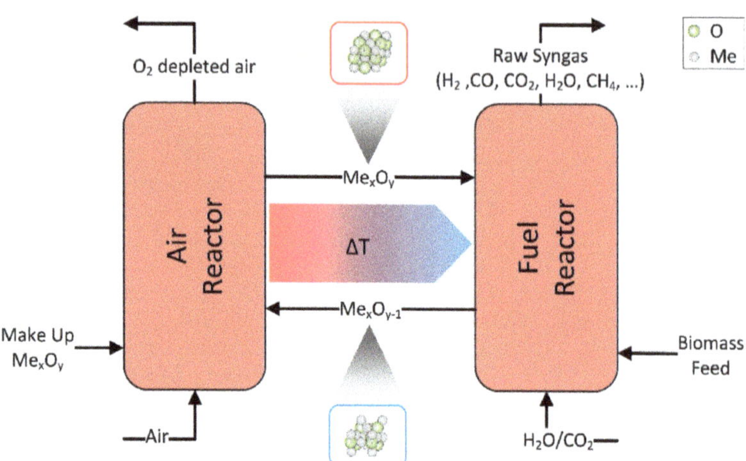

Figure 1. Schematic of the chemical looping gasification process.

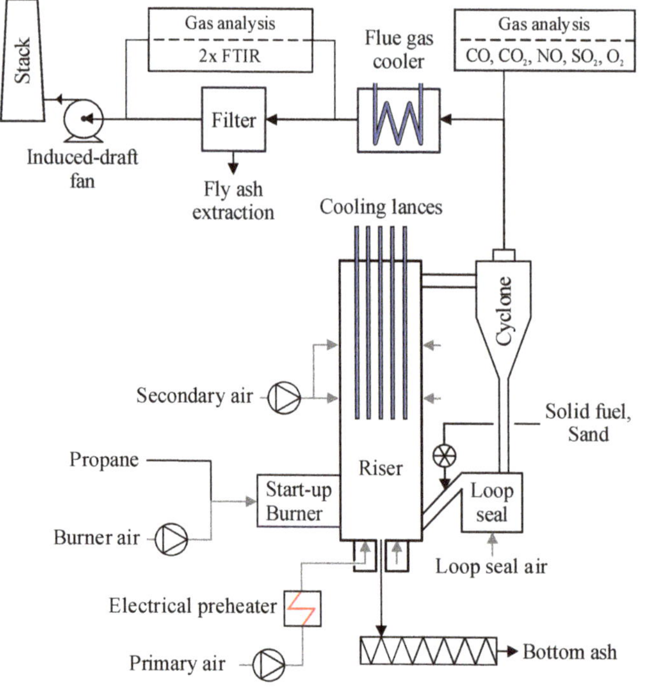

Figure 2. Simplified flow diagram of the 1 MW$_{th}$ pilot plant at the Technical University of Darmstadt.

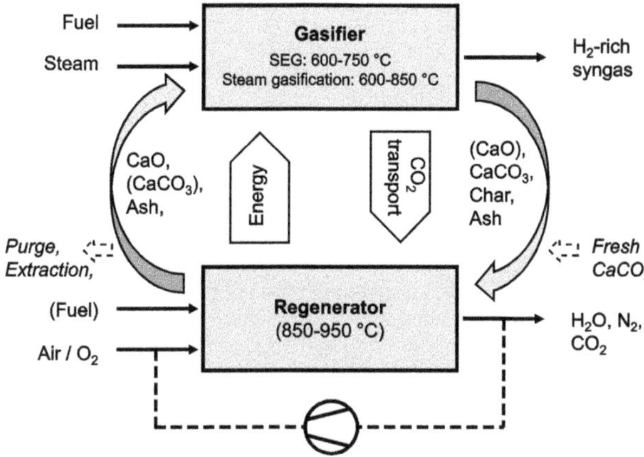

Figure 3. Schematic diagram of sorption enhanced gasification (SEG) process (up to 750 °C) and extended steam gasification mode (up to 850 °C); option of oxy-fuel operation.

3. Conclusions

The editors of this Special Issue are pleased to bring the recent advancements in thermochemical conversion processes for solid fuels and renewable energies to the scientific community. In this Editorial, the majority of published papers (in total four studies) was related to the gasification of low-rank solid fuels (e.g., biomass and lignite), subjected at the early stage of development to a single fluidized bed gasifier and recently to dual fluidized bed gasification systems, such as the sorption enhanced gasification and the chemical looping gasification. Three published papers focused on the evaluation of recent absorption and adsorption technologies for carbon capture. Two published papers were related to the most abundant renewable energy source available "Solar Energy". The solar energy in the first manuscript was used to operate a solar driven-absorption cooling system, while in the second manuscript it was converted into electrical power in an integrated solar combined cycle. The last paper discussed the operational flexibility of a circulating fluidized bed boiler, subjected to a typical operation during fluctuating electricity generation by renewables.

We hope the information collected in this Special Issue, involving new results on thermochemical conversion technologies, will benefit the readers of *Applied Sciences*. All papers were published online, free of cost or access barriers. We also look forward to more submissions to the second volume of this Special Issue "Thermochemical Conversion Processes for Solid Fuels and Renewable Energies: Volume II"—in particular, studies of high-quality, excellence, and clarity that can make a difference in this field of research.

Author Contributions: F.A. was responsible for writing, review, and editing this Editorial. J.S. has read and reviewed this Editorial. Both authors have read and agreed to the published version of the manuscript.

Funding: This research received no external funding.

Institutional Review Board Statement: Not applicable.

Informed Consent Statement: Not applicable.

Data Availability Statement: Not applicable.

Acknowledgments: The editors would like to thank the Assistant Editor Tamia Qing and her team for the technical support and organizational management of this Special Issue. Furthermore, many thanks go to all the reviewers for reviewing the above-referenced manuscripts. Their professional reviews have significantly helped to improve the quality of the submitted papers to this Special Issue.

Conflicts of Interest: The author declares no conflict of interest.

References

1. Masson-Delmotte, V.; Zhai, P.; Pörtner, H.-O.; Roberts, D.; Skea, J.; Shukla, P.R.; Pirani, A.; Moufouma-Okia, W.; Péan, C.; Pidcock, R. Global warming of 1.5 °C. *IPCC Spec. Rep. Impacts Glob. Warm.* **2018**, *1*, 1–9.
2. Nguyen, N.M.; Alobaid, F.; May, J.; Peters, J.; Epple, B. Experimental study on steam gasification of torrefied woodchips in a bubbling fluidized bed reactor. *Energy* **2020**, *202*, 117744. [CrossRef]
3. Alobaid, F.; Busch, J.-P.; Stroh, A.; Ströhle, J.; Epple, B. Experimental measurements for torrefied biomass Co-combustion in a 1 MWth pulverized coal-fired furnace. *J. Energy Inst.* **2020**, *93*, 833–846. [CrossRef]
4. Bui, M.; Mac Dowell, N. *Carbon Capture and Storage*; Royal Society of Chemistry: London, UK, 2019; Volume 26.
5. Araújo, O.d.Q.F.; de Medeiros, J.L. Carbon capture and storage technologies: Present scenario and drivers of innovation. *Curr. Opin. Chem. Eng.* **2017**, *17*, 22–34. [CrossRef]
6. Østergaard, P.A.; Duic, N.; Noorollahi, Y.; Mikulcic, H.; Kalogirou, S. Sustainable development using renewable energy technology. *Renew. Energy* **2020**, *146*, 2430–2437. [CrossRef]
7. Sinsel, S.R.; Riemke, R.L.; Hoffmann, V.H. Challenges and solution technologies for the integration of variable renewable energy sources—A review. *Renew. Energy* **2020**, *145*, 2271–2285. [CrossRef]
8. Alobaid, F.; Mertens, N.; Starkloff, R.; Lanz, T.; Heinze, C.; Epple, B. Progress in dynamic simulation of thermal power plants. *Prog. Energy Combust. Sci.* **2017**, *59*, 79–162. [CrossRef]
9. Rackley, S.A. *Carbon Capture and Storage*; Butterworth-Heinemann: Oxford, UK, 2017.
10. Bouillon, P.-A.; Hennes, S.; Mahieux, C. ECO2: Post-combustion or Oxyfuel–A comparison between coal power plants with integrated CO2 capture. *Energy Procedia* **2009**, *1*, 4015–4022. [CrossRef]
11. May, J.; Alobaid, F.; Ohlemueller, P.; Stroh, A.; Stroehle, J.; Epple, B. Reactive two–fluid model for chemical–looping combustion–Simulation of fuel and air reactors. *Int. J. Greenh. Gas Control* **2018**, *76*, 175–192. [CrossRef]
12. Ohlemüller, P.; Alobaid, F.; Abad, A.; Adanez, J.; Ströhle, J.; Epple, B. Development and validation of a 1D process model with autothermal operation of a 1 MWth chemical looping pilot plant. *Int. J. Greenh. Gas Control* **2018**, *73*, 29–41. [CrossRef]
13. Heinze, C.; May, J.; Peters, J.; Ströhle, J.; Epple, B. Techno-economic assessment of polygeneration based on fluidized bed gasification. *Fuel* **2019**, *250*, 285–291. [CrossRef]
14. May, J.; Alobaid, F.; Stroh, A.; Daikeler, A.; Ströhle, J.; Epple, B. Euler-Lagrange Model for the Simulation of Carbonate Looping Process. *Chem. Ing. Tech.* **2020**, *92*, 648–658. [CrossRef]
15. Pan, S.-Y.; Du, M.A.; Huang, I.-T.; Liu, I.-H.; Chang, E.; Chiang, P.-C. Strategies on implementation of waste-to-energy (WTE) supply chain for circular economy system: A review. *J. Clean. Prod.* **2015**, *108*, 409–421. [CrossRef]
16. Alobaid, F. *Numerical Simulation for Next Generation Thermal Power Plants*; Springer: Berlin, Germany, 2018.
17. Gallucci, K.; Taglieri, L.; Papa, A.A.; Di Lauro, F.; Ahmad, Z.; Gallifuoco, A. Non-Energy Valorization of Residual Biomasses via HTC: CO2 Capture onto Activated Hydrochars. *Appl. Sci.* **2020**, *10*, 1879. [CrossRef]
18. Heinze, C.; Langner, E.; May, J.; Epple, B. Determination of a Complete Conversion Model for Gasification of Lignite Char. *Appl. Sci.* **2020**, *10*, 1916. [CrossRef]
19. Almoslh, A.; Alobaid, F.; Heinze, C.; Epple, B. Comparison of Equilibrium-Stage and Rate-Based Models of a Packed Column for Tar Absorption Using Vegetable Oil. *Appl. Sci.* **2020**, *10*, 2362. [CrossRef]
20. Bhoi, P.R. Wet scrubbing of biomass producer gas tars using vegetable oil. Ph.D. Thesis, Oklahoma State University, Stillwater, OK, USA, 2014.
21. Savuto, E.; May, J.; Di Carlo, A.; Gallucci, K.; Di Giuliano, A.; Rapagnà, S. Steam gasification of lignite in a bench-scale fluidized-bed gasifier using olivine as bed material. *Appl. Sci.* **2020**, *10*, 2931. [CrossRef]
22. Al-Falahi, A.; Alobaid, F.; Epple, B. A new design of an integrated solar absorption cooling system driven by an evacuated tube collector: A case study for Baghdad, Iraq. *Appl. Sci.* **2020**, *10*, 3622. [CrossRef]
23. Dieringer, P.; Marx, F.; Alobaid, F.; Ströhle, J.; Epple, B. Process Control Strategies in Chemical Looping Gasification—A Novel Process for the Production of Biofuels Allowing for Net Negative CO2 Emissions. *Appl. Sci.* **2020**, *10*, 4271. [CrossRef]
24. Almoslh, A.; Alobaid, F.; Heinze, C.; Epple, B. Influence of Pressure on Gas/Liquid Interfacial Area in a Tray Column. *Appl. Sci.* **2020**, *10*, 4617. [CrossRef]
25. Temraz, A.; Rashad, A.; Elweteedy, A.; Alobaid, F.; Epple, B. Energy and Exergy Analyses of an Existing Solar-Assisted Combined Cycle Power Plant. *Appl. Sci.* **2020**, *10*, 4980. [CrossRef]
26. Peters, J.; Alobaid, F.; Epple, B. Operational Flexibility of a CFB Furnace during Fast Load Change—Experimental Measurements and Dynamic Model. *Appl. Sci.* **2020**, *10*, 5972. [CrossRef]
27. Alobaid, F.; Peters, J.; Amro, R.; Epple, B. Dynamic process simulation for Polish lignite combustion in a 1 MWth circulating fluidized bed during load changes. *Appl. Sci.* **2020**, *278*, 115662. [CrossRef]
28. Beirow, M.; Parvez, A.M.; Schmid, M.; Scheffknecht, G. A Detailed One-Dimensional Hydrodynamic and Kinetic Model for Sorption Enhanced Gasification. *Appl. Sci.* **2020**, *10*, 6136. [CrossRef]

Article

Non-Energy Valorization of Residual Biomasses via HTC: CO$_2$ Capture onto Activated Hydrochars

Katia Gallucci *, Luca Taglieri, Alessandro Antonio Papa, Francesco Di Lauro, Zaheer Ahmad and Alberto Gallifuoco

Industrial Engineering Department, University of L'Aquila, Monteluco di Roio—67100 L'Aquila, Italy;
luca.taglieri@univaq.it (L.T.); alessandroantonio.papa@graduate.univaq.it (A.A.P.);
francesco.dilauro@student.univaq.it (F.D.L.); zaheer.ahmad@graduate.univaq.it (Z.A.);
alberto.gallifuoco@univaq.it (A.G.)
* Correspondence: katia.gallucci@univaq.it

Received: 13 February 2020; Accepted: 5 March 2020; Published: 10 March 2020

Abstract: This study aims to investigate the CO$_2$ sorption capacity of hydrochar, obtained via hydrothermal carbonization (HTC). Silver fir sawdust was used as a model material. The batch runs went at 200 °C and up to 120 min. The hydrochar was activated with potassium hydroxide impregnation and subsequent thermal treatment (600 °C, 1 h). CO$_2$ capture was assayed using a pressure swing adsorption (PSA) process. The morphology and porosity of hydrochar, characterized through Brunauer-Emmett-Teller, Barrett-Joyner-Halenda (BET-BJH) and scanning electron microscopy (SEM) analyses, were reported and the sorbent capacity was compared with traditional sorbents. The hydrochar recovered immediately after the warm-up of the HTC reactor had better performances. The Langmuir equilibrium isotherm fits the experimental data satisfactorily. Selectivity tests performed with a model biogas mixture indicated a possible use of hydrochar for sustainable upgrading of biogas to bio-methane. It is conceivably a new, feasible, and promising option for CO$_2$ capture with low cost, environmentally friendly materials.

Keywords: hydrochar; hydrothermal carbonization; biogas upgrading; CO$_2$ capture; pressure swing adsorption

---

1. Introduction

Growing population and economy race are the two main concerns regarding the increase in emissions of greenhouse gases (GHG) and, subsequently, the accelerating climate change and global warming. Carbon dioxide is the first responsible for these unwanted effects [1]. Nowadays, reducing CO$_2$ emissions is one of the most challenging issues facing humanity [2]. Global climate reports show that the average world temperature is steadily increasing with respect to 20th century data [3]. Besides, worldwide researchers recognize the need for a sustainable development and more equity, including poverty eradication, and provide an ethical foundation for limiting the effects of climate changes. The rate of atmospheric accumulation of GHGs and the capacity to address its mitigation and adaptation differ for each nation. Often, many of the countries most vulnerable to climate change contribute little to GHG emissions.

Intergovernmental panel on climate change (IPCC) recently reported on the impacts of global warming of 1.5 °C above pre-industrial levels and the related global greenhouse gas emission pathways [4]. The attention is on the decision to adopt the Paris Agreement [5,6].

Comprehensive and sustainable strategies in response to climate change should consider the co-benefits, adverse side-effects, and risks inherent to both adaptation and mitigation options. European policy options emerge in the European Union (EU) Emission Trading System (ETS) that involve the promotion of investment on clean and low carbon technologies in power and heat generation, energy-intensive industry sectors, materials, and chemicals' production. The legislative framework of the EU Emission Trading System (ETS) achieved the EU's 2030 emission reduction targets [7] and contributed to the 2015 Paris Agreement. Now, the goals include keeping the rise in temperature below 2 °C, that is, CO_2 below 450 ppm [8].

Practical ways to meet these goals could be planting trees, caring the existing forests, rebuilding of soils, and developing the biomass-to-energy chain with coupled with carbon capture and storage.

Carbon dioxide capture/separation onto solid matrices is broadly studied for reducing greenhouse gas discharge [9]. Also, biofuels and bio-methane are strategic for fuel transition to sustainability, or as a reagent in the steam and dry reforming and catalytic processes. A paradigm is biogas production by anaerobic digestion of organic matter. The mixture mainly consists of 40 vol% to 75 vol% methane and 15 vol% to 60 vol% carbon dioxide [10,11].

Biogas cleaning and upgrading to bio-methane require to remove water, foam, dust H_2S, and trace components, as well as CO_2 [11] and bio-methane production, in Europe is spreading progressively [12]. In Italy, such technology struggles to spread with only five operating plants and a total upgrading capacity of 500 Nm^3/h. The delay is mainly owing to the imprecise legislative references regarding grid injection techniques. As a mean to overcome these limits, an effort was made toward the development of implementing rules and guidelines for accessing the incentives.

From a purely technological point of view, several techniques are available for commercial use, at laboratory and industrial scale [13–18]. CO_2 capture technologies, like absorption and adsorption, are proliferating as commonly used capture techniques worldwide, as proven by the several patent applications and published articles [19]. Adsorbent media include activated carbon, alumina, metallic oxides, and zeolites. The regeneration of the adsorbent is carried out by temperature or pressure swing adsorption (TSA or PSA) [20]. For TSA, solid adsorbents with a lower heat capacity are claimed for reducing the energy required for regeneration [21–23].

The most important property of an adsorbent is its CO_2 adsorption capacity, which depends strongly on the pore structure, the surface area, and the type of functionalization. Besides, the capacity depends on the partial pressure of CO_2, temperature, and humidity [24,25]. Absorption proved to not be economical for treating flue gas streams with CO_2 partial pressures lower than 15 vol% [26]. Chemical absorption has relatively high selectivity, but also high energy consumption for regeneration, chemicals' make-up, and high environmental impact [27,28]. Taking into account the potential role of porous carbon materials for CO_2 capture, the authors propose the production of these materials starting from renewable sources, mainly residual biomasses.

Hydrothermal carbonization (HTC) is an alternative synthesis method to produce precursors for high value-added renewable carbon materials from residual lignocellulosic biomasses [29]. This thermochemical conversion occurs in hot compressed water, at a relatively low temperature under autogenous pressure [30]. HTC optimization was studied considering the re-use of the liquid phases in the process itself [31,32], the online monitoring of the carbonization time-course [33,34], and the recovery of valuable platform chemicals from the liquid [35,36].

Hydrochar attracts the attention for several potential applications, mainly in support of the energy production chain [37,38]. Nevertheless, owing to its coal-like structure, hydrochar could be favorably transformed into porous activated carbon with tunable morphologies and porosities, which is hard to achieve using traditional pyrolytic methods, with a surface area up to 3000 m^2g^{-1} and pore volume in the range of 0.5–1.5 cm^3g^{-1} [39,40].

The focus of this paper is on the separation of CO_2 for the upgrading of biogas to bio-methane by PSA onto HTC from silver fir sawdust, a waste product available in Central Europe and *Abies* species available worldwide [41].

2. Materials and Methods

2.1. Synthesis of Carbon Porous Sorbent

2.1.1. Materials

The tested biomass was silver fir sawdust. All HTC experiments were performed using demineralized water (σ = 0.005 mS/cm). Potassium hydroxide (Sigma-Aldrich Comporation US, grade ACS reagent, Reg. Ph. Eur.) and HCl (Sigma-Aldrich grade puriss, 24.5%–26.0%) were used for the activation of the hydrochar.

2.1.2. Experimental Apparatus

The HTC apparatus was designed and realized on purpose. Details of the piping and control instrumentation are reported elsewhere [34]. The 316-stainless steel batch reactor has an internal volume of about 200 mL and is equipped with a valve for the air initial evacuation and gaseous product withdrawal. A thermocouple and a pressure-meter allow monitoring and controlling the process temperature and pressure, respectively.

2.1.3. Experimental Procedure

The HTC experiments were performed at 200 °C, with a constant water biomass ratio 7:1, and residence time equal to 0 and 120 min. The substrate was ground in a mortar and sieved. The fraction below 1 mm was recovered and used for the reaction after drying until constant weight in an oven, at 105 °C. Demineralized water and dry biomass were loaded into the reactor in the chosen weight ratio. Then, the reactor was sealed, evacuated for about 15 min, and heated up to the process temperature. The warm-up lasted for about 20–25 min. Residence time was measured once it reached the temperature set point. At the end of the run, the reactor was quenched with air until 160 °C, and then with water until room temperature.

After the quenching step, the gas phase was recovered and weighed. Solid and liquid products were recovered and separated by filtration. The filter and solid phases were put in an oven at 105 °C for at least 24 h for dry weight determination. After drying, the hydrochar was milled and sieved (certified ISO 3310–1 & 2 and ASTM E11, Endecott sieves). The fraction between 106 and 355 μm was recovered and stored in vials before the CHNS characterization (data reported in Supplementary Material) and the activation step. Then, 15 grams of 1:2 solid mixture of hydrochar and KOH was put in an oven under nitrogen gas flow and warmed up to 600 °C (3 °C/min). The dwell step lasted 1 h [40]. The samples were washed with 10 wt% HCl to remove any inorganic salts, and then with demineralized water until neutral pH. Finally, the samples were dried in an oven at 105 °C for 24 h.

2.1.4. Characterization Methods

The nitrogen sorption isotherms of activated and non-activated hydrochars were determined with a high-speed surface area and pore size analyzer NOVA 1200e Alfates Quartachrome. Brunauer-Emmett-Teller (BET) [42] and Barrett-Joyner-Halenda (BJH) [43] methods were utilized for a standard determination of surface area, pore volume, and pore size distribution and the average pore diameter [44]. Each sample was outgassed for 3 h at 100 °C.

Textural properties were investigated by means of scanning electron microscopy and energy dispersive X-ray spectroscopy (SEM-EDS). The analyses were performed with a Field Emission Zeiss Gemini500SEM analyser (acceleration voltage from 0.02 to 30 KV—store resolution up to 32 k × 24 k pixels—0.6 nm at 15 KV, in back-scattering electron mode—pressure from 5 to 500 Pa), equipped with EDS (energy dispersive spectroscopy) microanalyzer OXFORD Aztec Energy with INCA X-ACT PELTIER COOLED ADD detector. All samples were coated with 5 nm of Cr in an automatic sputter coater Q 150T ES.

2.2. CO_2 Adsorption Test

2.2.1. Materials

Three hydrochar types were tested, namely, HC_200_0 (non-activated, residence time 0 min), HCA_200_0 (activated, residence time 0 minute), and HCA_200_120 (activated, residence time 120 min). Grain size ranged from 106 μm to 355 μm. Nitrogen (grade 5.5), methane (grade 4.5), and carbon dioxide (grade 4.0) were used for dynamic adsorption tests (Rivoira Spa). Blank tests were performed, packing the column with glass beads (2500 kg/m^3, 106–355 μm).

2.2.2. Experimental Apparatus

The experimental apparatus is shown in Figure 1.

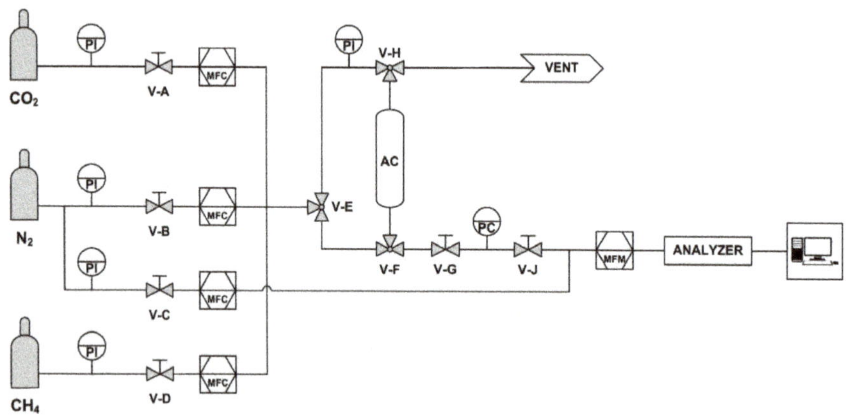

Figure 1. Flowsheet of the experimental apparatus: PI—pressure indicator; V-i—valves; MFC—mass flow controller; MFM—mass flow meter; PC—pressure controller; AC—adsorption column; analyzer—URAS ABB.

Details of the piping and control instrumentation are reported in a previous paper [22].

2.2.3. Experimental Procedure

Bed length was kept constant packing the column with the same HC volume same of the inert material volume used for the blank test.

Adsorption experiments were performed with two different feed mixtures ($CO_2 + N_2$ and $CO_2 + CH_4$). The PSA operating pressure varied up to 5 bars.

Tests with $CO_2 + N_2$ were carried out at 2, 3, 4, and 5 bar with 100 NmL/min of CO_2 and 80 NmL/min of N_2 (carrier gas). The regeneration step was performed with N_2 at 280 NmL/min. A mass flow meter (MFM) quantified the total amount of the gas mixture leaving the column. During the adsorption tests, V-A, V-B, and V-C valves were kept open. The configuration for regeneration after each run was as follows: V-A and V-G closed, V-H switched for a quick depressurization and then repositioned, V-G re-opened, and N_2 injected until no CO_2 was detected. The blank tests were carried out in the same way.

Tests with $CO_2 + CH_4$ mixture were carried out at 2 and 5 bar with 76 NmL/min of CO_2 and 76 NmL/min of CH_4. The flow rate of N_2 during the regeneration step was 280 NmL/min. The experiments were carried out as previously described, with CH_4 in the inlet flow rate instead of N_2. V-D and V-A operated simultaneously. The regeneration step lasted until no CH_4 was detected at the column outlet.

All tests were repeated almost five times. The synopsis is reported in Table 1.

Table 1. Synopsis of tests conditions.

Test	Hydrothermal Conditions	Adsorption Conditions		Activation
	Time [min]	P [bar]	W_{BED} [g]	
CO_2/N_2				
HC_200_120_2	120	2	1.135	no
HC_200_120_3	120	3	1.135	no
HC_200_120_4	120	4	1.135	no
HC_200_120_5	120	5	1.135	no
HCA_200_0_2	0	2	0.695	yes
HCA_200_0_3	0	3	0.695	yes
HCA_200_0_4	0	4	0.695	yes
HCA_200_0_5	0	5	0.695	yes
HCA_200_120_2	120	2	1.135	yes
HCA_200_120_3	120	3	1.135	yes
HCA_200_120_4	120	4	1.135	yes
HCA_200_120_5	120	5	1.135	yes
CO_2/CH_4				
HCA_200_0_2	0	2	0.695	yes
HCA_200_0_5	0	5	0.695	yes

2.2.4. Mathematical Model

The mathematical model for determining the amount of the adsorbed CO_2 was extensively described in Di Felice et al. [45]. The blank test highlighted the domination in the response curve of a flow-mixing in the ancillary equipment and provided a means of confronting the gas outlet responses to the CO_2 capture characteristics of the bed itself. A simple descriptive FOPDT model (first order plus dead time model) of the gas phase allows extracting the response of the particle phase from the measured gas phase response of the entire system under CO_2 capture conditions. The response curve gives the molar total gas holdup, viz, the product of the total molar throughput of gas and the area enclosed by the response curve and normalized ordinate 1 (Figures 2 and 3).

The response curve may be used to evaluate the total amount of CO_2 present in the whole system as a function of time (i.e., its holdup: the moles of CO_2 that entered the system at time t minus those that left). The CO_2 holdup in the solid phase of the bed (i.e., the CO_2 captured by the hydrochar) is merely the total holdup of the entire system minus the holdup of the gas phase of the entire system [45]. Adsorption performance in CO_2/CH_4 tests was evaluated estimating recovery, purity, and selectivity [22]. The gas recovery rate is the ratio of the quantity of gas recovered after the column to the fed quantity:

$$Recovery(i) = \left(\int_0^t (y_i\, q)_{OUT}\, dt \right) / \left(\int_0^t (y_i\, q)_{BLANK}\, dt \right) \tag{1}$$

The purity is defined by the quantity of the gas during the saturation phase divided by total gas leaving the column at the same time.

$$Purity(i) = \left(\int_0^t (y_i\, q)_{OUT}\, dt \right) / \left(\int_0^t q_{OUT}\, dt \right) \tag{2}$$

where y_i is the molar fraction of the specific gas and q is the total gas flow.

The selectivity is expressed as follows:

$$Selectivity = \frac{mol_{CO_2}/kg_{adsorbent}}{mol_{CH_4}/kg_{adsorbent}} \tag{3}$$

3. Results and Discussions

The assessment of new porous carbon material should be based on its adsorption properties, evaluated through the well-established procedures. Table 2 lists the results of the BET analysis.

Table 2. BET and BJH results of all materials tested.

Sample Parameters	Surface Area (A_{BET}) [m^2/g]	BJH Desorption Pore Volume (V_{BJH}) [cm^3/g]	Average Diameter of the Pores ($D_{av,BJH}$) [Å]
HC_200_0	1.13	0.004	142
HCA_200_0	881	0.241	11
HC_200_120	1.33	0.008	241
HCA_200_120	284	0.109	15

The surface area and the average diameter of the pore $D_{av,BJH}$ (obtained as $4 \cdot V_{BJHa}/A_{BET}$, where V_{BJHa} is the BJH desorption pore volume and A_{BET} is the BET surface area) are critical parameters for physical adsorption. In general, the higher the superficial area, the better the sorbent capacity.

The increase of porosity is also highlighted by SEM-EDS images (Figure 2), where, in the left-hand side (Figure 2a,c,e), the original wood structure is always present after hydrothermal treatment with several typical pits [46] and, at higher magnification (3000 and 10,000×), 10 µm macropores are also visible. On the right-hand side (Figure 2b,d,f), after the activation procedure, as expected, a well-developed sponge structure is evident at increasing magnification (400, 2000, and 10,000×).

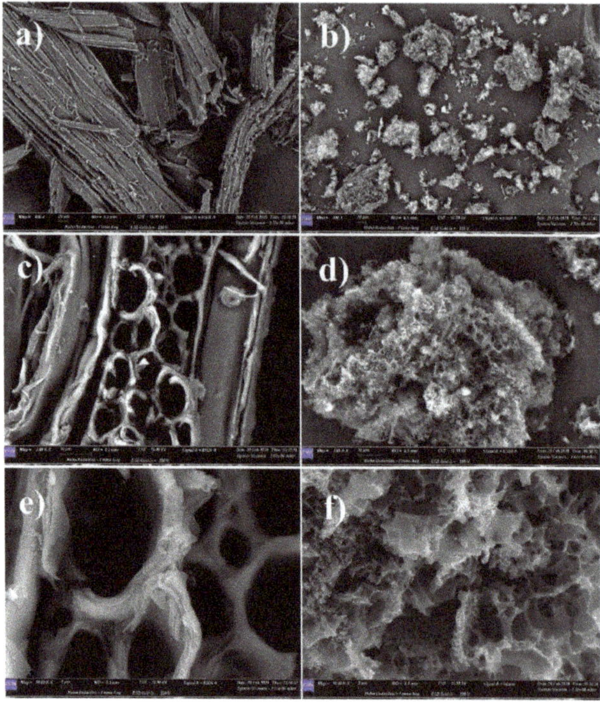

Figure 2. Scanning electron microscopy (SEM) images of HC_200_0 (left-hand side: (**a,c,d**) and HCA_200_0 (right-hand side: (**b**), (**d,f**) at different magnification: (**a,b**) 400×; (**c**) 3000×; (**d**) 2000×; (**e,f**) 10,000×.

The EDS analysis, not reported here, shows the carbon as the main element with a distributed trace of chlorine, owing to the HCl washing step of the activation method.

As far as the time course of the outflowing CO_2 and CH_4 concentration is concerned, a sigmoidal behavior was observed for all tests and at all values of the operating pressure, as shown in Figure 3. The recorded signals are reported as a normalized value, that is, divided by the inlet concentration.

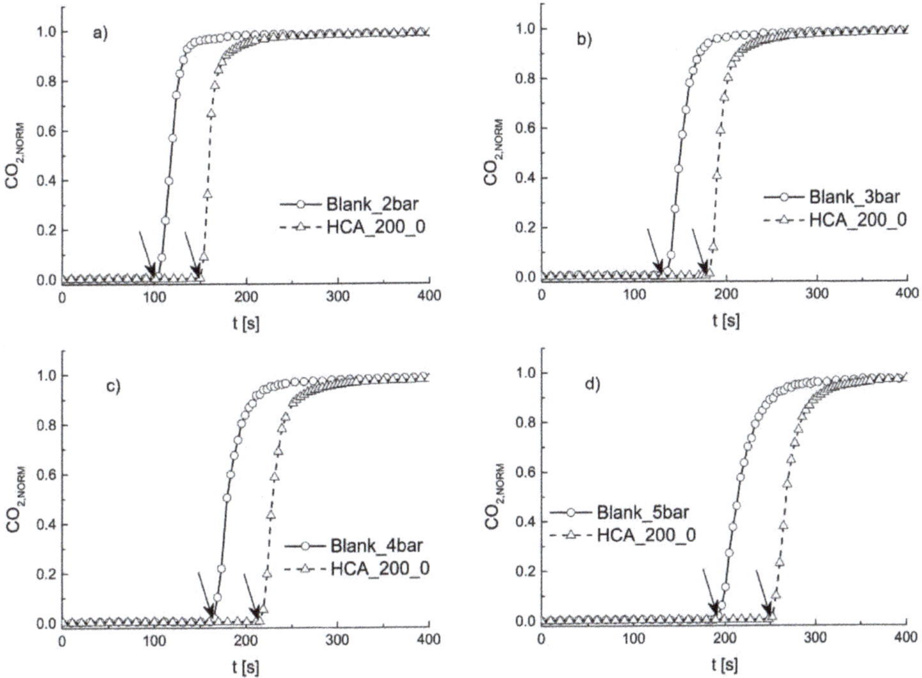

Figure 3. CO_2/N_2 adsorption curves at (**a**) 2 bar; (**b**) 3 bar; (**c**) 4 bar; and (**d**) 5 bar.

By inspection of Figure 3, the time of first detection (arrows) depends evidently on the operating pressure. The translation of blank curves is inherent in the operating modality of the apparatus and depends linearly on the pressure (data not reported for the sake of brevity). The translation of the response curves suffers from the superimposition owing to the presence of the adsorbent bed. The difference between the two delays increases steadily. Table 3 is a synopsis of all the results.

The sorbent capacity of the activated hydrochar obtained at 200 °C and 0 min is higher than that of the corresponding material recovered after 120 min of retention in the HTC reactor. The results of the BET analysis confirm this finding, where the 0 minute sample shows a specific area as high as 210% (3.1 time bigger) and an average pore diameter decreased by 27% in comparison with sample HCA_200_120. These results warn that the HTC reaction time is a crucial parameter and that the existence of a possible optimal retention time will be worth study, and in the further developments, it will be coupled with a cost optimization aimed at industrial exploitation of the results.

On the other hand, the performances of non-activated hydrochar denounce of an adsorbent capacity of an order of magnitude lower, and are thus not acceptable for industrial applications.

Table 3. CO_2 sorbent capacity.

Sample	TEST	P [bar]	Sorbent Capacity * [mmol/g]
HC_200_120	CO_2/N_2	2	0.335 ± 0.129
		3	0.425 ± 0.181
		4	0.688 ± 0.210
		5	0.692 ± 0.228
HCA_200_120	CO_2/N_2	2	2.651 ± 0.225
		3	3.366 ± 0.292
		4	3.475 ± 0.368
		5	3.635 ± 0.268
HCA_200_0	CO_2/N_2	2	4.957 ± 0.952
		3	5.658 ± 0.070
		4	6.199 ± 0.081
		5	6.569 ± 0.119

* Mean ± standard deviation.

Figure 4 shows typical adsorption curves obtained with the activated hydrochar at 2 and 5 bar (a and b, respectively) and the mixture CH_4/CO_2. Blank runs are reported as a reference. Arrows signal the time of the first detection. Both diagrams show that, in the blank runs, the CO_2 and CH_4 signals are indistinguishable. On the contrary, in the presence of a hydrochar bed, a selectivity appears evident, as proven by the temporal separation of the arrows. Methane appears first in the column outlet regardless of the operating pressure. The delay between the two signals is an increasing function of the operating pressure, and in any case, it is sufficiently broad for envisaging the development of an industrial process. All of this evidence proves that hydrochar is a suitable medium for separating the mixture by selective adsorption.

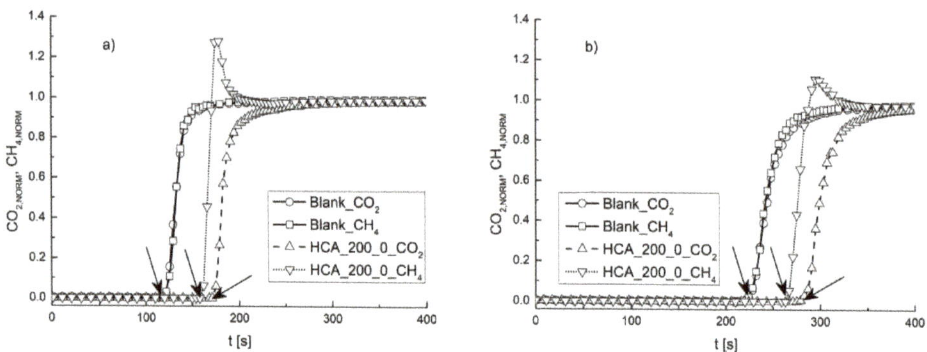

Figure 4. CO_2/CH_4 adsorption curves at (**a**) 2 bar and (**b**) 5 bar.

Table 4 quotes the obtained selectivities and recoveries calculated using Equations (1)–(3). The obtained selectivities confirm that CO_2 is preferred to CH_4. The performance of hydrochar is worse than that obtainable with commercial porous sorbents such as zeolites or activated carbons [22]. As expected, the methane recovery is relatively low with a purity of 95% (Italian regulation for network injection) regardless of the investigated operating pressure. The doubling of recovery obtained with a purity of 70% is a valuable result because of energetic applications.

Table 4. Summary of the results of the adsorption tests (mixture CO_2/CH_4).

Sample	TEST	P [bar]	Recovery (Purity = 95%) [%]	Recovery (Purity = 70%) [%]	Selectivity
HCA_200_0	CO_2/CH_4	2	31.98 ± 0.13	60.12 ± 0.35	1.88 ± 0.02
		5	36.09 ± 0.87	68.11 ± 0.10	2.10 ± 0.02

Figure 5 reports the averaged CO_2 sorbent capacity for the two samples of activated hydrochar as a function of the corresponding partial pressure in the gas phase. Data fit well to the Langmuir equation [47].

$$CO_{2Sorbent\ capacity} = \left(C_{Max} \cdot p_{CO_2}\right)/\left(K + p_{CO_2}\right) \tag{4}$$

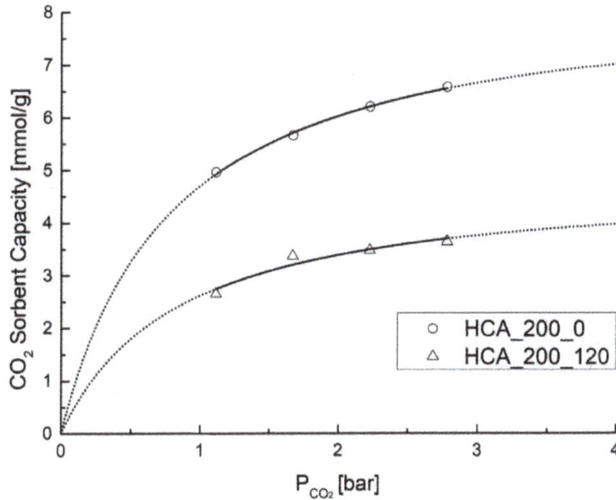

Figure 5. CO_2 sorbent capacity vs. p_{CO2}.

The regressions are reported as solid lines in the explored range and as dotted lines in the extrapolated ones. A wider partial pressure range should be explored to ascertain the correct equilibrium law. For the present paper, this preliminary investigation gives valuable information for steering future studies.

The regression parameters are reported in Table 5.

Table 5. Synopsis of regression parameters for Langmuir equation.

	C_{Max}	K	R^2
HCA_200_0	8.38 ± 0.15	0.77 ± 0.05	0.99
HCA_200_120	4.79 ± 0.45	0.83 ± 0.26	0.90

Figures 6 and 7 compare the performances of the hydrochars to those of traditional solid materials and those of some innovative materials, as reported in the literature [22,40,48–54]. As a general finding, data on PSA at 1 bar appear in the literature sparingly, even for traditional sorption media. Figure 6 reports a possible comparison of the hydrochar sorption capacity. It appears that, despite the different capture techniques, the data of "HCA_200_0_calc" at 1 bar (calculated by Equation (4)) are comparable with those of the literature. On the other hand, the comparison with PSA experiments conducted on zeolites, activated carbon, and fly ash shows that, at the same operating conditions, the HCA exhibits much better results in terms of CO_2 capture capacity.

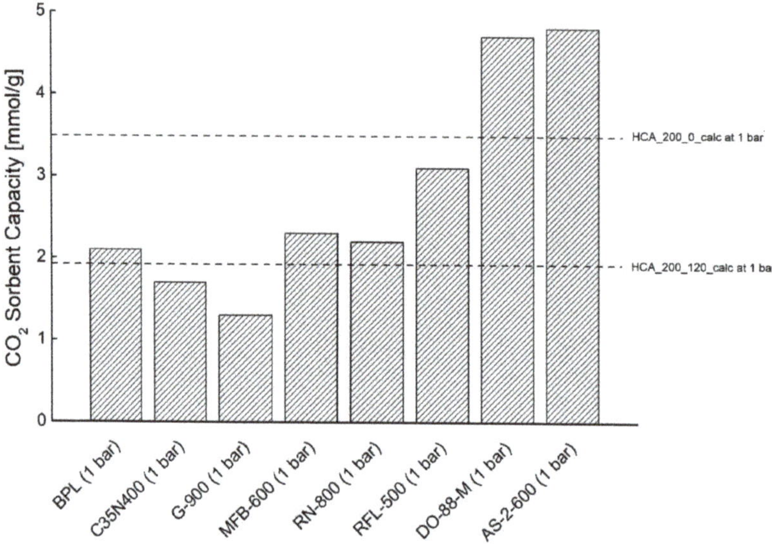

Figure 6. Comparison of CO_2 sorbent capacities with literature data obtained with batch equilibrium method (BPL: commercial activated carbon [48]; C35N400: ammonia-treated activated carbon [49]; G-900: activated graphite fibers [50]; MFB-600: N-doped activated carbon [51]; RN-800: ammonia-treated activated carbon [52]; RFL-500: N-doped porous carbon [53]; DO-88-M: activated carbon from petroleum pitch [54]; AS-2-600: sawdust-based porous carbon [40]).

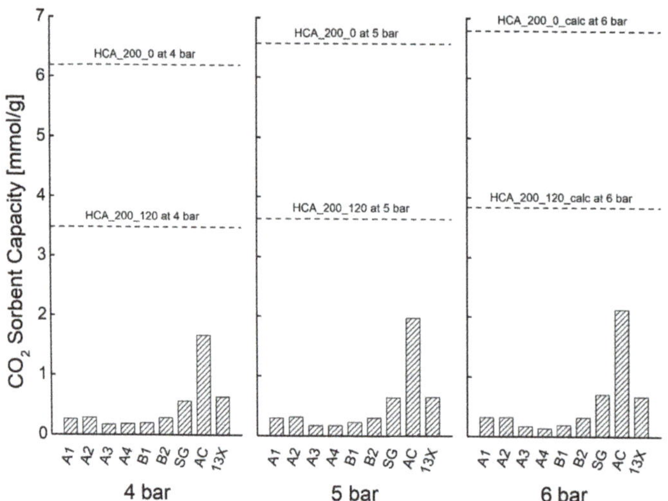

Figure 7. Comparison of CO_2 sorbent capacities with literature data obtained with the dynamic experimental method (pressure swing adsorption (PSA)) (A1, A2, A3, A4, B1, B2: zeolites from fly ash; SG: commercial silica gel; AC: commercial activated carbon; 13X: commercial zeolite [22]).

Figure 7 shows that the HCA_200_0 has the best CO_2 sorbent capacity: 6.569 mmol/g, threefold concerning the best performance of traditional sorbents [22]. Another significant result appears in Figure 7. Hydrochars prepared after 120 min of HTC reaction halve their performance, even though remaining well above commercial zeolites and similar materials. This suggests investing in further research aimed to ascertain if a reaction time exists, which maximizes the sorption capacity. This more-in-depth investigation is of the utmost importance for industrial-scale process optimization.

All of the results here reported highlight the concrete possibility of exploiting the residual biomass as an adsorption medium for biogas upgrading and encourages continuing research in this way.

4. Conclusions

Hydrochar from lignocellulosic residual biomass furnishes porous carbon materials via usual activation. The best reaction conditions were 200 °C and 0 min. The best results were BET area of 881 m^2/g, average pore diameter of 11 Å, and sorbent capacity of 6.569 mmol/g at 5 bar. The hydrochar exhibits higher CO_2 adsorption than that of some traditional sorbents. The material possesses adequate capacity to adsorb CO_2 from mixtures with CH_4 selectively. This residual material could be exploited in a sustainable biogas upgrading process, contextually reducing CO_2 emissions and the related environmental impact.

Supplementary Materials: The following are available online at http://www.mdpi.com/2076-3417/10/5/1879/s1.

Author Contributions: Funding acquisition, K.G.; Investigation, L.T., A.A.P., F.D.L. and Z.A.; Methodology, K.G.; Project administration, K.G.; Supervision, A.G.; Writing—original draft and revision K.G., L.T. and A.A.P. All authors have read and approved the manuscript for publication.

Funding: This research was funded by European Regional Development Fund 2014–2020 grant number PON—R&I E12H18000230001 for Industrial PhD scholarship (http://www.ponricerca.gov.it/) and the APC was funded by the dedicated budget for PhD research activities.

Conflicts of Interest: The authors declare no conflict of interest.

References

1. Vitillo, J.G.; Smit, B.; Gagliardi, L. Introduction: Carbon Capture and Separation. *Energy Convers. Manag.* **2002**, *43*, 63–75. [CrossRef]
2. IEA International Energy Agency (Paris (France)). *Technology Roadmap: Carbon Capture and Storage.* 2013. Available online: https://insideclimatenews.org/sites/default/files/IEA-CCS%20Roadmap.pdf (accessed on 8 March 2020).
3. NOAA National Centers for Environmental Information. State of the Climate: Global Climate Report for September 2018. 2018. Available online: https://www.ncdc.noaa.gov/sotc/global/201809 (accessed on 8 March 2020).
4. Hoegh-Guldberg, O.; Jacob, D.; Taylor, M.; Bindi, M.; Brown, S.; Camilloni, I.; Diedhiou, A.; Djalante, R.; Ebi, K.L.; Engelbrecht, F.; et al. Impacts of 1.5 °C Global Warming on Natural and Human Systems. In *Global Warming of 1.5 °C. An IPCC Special Report on the Impacts of Global Warming of 1.5 °C above Pre-Industrial Levels and Related Global Greenhouse Gas Emission Pathways, in the Context of Strengthening the Global Response to the Threat of Climate Change, Sustainable Development, and Efforts to Eradicate Poverty*; IPCC Secretariat: Geneva, Switzerland, 2018.
5. IPCC. Annex II: Glossary. In *Climate Change 2014: Synthesis Report. Contribution of Working Groups I, II and III to the Fifth Assessment Report of the Intergovernmental Panel on Climate Change*; Core Writing Team, Pachauri, R.K., Meyer, L.A., Eds.; IPCC: Geneva, Switzerland, 2014; pp. 117–130.
6. WHO. *COP24 Special Report: Health and Climate Change*; World Health Organization: Geneva, Switzerland, 2018; Available online: http://apps.who.int/iris (accessed on 9 March 2020).
7. European Commission. 2030 Climate & Energy Framework. 2014. Available online: https://ec.europa.eu/clima/policies/strategies/2030_en (accessed on 8 March 2020).

8. European Commission. Paris Agreement 2015. Available online: https://ec.europa.eu/clima/policies/international/negotiations/paris_en (accessed on 8 March 2020).

9. Harrison, D.P. Sorption-Enhanced Hydrogen Production: A Review. *Ind. Eng. Chem. Res.* **2008**, *47*, 6486–6501. [CrossRef]

10. Jain, S.; Jain, S.; Wolf, I.T.; Lee, J.; Tong, Y.W. A comprehensive review on operating parameters and different pretreatment methodologies for anaerobic digestion of municipal solid waste. *Renew. Sustain. Energy Rev.* **2015**, *52*, 142–154. [CrossRef]

11. Ryckebosch, E.; Drouillon, M.; Vervaeren, H. Techniques for transformation of biogas to biomethane. *Biomass Bioenergy* **2011**, *35*, 1633–1645. [CrossRef]

12. EBA—European Biomass Association. EBA Biomethane & Biogas Report 2015. 2015. Available online: http://european-biogas.eu/2015/12/16/biogasreport2015/ (accessed on 8 March 2020).

13. Favre, E. Carbon dioxide recovery from post-combustion processes: Can gas permeation membranes compete with absorption? *J. Memb. Sci.* **2007**, *294*, 50–59. [CrossRef]

14. Khan, I.U.; Othman, M.H.D.; Hashim, H.; Matsuura, T.; Ismail, A.F.; Rezaei-DashtArzhandi, M.; Azelee, I.W. Biogas as a renewable energy fuel—A review of biogas upgrading, utilisation and storage. *Energy Convers. Manag.* **2017**, *150*, 277–294. [CrossRef]

15. Tuinier, M.J.; van Sint Annaland, M.; Kramer, G.J.; Kuipers, J.A.M. Cryogenic CO_2 capture using dynamically operated packed beds. *Chem. Eng. Sci.* **2010**, *65*, 114–119. [CrossRef]

16. Olajire, A.A. CO_2 capture and separation technologies for end-of-pipe applications—A review. *Energy* **2010**, *35*, 2610–2628. [CrossRef]

17. Mondal, M.K.; Balsora, H.K.; Varshney, P. Progress and trends in CO_2 capture/separation technologies: A review. *Energy* **2012**, *46*, 431–441. [CrossRef]

18. Haefeli, S.; Bosi, M.; Philibert, C. Carbon Dioxide Capture and Storage Issues. Accounting and Baselines under the United Nations Framework Convention on Climate Change (UNFCCC). IEA Information Paper. Available online: https://www.osti.gov/etdeweb/biblio/20543208 (accessed on 8 March 2020).

19. Quintella, C.M.; Hatimondi, S.A.; Musse, A.P.S.; Miyazaki, S.F.; Cerqueira, G.S.; De Araujo Moreira, A. CO_2 capture technologies: An overview with technology assessment based on patents and articles. *Energy Procedia* **2011**, *4*, 2050–2057. [CrossRef]

20. Zhao, Z.; Cui, X.; Ma, J.; Li, R. Adsorption of carbon dioxide on alkali-modified zeolite 13X adsorbents. *Int. J. Greenh. Gas. Control* **2007**, *1*, 355–359. [CrossRef]

21. Liu, Y.; Lin, X.; Wu, X.; Liu, M.; Shi, R.; Yu, X. Pentaethylenehexamine loaded SBA-16 for CO_2 capture from simulated flue gas. *Powder Technol.* **2017**, *318*, 186–192. [CrossRef]

22. Ferella, F.; Puca, A.; Taglieri, G.; Rossi, L.; Gallucci, K. Separation of carbon dioxide for biogas upgrading to biomethane. *J. Clean. Prod.* **2017**, *164*, 1205–1218. [CrossRef]

23. Dawson, R.; Cooper, A.I.; Adams, D.J. Chemical functionalization strategies for carbon dioxide capture in microporous organic polymers. *Polym. Int.* **2013**, *62*, 345–352. [CrossRef]

24. Kenarsari, S.D.; Yang, D.; Jiang, G.; Zhang, S.; Wang, J.; Russell, A.G.; Weif, Q.; Fan, M. Review of recent advances in carbon dioxide separation and capture. *RSC Adv.* **2013**, *3*, 22739–22773. [CrossRef]

25. Philibert, C. Technology Penetration and Capital Stock Turnover. Lessons from IEA Scenario Analysis. France. 2007. Available online: http://www.oecd.org/dataoecd/55/52/38523883.pdf (accessed on 8 March 2020).

26. Chakravarti, S.; Gupta, A.; Hunek, B. Advanced Technology for the Capture of Carbon Dioxide from Flue Gases. In Proceedings of the First National Conference on Carbon Sequestration, Washington, DC, USA, 14–17 May 2001; U.S. Department of Energy National Energy Technology Laboratory: Washington, DC, USA, 2001; pp. 1–11. [CrossRef]

27. IPCC. *Carbon Dioxide Capture and Storage*; Metz, B., Davidson, O., de Coninck, H., Loos, M., Meyer, L., Eds.; Cambridge University Press: New York, NY, USA, 2005; 431p, Available online: https://archive.ipcc.ch/pdf/special-reports/srccs/srccs_wholereport.pdf (accessed on 8 March 2020).

28. Wang, M.; Lawal, A.; Stephenson, P.; Sidders, J.; Ramshaw, C. Post-combustion CO_2 capture with chemical absorption: A state-of-the-art review. *Chem. Eng. Res. Des.* **2011**, *89*, 1609–1624. [CrossRef]

29. Titirici, M.M.; White, R.J.; Falco, C.; Sevilla, M. Black perspectives for a green future: Hydrothermal carbons for environment protection and energy storage. *Energy Environ. Sci.* **2012**, *5*, 6796–6822. [CrossRef]

30. Berge, N.D.; Ro, K.S.; Mao, J.; Flora, J.R.V.; Chappell, M.A.; Bae, S. Hydrothermal carbonization of municipal waste streams. *Environ Sci. Technol.* **2011**, *45*, 5696–5703. [CrossRef]

31. Uddin, H.M.; Reza, M.T.; Lynam, J.G.; Coronella, J.C. Effects of Water Recycling in Hydrothermal Carbonization of Loblolly Pine. *Environ. Prog. Sustain.* **2013**, *33*, 1309–1315. [CrossRef]

32. Stemann, J.; Ziegler, F. Hydrothermal Carbonisation (Htc): Recycling of Process Water. In Proceedings of the 19th European Biomass Conferences and Exhibition, Berlin, Germany, 6–10 June 2011.

33. Gallifuoco, A.; Taglieri, L.; Scimia, F.; Papa, A.A.A.; Di Giacomo, G. Hydrothermal carbonization of Biomass: New experimental procedures for improving the industrial Processes. *Bioresour Technol.* **2017**, *244*, 160–165. [CrossRef]

34. Gallifuoco, A.; Taglieri, L.; Papa, A.A.; Scimia, F.; Di Giacomo, G. Hydrothermal conversions of waste biomass: Assessment of kinetic models using liquid-phase electrical conductivity measurements. *Waste Manag.* **2018**, *77*, 586–592. [CrossRef] [PubMed]

35. Mariscal, R.; Maireles-Torres, P.; Ojeda, M.; Sádaba, I.; López Granados, M. Furfural: A renewable and versatile platform molecule for the synthesis of chemicals and fuels. *Energy Environ. Sci.* **2016**, *9*, 1144–1189. [CrossRef]

36. Centi, G.; Lanzafame, P.; Perathoner, S. Analysis of the alternative routes in the catalytic transformation of lignocellulosic materials. *Catal. Today* **2011**, *167*, 14–30. [CrossRef]

37. Reza, M.T.; Andert, J.; Wirth, B.; Busch, D.; Pielert, J.; Lynam, J.G.; Mumme, J. Hydrothermal Carbonization of Biomass for Energy and Crop Production. *Appl. Bioenergy* **2014**, *1*, 11–29. [CrossRef]

38. Kruse, A.; Funke, A.; Titirici, M.M. Hydrothermal conversion of biomass to fuels and energetic materials. *Curr. Opin. Chem. Biol.* **2013**, *17*, 515–521. [CrossRef] [PubMed]

39. Román, S.; Valente Nabais, J.M.; Ledesma, B.; González, J.F.; Laginhas, C.; Titirici, M.M. Production of low-cost adsorbents with tunable surface chemistry by conjunction of hydrothermal carbonization and activation processes. *Microporous Mesoporous Mater.* **2013**, *165*, 127–133. [CrossRef]

40. Sevilla, M.; Fuertes, A.B. Sustainable porous carbons with a superior performance for CO_2 capture. *Energy Environ. Sci.* **2011**, *4*, 1765–1771. [CrossRef]

41. Farjon, A. *Abies alba*. The IUCN Red List of Threatened Species 2017: E.T42270A83978869. Available online: https://www.iucnredlist.org/species/42270/83978869 (accessed on 2 March 2020).

42. Brunauer, S.; Emmett, P.H.; Teller, E. Adsorption of Gases in Multimolecular Layers. *J. Am. Chem. Soc.* **1938**, *60*, 309–319. [CrossRef]

43. Barrett, E.P.; Joyner, L.G.; Halenda, P.P. The Determination of Pore Volume and Area Distributions in Porous Substances. I. Computations from Nitrogen Isotherms. *J. Am. Chem. Soc.* **1951**, *73*, 373–380. [CrossRef]

44. Thommes, M.; Kaneko, K.; Neimark, A.V.; Olivier, J.P.; Rodriguez-Reinoso, F.; Rouquerol, J.; Sing, K.S.W. Physisorption of gases, with special reference to the evaluation of surface area and pore size distribution (IUPAC Technical Report). *Pure Appl. Chem.* **2015**, *87*, 1051–1069. [CrossRef]

45. Di Felice, L.; Foscolo, P.; Gibilaro, L.G. CO_2 capture by dolomite particles in a gas fluidized bed: Experimental Data and Numerical Simulations. *Int. J. Chem. React. Eng.* **2011**, *9*, 1542–6580. [CrossRef]

46. Ruayruay, W.; Khongtong, S. Impregnation of Natural Rubber into Rubber Wood: A Green Wood Composite. *BioResources* **2014**, *9*. [CrossRef]

47. Rackley, S.A. 7-Adsorption capture systems. In *Carbon Capture and Storage*, 2nd ed.; Rackley, S.A., Ed.; Butterworth-Heinemann: Oxford, UK, 2017; ISBN 9780128120415. [CrossRef]

48. Kikkinides, E.S.; Yang, R.T.; Cho, S.H. Concentration and recovery of carbon dioxide from flue gas by pressure swing adsorption. *Ind. Eng. Chem. Res.* **1993**, *32*, 2714–2720. [CrossRef]

49. Przepiórski, J.; Skrodzewicz, M.; Morawski, A. High temperature ammonia treatment of activated carbon for enhancement of CO_2 adsorption. *Appl. Surf. Sci.* **2004**, *225*, 235–242. [CrossRef]

50. Meng, L.-Y.; Park, S.-J. Effect of heat treatment on CO_2 adsorption of KOH-activated graphite nanofibers. *J. Colloid Interface Sci.* **2010**, *352*, 498–503. [CrossRef]

51. Pevida, C.; Drage, T.C.; Snape, C.E. Silica-templated melamine–formaldehyde resin derived adsorbents for CO_2 capture. *Carbon* **2008**, *46*, 1464–1474. [CrossRef]

52. Plaza, M.G.; Pevida, C.; Arias, B.; Fermoso, J.; Rubiera, F.; Pis, J.J. A comparison of two methods for producing CO_2 capture adsorbents. *Energy Procedia* **2009**, *1*, 1107–1113. [CrossRef]

Appl. Sci. **2020**, *10*, 1879

53. Hao, G.-P.; Li, W.-C.; Qian, D.; Lu, A.-H. Rapid Synthesis of Nitrogen-Doped Porous Carbon Monolith for CO_2 Capture. *Adv. Mater.* **2010**, *22*, 853–857. [CrossRef]

54. Wahby, A.; Ramos-Fernández, J.M.; Martínez-Escandell, M.; Sepúlveda-Escribano, A.; Silvestre-Albero, J.; Rodríguez-Reinoso, F. High-Surface-Area Carbon Molecular Sieves for Selective CO_2 Adsorption. *ChemSusChem* **2010**, *3*, 974–981. [CrossRef]

 *applied sciences*

Article

Determination of a Complete Conversion Model for Gasification of Lignite Char

Christian Heinze *, Eric Langner, Jan May and Bernd Epple

TU Darmstadt, Institute for Energy Systems and Technology, 64287 Darmstadt, Germany;
Eric.Langner@est.tu-darmstadt.de (E.L.); jan.may@est.tu-darmstadt.de (J.M.); bernd.epple@est.tu-darmstadt.de (B.E.)
* Correspondence: christian.heinze@est.tu-darmstadt.de; Tel.: +49-6151/16-23130; Fax: +49-(0)-6151/16-22690

Received: 14 February 2020; Accepted: 6 March 2020; Published: 11 March 2020

Abstract: The conversion of solid fuels via gasification is a viable method to produce valuable fuels and chemicals or electricity while also offering the option of carbon capture. Fluidized bed gasifiers are most suitable for abundantly available low-rank coal. The design of these gasifiers requires well-developed kinetic models of gasification. Numerous studies deal with single aspects of char gasification, like influence of gas compositions or pre-treatment. Nevertheless, no unified theory for the gasification mechanisms exists that is able to explain the reaction rate over the full range of possible temperatures, gas compositions, carbon conversion, etc. This study aims to demonstrate a rigorous methodology to provide a complete char gasification model for all conditions in a fluidized bed gasifier for one specific fuel. The non-isothermal thermogravimetric method was applied to steam and CO_2 gasification from 500 °C to 1100 °C. The inhibiting effect of product gases H_2 and CO was taken into account. All measurements were evaluated for their accuracy with the Allan variance. Two reaction models (i.e., Arrhenius and Langmuir–Hinshelwood) and four conversion models (i.e., volumetric model, grain model, random pore model and Johnson model) were fitted to the measurement results and assessed depending on their coefficient of determination. The results for the chosen char show that the Langmuir–Hinshelwood reaction model together with the Johnson conversion model is most suitable to describe the char conversion for both steam and CO_2 gasification of the tested lignite. The coefficient of determination is 98% and 95%, respectively.

Keywords: gasification; kinetic model; conversion model; reaction model; low-rank coal

---

1. Introduction

The electric power sector contributes to about a quarter of the total CO_2 emissions worldwide. Therefore, in most mitigation scenarios for climate change the share of low-carbon electricity supply (comprising renewable energy, nuclear and carbon capture and storage) increases from the current share of approximately 30% to more than 80% by 2050 [1].

A power plant based on integrated gasification combined cycle (IGCC) is a very suitable addition to any future power system, because it offers the possibility to capture CO_2 in a very efficient pre-combustion process. Furthermore, in a poly-generation configuration, this technology is able to accommodate the intermittent renewable power generation from wind and solar and operate the gasification island at full load by producing synthetic chemical products like hydrogen, SNG, methanol, and Fischer–Tropsch fuels. For high-ash and low-rank coals, fluidized bed gasifiers are especially suitable [2].

The rate of char gasification is the limiting step in gasifiers and most relevant for determining residence times of the particles and size of the reactors. Therefore, an understanding of the mechanics of char gasification for the chosen fuel is essential for the design of gasifiers. Considerable work has been done already in the field of char gasification processes. Irfan et al. [3] did a comprehensive review on

CO$_2$ gasification of coal regarding different factors of influence like coal rank, pressure, gas composition, temperature, and mineral matter. In this study, it was concluded that CO$_2$ gasification characteristics are hard to conclude with full authenticity and the researchers observed those differently for a variety of coals. Generally, the same is true for steam gasification [4]. Ye et al [5] investigated the kinetics and reactivity of two South Australian low-rank coals and quantified the reaction rate for steam and CO$_2$ gasification as well as the influence of mineral content and particle size. Nevertheless, the carbon conversion in the presented data never exceeded 70% and the inhibiting effect of the products has not been included in the model. Another study by Huang et al. [6] focused on the influence of H$_2$ and CO at different temperatures, but only worked with the reaction rate at 50% carbon conversion and omitted a comparison of different conversion models. Fermoso et al. [7] used non-isothermal experiments to determine a suitable conversion model and made statements on the errors of the models but again omitted any inhibiting influence of product gases. Everson et al. [8] assessed the gasification kinetics with steam and CO$_2$ including the influence of the product gases of an inertinite-rich coal with isothermal measurements in a temperature range of 150 K. They used data for almost the complete carbon conversion and validated their assumed kinetic and conversion model with the measurement results but neglected to test other possible models.

It can be stated that most existing work focuses on analysis of single aspects of the gasification, and is not suitable to describe the conversion process in a gasifier completely with its changing gas compositions, temperatures, and particles of varying carbon conversion. For correct prediction of the gasifier behavior with Computational Fluid Dynamic (CFD), a sound modelling of the reaction properties of the fuel is imperative. Therefore, this work aims at demonstrating a methodology to find a complete model for the char conversion during gasification of one specific lignite char that takes into account all relevant temperatures and gas compositions for the full range of char conversion.

2. Theory

2.1. Kinetic Models

Generally, external mass transport from the gas phase to the outer particle surface, the intra-particle diffusion and/or the chemical reaction at the char surface determine the rate of char–gas reactions, depending on temperature and particle properties. For temperatures below 1000 °C and particles in the order of magnitude of 0.1 mm, the reaction rate is controlled by the chemical reaction [9].

Equation (1) is a general expression for the chemical reaction rate, given by Lu et al. [10].

$$\frac{dX}{dt} = k\left(T, \bar{p}_g\right) f(X) \tag{1}$$

Here, k is the apparent reaction rate depending on temperature T and the partial pressures, of the gasifying agents and gas phase products, described by the vector $\bar{p}_g$, according to a reaction model. $f(X)$ describes the change in physical or chemical properties of the char with ongoing char conversion, X, according to a conversion model.

A simple representation of the apparent reaction rate during gasification is the Arrhenius reaction model, which only considers the partial pressure p_g of the gasifying agent and the temperature.

$$k_{Arr}\left(T, p_g\right) = p_g\, k_0\, e^{-\frac{E_a}{RT}} \tag{2}$$

The kinetic parameters for this model are the pre-exponential factor k_0 and the activation energy E_a. The inhibitive influence of the product, which has been observed in several studies [11,12], is considered when applying the Langmuir–Hinshelwood reaction model (L–H model) to the gasification mechanism [13]. Here, the rate-determining step is the formation of occupied sites on the carbon surface. Equations (3) and (4) apply to the CO_2 and H_2O gasification respectively.

$$k_{LH,CO_2}\left(T,\bar{p}_g\right) = \frac{p_{CO_2}\, k_1\, e^{-\frac{E_{a,1}}{RT}}}{1 + p_{CO}\, k_2\, e^{-\frac{E_{a,2}}{RT}} + p_{CO_2}\, k_3\, e^{-\frac{E_{a,3}}{RT}}} \tag{3}$$

$$k_{LH,H_2O}\left(T,\bar{p}_g\right) = \frac{p_{H_2O}\, k_1\, e^{-\frac{E_{a,1}}{RT}}}{1 + p_{H_2}\, k_2\, e^{-\frac{E_{a,2}}{RT}} + p_{H_2O}\, k_3\, e^{-\frac{E_{a,3}}{RT}}} \tag{4}$$

Here, the kinetic parameters are the three pre-exponential factors k_1, k_2 and k_3 and the three activation energies $E_{a,1}$, $E_{a,2}$ and $E_{a,3}$. They have to be determined separately for steam and CO_2 gasification.

In this work, four conversion models for the change in char properties with progressing char conversion are investigated with respect to their applicability for the fuel sample: the volumetric model (VM), the grain model (GM), the random pore model (RPM), which are the most common models used in gasification kinetics [14], and the Johnson model (JM). According to Equation (5), the VM assumes a decreasing reaction surface proportional to the remaining volume or mass of the particle.

$$\frac{dX}{dt} = k\left(T,\bar{p}_g\right)(1-X) \tag{5}$$

In Equation (6), the GM or shrinking core model considers the particles as an assembly of nonporous spheres with constant density and decreasing diameter [15]. The reaction only takes place at the surface.

$$\frac{dX}{dt} = k\left(T,\bar{p}_g\right)(1-X)^{\frac{2}{3}} \tag{6}$$

The RPM was proposed as an semi-empirical model by Bhatia and Perlmutter [16]. It considers arbitrary pore size distributions in the reacting solid and is able to predict a first increasing and then decreasing reaction rate due to the growth and later the coalescence of pores. In the according equation, Equation (7), ψ is a parameter related to the pore structure of the unreacted sample.

$$\frac{dX}{dt} = k\left(T,\bar{p}_g\right)(1-X)\left(1-\psi\ln(1-X)\right)^{0.5} \tag{7}$$

The JM is another semi-empirical approach by Johnson [17].

$$\frac{dX}{dt} = k\left(T,\bar{p}_g\right)(1-X)^{\frac{2}{3}} e^{\alpha X^2} \tag{8}$$

In Equation (8), the term $(1-X)^{\frac{2}{3}}$ is proportional to the effective surface area, as in the shrinking core model, and the term $e^{\alpha X^2}$ represents the relative reactivity of the effective surface area, which decreases with increasing conversion levels.

2.2. Mass Influence

It is commonly known that diffusional effects play a major role, when kinetic studies are performed in thermoscopes. Ollero et al. [18] have shown that the kinetic results of thermo-gravimetric analyzer (TGA) measurements depend on the geometry and the mass of the sample because of the influence on the local partial pressure distribution within the sample. On the other hand, they also showed that the assumption of a constant temperature throughout the sample is applicable without any significant influence on the results. Therefore, the diffusional effect should be incorporated into the kinetic model by only correcting the frequency factor of the reaction. In this work, the mass influence is considered through a correction factor $g(m_0)$, with m_0 as the sample mass. The general rate equation, Equation (1), extends to following form to model the reaction rate in the experiment.

$$\left(\frac{dX}{dt}\right)_{exp} = \frac{dX}{dt}\, g(m_0) \tag{9}$$

3. Experimental

3.1. Fuel Sample and Char Preparation

The raw material used in this work is abundantly available lignite from the Rhenish area, which was pre-processed and pre-dried for the use in a 2300 MW_{th} lignite power plant. The pre-drying was performed via a fluidized bed with internal waste heat utilization [19]. The ultimate and proximate analysis of the fuel sample is shown in Table 1. The mean Sauter diameter of the sample is about 140 μm.

Table 1. Proximate and ultimate analyses of the char samples, oxygen calculated by difference.

Proximate Analysis (wt%)				Ultimate Analysis (wt%, daf)				
Water	Ash	Volatile Matter	Fixed Carbon	C	H	O (Calculated)	N	S
15.3	17.05	37.1	30.55	70.1	4.84	23.12	0.75	1.19

The chars for the TGA experiments were prepared by devolatilizing the raw fuels in a muffle furnace at 900 °C for 7 min according to DIN 51720 or ISO 562:2010, respectively. Generally, the kinetics of the gasification strongly depend on the duration of the pyrolysis and the temperature gradient used during the heat up [20]. Devolatilizing the samples according to DIN 51720 leads to significantly higher heating rates than any possible pyrolysis in the TGA, but still is slower than heating rates one can expect in a fluidized bed gasifier.

3.2. Experimental Setup

The gasification tests were conducted in a TGA (Netzsch STA 449 F3 Jupiter) at atmospheric pressure, which allows the injection of two dry reaction gases and is fitted with a steam generator. The flow rate of gases is controlled with mass flow controllers (MFCs).

The crucibles used were plate-shaped and had a diameter of 17 mm. On these plates, 10 ± 2 mg of the fuel samples was evenly distributed for ideal gas exchange. The temperature of the sample was monitored with a thermocouple in the sample carrier. The systematic error of the mass measurement was mitigated by a correction run for every gas composition with an empty crucible. The correction measurement was then subtracted from the actual measurement.

In the first step, prior to the actual gasification tests, the samples were heated in the TGA to approximately 1100 °C with 20 K/min in a nitrogen atmosphere to ensure a complete drying and devolatilization of the samples. Then, the samples were cooled down to 500 °C and stabilized at this temperature for 30 min. During this time period, the reactive purge gases were injected into the oven to ensure enough time for gas mixing and gas distribution with the oven. Then, the experiments were performed under non-isothermal conditions with a heating rate of 20 K/min up to a temperature of 1100 °C. Finally, this temperature was held constant for 20 min for a complete reaction of the carbon.

The experiments were conducted for CO_2 gasification as well as steam gasification. For both sets of experiments, the gasifying agent was introduced into the oven with 20%, 25%, 33% and 50% of the gas flow. For each gas concentration, a configuration with and without gasifying product, H_2 or CO respectively, was tested. In total, 16 different atmospheres were used for the CO_2 and steam gasification. Tables 2 and 3 display the matrix of the experiment configurations. For each of the configurations, three separate experiments were performed to test the reproducibility of the results.

Table 2. Experiment configurations for CO_2 gasification; N_2 was added to yield a total gas flow of 100 mL/min.

Config. No.	Flow Rate CO_2 (mL/min)	Flow Rate CO (mL/min)
1	20	0
2	20	24
3	25	0
4	25	22.5
5	33	0
6	33	20
7	50	0
8	50	15

Table 3. Experiment configurations for steam gasification; N_2 was added to yield a total gas flow of 100 mL/min.

Config. No.	Flow Rate H_2O (mL/min)	Flow Rate H_2 (mL/min)
1	20	0
2	20	35
3	25	0
4	25	35
5	33	0
6	33	35
7	50	0
8	50	35

In order to define the type of function for the mass influence $g(m_0)$, an additional set of experiments was performed for a constant gas composition but varying sample masses. The experiments are listed in Table 4.

Table 4. Experiment configurations for determination of the sample mass influence; N_2 was added to yield a total gas flow of 100 mL/min.

Config. No.	Sample Mass (mg)	Flow Rate CO_2 (mL/min)
1	2.5	20
2	5	20
3	10	20
4	13	20
5	20	20
6	45	20

3.3. Data Preparation and Evaluation

For each measurement, weight and temperature were recorded with a sampling rate of 300 Hz. The char conversion is calculated depending on starting and final mass for every measurement according to Equation (10).

$$X(m) = \frac{m_0 - m}{m_0 - m_{ash}} \tag{10}$$

The char conversion during the experiment is represented over temperature and time, as well as dX/dt over temperature. To reduce the measurement noise, the signal is smoothed with a first order Savitzky–Golay-Filter [21]. The integration time for the Savitzky–Golay-Filter was obtained based on the method described by Werle et al. [22] through minimizing the Allan variance for each measurement. The standard deviation for the char reaction rate after filtering is in the range of 3.6×10^{-3} min^{-1} and 15.6×10^{-3} min^{-1} with the mean at 8.5×10^{-3} min^{-1}.

The models investigated in this paper were fitted to the measurement with the nonlinear least-squares method and the parameters of the models were calculated by minimizing the objective function OF, Equation (11).

$$OF = \sum_{i=1}^{n} \left(\left(\frac{dX}{dt}\right)_{exp,i} - \left(\frac{dX}{dt}\right)_{calc,i} \right)^2 \tag{11}$$

4. Results and Discussion

4.1. Mass Influence

To gain an understanding of the effect of varying masses in the TGA as explained in Section 2.2, the experiments listed in Table 4 were evaluated using the Arrhenius equation together with all four kinetic models (VM, GM, RPM, and JM). In Figure 1, the frequency factor is plotted over the sample mass for each conversion model. A correction term of an exponential type fits the results for every conversion model to a satisfactory degree ($R^2 \approx 95\%$), with b as the model parameter.

$$g(m_0) = e^{-b\,m_0} \tag{12}$$

With the addition of this correction term, a determination of the intrinsic gasification rate is possible. Therefore, all further experiments were evaluated with the following rate equation, Equation (13), and the value for b is obtained by optimization with Equation (11).

$$\left(\frac{dX}{dt}\right)_{exp} = k\left(T,\bar{p}_g\right) f(X)\, e^{-b\,m_0} \tag{13}$$

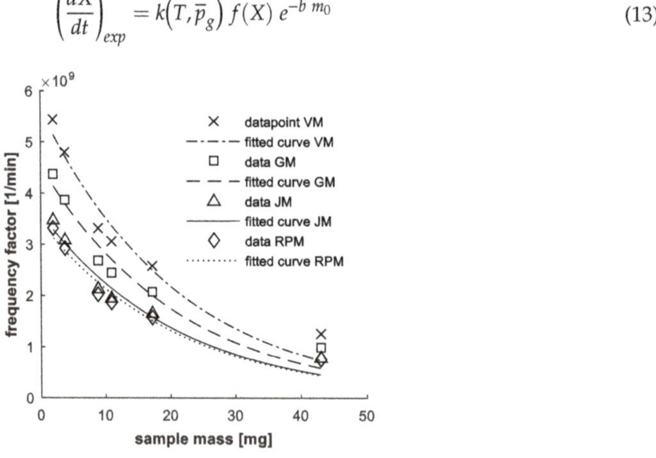

Figure 1. Frequency factor over sample mass for all four conversion models.

4.2. Confirmation of Reaction Model

The kinetic model for the gasification of the char has to describe sufficiently the change in reaction rate with ongoing carbon conversion. In Figure 2, the Arrhenius plot is shown for the CO_2 gasification in configurations No. 2 and 5 as well as for the steam gasification in configuration No. 3 and 8 for the volumetric model.

It can be seen that the gradient of the Arrhenius graph (i.e., activation energy) increases with increasing temperature. Generally, two possible explanations exist for the change in reactivity. Firstly, two separate reactions with different activation energies and frequency factors could determine the gasification of the char at different temperatures, e.g., a catalyzed and a non-catalyzed reaction. In this case, a suitable reaction model must be selected. Secondly, the reactivity of the char increases more with ongoing gasification than the volumetric model predicts. In this case, another conversion model should be used, like the GM, RPM or JM. Only additional runs with different heating rates can make the distinction between those two possible explanations, as stated by Miura et al. [23].

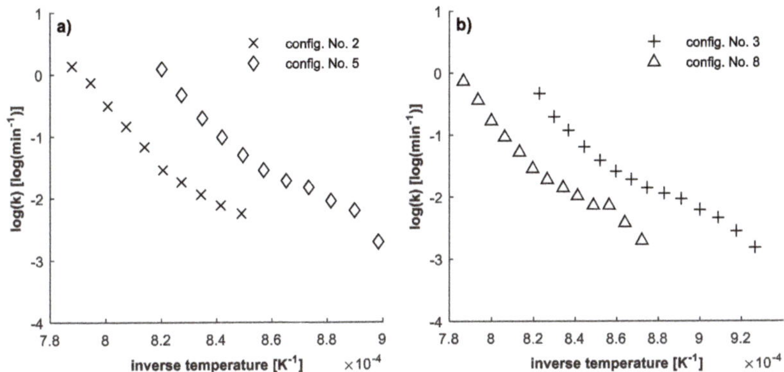

Figure 2. Arrhenius plot for (**a**) CO_2 gasification (configurations No. 2, 5) and (**b**) steam gasification (configuration No. 3, 8) (VM).

The test configuration No. 1 of the CO_2 gasification was repeated with heating rates of 20 K/min and 40 K/min in another oven of the TGA. In Figure 3, one representative run for each heating rate is plotted. The solid lines represent the reaction rate according to the Arrhenius model for both runs. The dashed lines show the reaction rate according to the Arrhenius model, when only inverse temperatures of more than 8.2×10^{-4} K^{-1} are considered. It is apparent that the change in reaction rate happens at different temperatures.

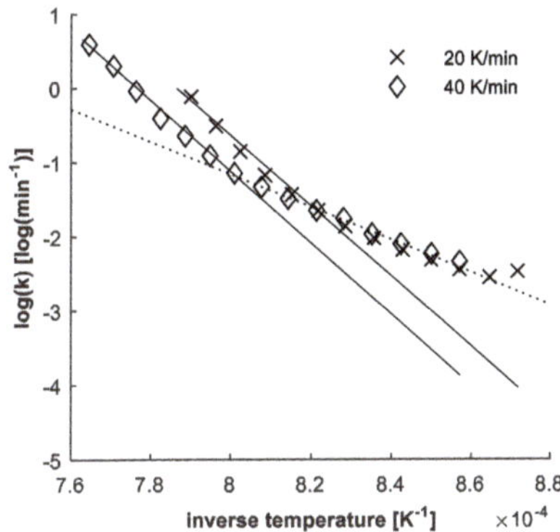

Figure 3. Arrhenius plot for CO_2 gasification in configuration No. 1 with 20 K/min and 40 K/min heating rate.

Additionally, the samples were acid washed according to ISO 602:2015 to remove the mineral content. The acid washed samples were used in CO_2 gasification experiments (Table 2) of the configurations No. 1, 3, 5, and 7. Compared to the non-acid washed samples, the Arrhenius plots have the same shape, but exhibit a shift to lower frequency factors by about 0.15–0.4 log(min^{-1}). This indicates the general catalytic effect of the mineral matter in the ash but without any temperature dependency.

Therefore, it can be concluded that the change in char reactivity has to be explained by a suitable conversion model.

4.3. Determination of Kinetic Parameters

For the determination of the kinetic parameters, the objective function (11) was minimized for the Arrhenius equation and the L–H model in combination with all four kinetic models (VM, RM, RPM and JM) for both the CO_2 gasification as well as the steam gasification. The coefficients of determination R^2 are listed in Table 5. The L–H model generally leads to significantly better results than the Arrhenius model with the coefficient of determination being larger for any given conversion model except from the GM for CO_2 gasification. In this case, R^2 is very similar for the Arrhenius and the L–H model. Regarding the conversion models, the JM is the most suitable for the selected char samples.

Table 5. Coefficient of determination R^2 for all model combinations.

Gasifying Agent	Carbon Dioxide				Steam			
Conversion Model	VM	GM	RPM	JM	VM	GM	RPM	JM
R^2 for Arrhenius Model [%]	57.5	72.1	79.3	78.4	32.7	41.8	53.6	57.3
R^2 for L–H Model [%]	85.1	71.8	90.6	94.8	86.1	92.4	89.4	98.4

Table 6 shows the kinetic parameters for the L–H model, the parameter α of the JM and the parameter b for the mass influence.

Table 6. Results for the L–H reaction model with the Johnson conversion model.

k_1	k_2	k_3	$E_{a,1}$	$E_{a,2}$	$E_{a,3}$	α	b
[kPa^{-1} min^{-1}]	[kPa^{-1}]			[kJ/mol]		[-]	[µg^{-1}]
CO$_2$ Gasification							
3.70×10^8	4.04×10^{-6}	8.73×10^9	236.1	−87.8	256.7	1.41	7.7
Steam Gasification							
1.39×10^{12}	5.14×10^{-3}	3.25×10^{12}	298.5	−39.8	287.5	1.89	10.8

A closer look has to be taken at the activation energy $E_{a,2}$, which is negative for both CO_2 and steam gasification. Negative activation energies have been observed for gasification before [6] and are a hint that the L–H model does not completely describe the reaction mechanisms during gasification. In the case of steam gasification, a possible reason is the hydrogen inhibition through the irreversible adsorption of hydrogen on the active char sites, described by Hüttinger et al. [24].

For both, steam and CO_2 gasification, the influence of ash acting as a catalyst is another explanation. Still, with a coefficient of determination at about 98% and 95% for steam and CO_2 gasification respectively, the confidence intervals are narrow enough for practical use.

Figure 4 exemplarily shows the results of a steam gasification run in test configuration No. 2 with 20 mL/min steam and 35 mL/min hydrogen for a selection of temperatures with their error bar according to the Allan deviation. Additionally, the predicted reaction rates of the L–H model together with all four conversion models are plotted. For the JM, the 95% confidence interval is marked in the plot, too. The measurements show a digressive change in reaction rate between 900 °C and 970 °C that was observed in all measurements with steam. Only the JM satisfactorily models this characteristic, leading to the very high coefficient of determination. The results are in good agreement with the JM model, especially in the relevant range from 750 °C to 950 °C.

Figure 5 shows the respective information for a trial with 33% CO_2 and 20% CO for the L–H reaction model together with all conversion models. Here, the L–H model emulates the measurements for the CO_2 gasification best, too. In addition, similarly to Figure 4, the change in reaction rate decreases between 900 °C and 950 °C. However, for the CO_2 gasification, the JM is not able to model this effect correctly. For very high conversions and temperatures, all models overestimate the reaction rate. These deviations are observed for most CO_2 measurements and lead to a smaller coefficient of determination. Still, the results are satisfying in the relevant temperature range for practical use in fluidized bed gasification.

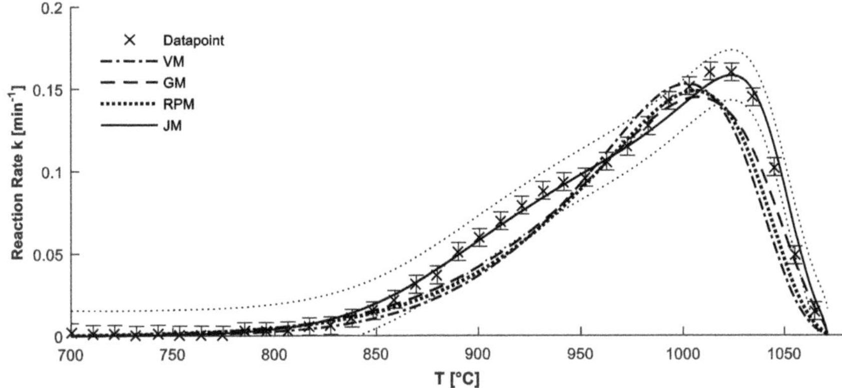

Figure 4. Measurement and L–H model results for steam gasification in configuration No. 2 (20 mL/min H_2O, 35 mL/min H_2); 95% confidence interval for JM is marked with a thin dotted line.

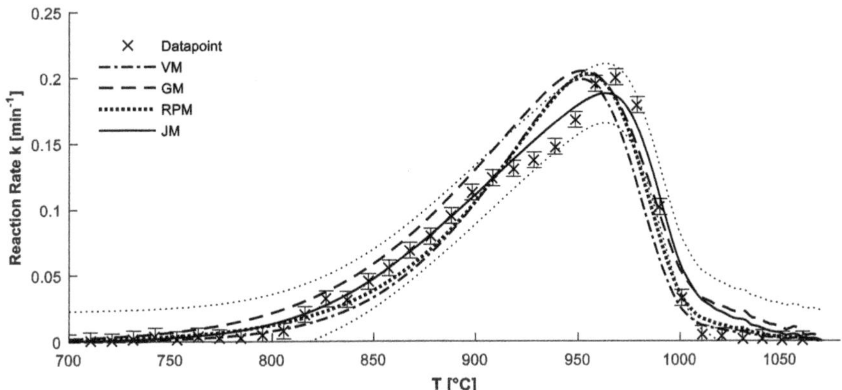

Figure 5. Measurement and L–H model results for CO_2 gasification in configuration No. 6 (33 mL/min CO_2, 20 mL/min CO); 95% confidence interval for JM is marked with a thin dotted line.

5. Conclusions

Within this work, it was possible to demonstrate a suitable approach for determining gasification kinetics of one char with thermogravimetric analysis by the followings steps: determination of the mass influence; determination of possible influence of mineral matter or conversion model; non-isothermal measurement of a representative set of gas compositions; evaluation of measurement error; fitting results to possible models and assessment of model quality.

Hence, the char of a Rhenish lignite was gasified in a TGA at atmospheric pressure under non-isothermal conditions in order to determine the gasification kinetics for steam and CO_2 gasification with the inhibiting effect of H_2 and CO respectively. The measurement data were filtered with a first order Savitzky–Golay-Filter and an optimal integration time determined by means of the Allan variance.

Two reactions models, Arrhenius and Langmuir–Hinshelwood, and four conversion models, the volumetric model (VM), the grain model (GM), the random pore model (RPM) and the Johnson model (JM), were investigated. Furthermore, a model for the influence of the sample mass in a TGA was incorporated to account for mass transfer effects.

With this rigorous approach, it was found that the reaction is best described by the L–H rate equation together with the JM as the conversion model. For both, steam and CO_2 gasification, the activation energies of the reverse reactions are negative. This is a hint that the L–H model does not completely describe the underlying reaction mechanism. Still, for typical environments of fluidized bed gasifiers, the L–H model can be used to predict the reactions rates.

Author Contributions: C.H. is responsible for administration, conceptualization, the original draft, and developed the applied methodology. The experimental investigation were conducted by E.L. J.M. supported the writing process with his reviews and edits. B.E. supervised the research progress and the presented work. All authors have read and agreed to the published version of the manuscript.

Funding: The research leading to these results has received funding from the COORETEC initiative as part of the 6th Program on Energy Research of the Federal Ministry for Economic Affairs and Energy of Germany under grant agreement No. 03ET7048A (FABIENE; Flexible Supply of Electricity and Fuels from Gasification of Lignite in a Fluidized Bed). The funding is gratefully acknowledged.

Conflicts of Interest: The authors declare no conflict of interest.

Abbreviations

IGCC	Integrated Gasification Combined Cycle
CFD	Computational Fluid Dynamic
L–H Model	Langmuir–Hinshelwood Reaction Model
VM	Volumetric Model
GM	Grain Model
RPM	Random Pore Model
JM	Johnson Model
TGA	Thermo-Gravimetric Analyzer

References

1. Edenhofer, O.; Pichs-Madruga, R.; Sokona, Y.; Farahani, E.; Kadner, S.; Seyboth, K.; Adler, A.; Baum, I.; Brunner, S.; Eickemeier, P. IPCC, 2014: Summary for policymakers—Sectoral and cross-sectoral mitigation pathways and measures. *Clim. Chang.* **2014**, 17–26.
2. Heinze, C.; Peters, J.; Ströhle, J.; Epple, B.; Hannes, J.; Ullrich, N.; Wulcko, I. Polygeneration von Strom und Kraftstoffen basierend auf der Wirbelschichtvergasung von Braunkohle. *DGMK Tag.* **2016**, *2*, 205–213.
3. Irfan, M.F.; Usman, M.R.; Kusakabe, K. Coal gasification in CO_2 atmosphere and its kinetics since 1948: Abrief review. *Energy* **2011**, *36*, 12–40. [CrossRef]
4. Takarada, T.; Tamai, Y.; Tomita, A. Reactivities of 34 coals under steam gasification. *Fuel* **1985**, *64*, 1438–1442. [CrossRef]
5. Ye, D.; Agnew, J.; Zhang, D. Gasification of a South Australian low-rank coal with carbon dioxide and steam: kinetics and reactivity studies. *Fuel* **1998**, *77*, 1209–1219. [CrossRef]
6. Huang, Z.; Zhang, J.; Zhao, Y.; Zhang, H.; Yue, G.; Suda, T.; Narukawa, M. Kinetic studies of char gasification by steam and CO_2 in the presence of H_2 and CO. *Fuel Process. Technol.* **2010**, *91*, 843–847. [CrossRef]
7. Fermoso, J.; Gil, M.V.; Pevida, C.; Pis, J.; Rubiera, F. Kinetic models comparison for non-isothermal steam gasification of coal–biomass blend chars. *Chem. Eng. J.* **2010**, *161*, 276–284. [CrossRef]
8. Everson, R.C.; Neomagus, H.W.; Kasaini, H.; Njapha, D. Reaction kinetics of pulverized coal-chars derived from inertinite-rich coal discards: gasification with carbon dioxide and steam. *Fuel* **2006**, *85*, 1076–1082. [CrossRef]
9. Laurendeau, N.M. Heterogeneous kinetics of coal char gasification and combustion. *Prog. Energy Combust. Sci.* **1978**, *4*, 221–270. [CrossRef]
10. Lu, G.; Do, D. Comparison of structural models for high-ash char gasification. *Carbon* **1994**, *32*, 247–263. [CrossRef]

11. Barrio, M.; Hustad, J. CO$_2$ gasification of birch char and the effect of CO inhibition on the calculation of chemical kinetics. *Prog. Thermochem. Biomass Convers.* **2008**, 47–60.

12. Lussier, M.; Zhang, Z.; Miller, D.J. Characterizing rate inhibition in steam/hydrogen gasification via analysis of adsorbed hydrogen. *Carbon* **1998**, *36*, 1361–1369. [CrossRef]

13. Ergun, S. *Kinetics of the Reactions of Carbon Dioxide and Steam with Coke*; US Government Printing Office: Washington, DC USA, 1962.

14. Molina, A.; Mondragon, F. Reactivity of coal gasification with steam and CO$_2$. *Fuel* **1998**, *77*, 1831–1839. [CrossRef]

15. Szekely, J.; Evans, J. A structural model for gas—Solid reactions with a moving boundary. *Chem. Eng. Sci.* **1970**, *25*, 1091–1107. [CrossRef]

16. Bhatia, S.K.; Perlmutter, D. A random pore model for fluid-solid reactions: I. Isothermal, kinetic control. *AIChE J.* **1980**, *26*, 379–386. [CrossRef]

17. Johnson, J. Kinetics of bituminous coal char gasification with gases containing steam and hydrogen. *Am. Chem. Soc. Coal Gasification (USA)* **1974**, *10*, 145–178.

18. Ollero, P.; Serrera, A.; Arjona, R.; Alcantarilla, S. Diffusional effects in TGA gasification experiments for kinetic determination. *Fuel* **2002**, *81*, 1989–2000. [CrossRef]

19. Klutz, H.-J.; Moser, C.; Block, D. Development status of WTA fluidized-bed drying for lignite at RWE Power AG. *Power Plant Technol. Secure Sustain. Energy Supply* **2010**, *2*, 427–444.

20. Liu, H.; Kaneko, M.; Luo, C.; Kato, S.; Kojima, T. Effect of pyrolysis time on the gasification reactivity of char with CO$_2$ at elevated temperatures. *Fuel* **2004**, *83*, 1055–1061. [CrossRef]

21. Savitzky, A.; Golay, M.J. Smoothing and differentiation of data by simplified least squares procedures. *Anal. Chem.* **1964**, *36*, 1627–1639. [CrossRef]

22. Werle, P.; Mücke, R.; Slemr, F. The limits of signal averaging in atmospheric trace-gas monitoring by tunable diode-laser absorption spectroscopy (TDLAS). *Appl. Phys. B Lasers Opt.* **1993**, *57*, 131–139. [CrossRef]

23. Miura, K.; Silveston, P.L. Analysis of gas-solid reactions by use of a temperature-programmed reaction technique. *Energy Fuels* **1989**, *3*, 243–249. [CrossRef]

24. Hüttinger, K.J.; Merdes, W.F. The carbon-steam reaction at elevated pressure: formations of product gases and hydrogen inhibitions. *Carbon* **1992**, *30*, 883–894. [CrossRef]

 *applied sciences*

Article

Comparison of Equilibrium-Stage and Rate-Based Models of a Packed Column for Tar Absorption Using Vegetable Oil

Adel Almoslh *, Falah Alobaid, Christian Heinze and Bernd Epple

Institut Energiesysteme und Energietechnik, Technische Universität Darmstadt, Otto-Berndt-Straße 2, 64287 Darmstadt, Germany; falah.alobaid@est.tu-darmstadt.de (F.A.); christian.heinze@est.tu-darmstadt.de (C.H.); bernd.epple@est.tu-darmstadt.de (B.E.)
* Correspondence: adel.almoslh@est.tu-darmstadt.de; Tel.: +49-6151/16-23004; Fax: +49-(0)-6151/16-22690

Received: 18 February 2020; Accepted: 18 March 2020; Published: 30 March 2020

Abstract: In this study two mathematical models, rate-based and equilibrium-stage models in Aspen Plus process simulator, were used to simulate the tar absorption processes using soybean oil as a solvent in a research lab-scale experiment. The matching between simulation results and experimental data shows a good agreement. The simulation results predicted by the rate-based model show a higher level of agreement than the equilibrium model compared with the experimental data. Analysis study of tar absorption process was carried out which revealed the effect of temperature and flow rate on the soybean oil, and height-packed bed on tar removal efficiency. The methodology of selecting the optimum (most economical) operation conditions has also been performed in this study.

Keywords: gasification; tar absorption; process simulation; validation study; sensitivity analyses

1. Introduction

Gasification technology makes the biomass a vital source of energy. It converts the biomass into raw syngas that is a fuel gas mixture consisting primarily of hydrogen, carbon monoxide, tar, and other gases which are considered contaminants such as hydrogen chloride, carbon dioxide, hydrogen sulfide. The tar in gasifier may create fouling and soot accumulation in downstream processes, moreover, tar solubility in the water may generate wastewater difficulties [1]. In literature, many definitions for tar have been reported. All the definitions seek to present a view about the nature of the tar. Besides, these definitions are influenced by the gas quality specifications required for a particular end-use application and how the tar is assembled and analyzed [2]. One of the definitions of the tar was described by Milne et al. [3] as follows: It is the organics components that are created under thermal or partial-oxidation process (gasification) and are supposed to be mostly aromatic. Devi et al. [4] defined the tar as a complex blend of condensable hydrocarbons, which comprises single to multiple ring aromatic compounds along with other oxygen-containing hydrocarbons and complex polycyclic aromatic hydrocarbons. The Energy Research Centre of the Netherlands (ECN) considers that the tar consists of organic molecules, which have a higher molecular weight than benzene (benzene is not considered to be tar) [5]. According to Unger et al. [6], tar is a mix of the hydrocarbons that can form liquid or highly viscous to solid accumulation by dropping the temperature of the gaseous phase down to ambient temperature; it consists of carbon, hydrogen, and other organic linked elements such as oxygen (O), nitrogen (N), or sulfur (S). The tar can be classified based on different criteria. Li et al. [7] listed tar into five classes: GC-undetectable, heterocyclic aromatics, light aromatic (one ring), light PAH compounds (two to three rings), and heavy PAH compounds (four to seven rings).

Several studies [3,8–10] published that the tar can also be classified into primary, secondary, and tertiary tar. Wolfesberger et al. [10] described how the tar components are created and what is the effect of temperature on the nature of the formed tar components. The primary tar components begin to appear during the pyrolysis process, the complex polymers that make the main parts of biomass (cellulose, hemicellulose and lignin) are broken down from cellulose and hemicellulose, tar components like alcohols, ketones, aldehydes, or carbon acids are formed, whereas bi-and trifunctional monoaromatics mostly substituted phenols are derived from lignin [10]. By growing temperature and attendance of an oxidant, a portion of the cellulose-contributed primary tars convert to small gaseous molecules, the remaining primary tar creates secondary tar, examples for secondary tar components are like alkylated mono- and diaromatics including heteroaromatics such as pyridine, furan, dioxin, and thiophene [10]. At a temperature above 800 °C, components such as benzene, naphthalene, phenanthrene, pyrene, and benzopyrene (polynuclear aromatic hydrocarbons (PAH)) are created; these components form the tertiary tars components [10].

1.1. Tar Treatments

The methods of tar removal can be categorized into primary and secondary measures based on the place where tar is removed [11]. With the primary methods, the tar is removed by applying processes such as thermal or catalytic cracking in the gasifier itself, while the secondary methods, the tar is separated outside the gasifier [2]. Although the primary methods have some disadvantages such as the complex construction of the gasifier and the limited flexibility of feedstock, it promises high tar removal efficiency by promoting this technology with time [2]. However, Figure 1 shows that a combination of both methods can only achieve high tar removal efficiency. Currently, secondary methods are fitting for tar separation from the produced syngas because of their low cost and simple measures [12]. Wet scrubbing process is one of the secondary methods, which applied an absorber to exclude the tar. The absorber can be a plate or packed column. It is recommended to use the packed absorber because of its high capacity [13].

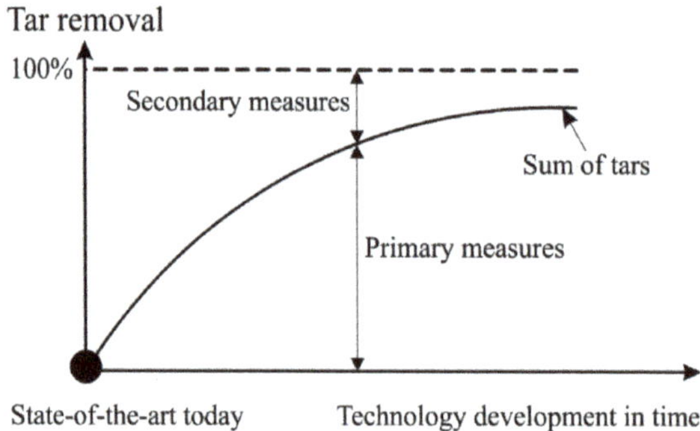

Figure 1. Illustration of the need for primary and secondary measures [2].

Furthermore, the packed absorber can operate with lower overall pressure drops than the tray columns [14]. The packed absorber materials are categorized into random or structured packing. The modern random packings had a different wide range of geometries and shapes and are made from ceramic, metal, or plastics. The structured packings are ideal for lower pressures (i.e., less than 2 bar) and lower liquid rates (i.e., less than 50 m^3/m^2·h) [14].

A suitable solvent must be appropriately selected for the absorption process since the solvent type has a significant influence on equipment sizing and operating costs [15]. Phuphuakrat et al. [16] summarized that the absorption process should concentrate on separating the components of the tar that cause the fouling problem. These components as per tar classification of Bergman et al. [17] are heterocyclic compounds, light polycyclic aromatic hydrocarbons such as naphthalene, and the heavier hydrocarbons that condensate easily. According to Phuphuakrat et al. [16], light aromatic hydrocarbon tares (one ring aromatic hydrocarbon) are not the reason for blocking and fouling problems, they studied some scrubbing liquids such as diesel fuel, vegetable oil, engine oil, and water as a solvent to remove the tar. The removal efficiencies by using these solvents are shown in Table 1.

Table 1. Absorption efficiencies of tar components by different solvents (%) [16].

Absorbent	Water	Diesel Fuel	Biodiesel Fuel	Vegetable Oil	Engine Oil
Benzene	24.1	77.0	86.1	77.6	61.7
Toluene	22.5	63.2	94.7	91.1	82.3
Xylene	22.1	−730.1	97.8	96.4	90.7
Styrene	23.5	57.7	98.1	97.1	91.1
Phenol	92.8	−111.1	99.9	99.7	97.7
Indene	28.2	97.9	97.2	97.6	88.7
Naphthalene	38.9	97.4	90.3	93.5	76.2

From Table 1, it can be observed that diesel fuel is the most effective solvent used to remove naphthalene. However, diesel is considered an uneconomic solvent because of its simple evaporation, which raises the losses of the solvent [16]. The vegetable oil has proven to be efficient to separate naphthalene [12]. The water has comparatively high removal efficiency for phenol because the phenol is a hydrophilic component and it can lose H+ (ion) from a hydroxyl group, whereas the other components are nonpolar substances [12]. Applying water as a solvent to remove the tar achieves removal efficiency of about 31.8%, however water is not as effective solvent since the tar has a low solubility in water and the separation of the tar from the water is difficult and expensive [12]. Phuphuakrat et al. [16] placed the efficiency of the solvent as: vegetable oil > engine oil > water > diesel fuel. Paethanom et al. [18] published the tar removal efficiency of vegetable oil as 89.8% and cooking oil as 81.4%. Bhoi [19] investigated the effect of two kinds of vegetable oils, namely soybean and canola oil to separate the tar. The author summarized that there is no significant difference between the soybean and canola oils for all the conditions of absorbent like temperatures and volumetric flow rates. Ozturk et al. [20] analyzed the relationship between the operating time and removal efficiency of some oily solvents like benzene and toluene. They concluded that the removal efficiency declines with time because of increasing the tar concentration in the absorbent.

1.2. Modelling Approaches for the Packed Column Used for Absorption Processes

Mathematical models contribute to a better understanding of the process and play an essential role in enhancing plant efficiency. A recent literature review tells that there are several studies concerning modelling the absorption process. These studies are based on two standard models: the equilibrium model and the rate-based model. The equilibrium-stage model is developed by Mofarahi et al. [21], whereas the rate-based model depends on the early work published by Pandya [22], who presented a model for rate-based CO_2 absorption. Recently the rate-based model and equilibrium model of CO_2 capture with Amin solvent in a packed column have been investigated by several authors. Many studies [23–25] applied the rate-based model for studying CO_2 absorption. Afkhamipour et al. [26] compared the rate-based and equilibrium models for CO_2 capturing with AMP solution in a packed column. Bhoi [19] employed equilibrium model to explain the experimental data for absorption tar by vegetable oil. The author tested two vegetable oils as absolvents namely soybean oil and canola oil.

To the best of our knowledge, few studies have been published regarding the modelling of tar absorption with vegetable oil in a packed column. Most of these studies have concentrated on modelling absorption process for CO_2 capture, while the modelling of the tar absorption process is rarely presented. The objectives of this study are as follows:

Assembling a property package for tar-soybean oil and build a rate-based model as well as equilibrium model by applied Aspen Plus software for simulation of tar absorption using soybean oil as a solvent. Aspen Plus software is used because of its extensive property databanks and rigorous equation solvers.

Validation of both the mathematical models (rate-based and equilibrium stage) against experimental data is carried out at different operation points. The experimental data reported by Bhoi [19] is used for validation of the models.

The accuracy of the results predicted by the two mathematical models (rate-based and equilibrium stage) is compared with the experimental data.

Analysis of tar absorption process is essential to study the process parameters on tar removal efficiencies such as the solvent temperature, flow rate of solvent, and height packed bed.

This study presents a methodology for selecting the optimum (most economical) operating conditions.

This work is a contribution to the knowledge available for modelling studies for tar absorption using vegetable oil as a solvent in the wet packed column.

2. Description of the Experiment

The experimental research accomplished by Bhoi [19] was used as a tar removal unit, illustrated in Figure 2. From Figure 2 the pilot plant comprises two major sections, namely the gas mixing section and an absorption column, the gas mixing section consists of a sequence of instruments, which have different functions. The gas mixing section prepares a simulated gas from air and tar with a specific temperature, pressure, and volumetric flow rate [19]. The air as tar holder is heated to 350 °C to make sure that when the tar injected into the heated air, the liquid tar components are evaporated immediately and carried by the air stream [19]. The wet packed bed scrubbing system consists of a stainless steel column, water bath heater to heat the solvent soybean oil or canola to a specific temperature, and a peristaltic pump to recycle the solvent to the absorption column [19]. The designed internal diameter of the column is 50 mm, and the height of the column is 150 cm [19]. The selected packing materials were from kind metal Raschig rings with size 6-mm, a metal material to provide better strength and wettability compared to ceramic and plastic packings [19]. Raschig rings used in the experiment are of size (diameter × length × thickness) 6 × 6 × 0.3 mm respectively, density is 900 kg/m^3, the specific surface area is 900 m^2/m^3, packing factor is 2297 1/m, and the void fraction is 89% [19].

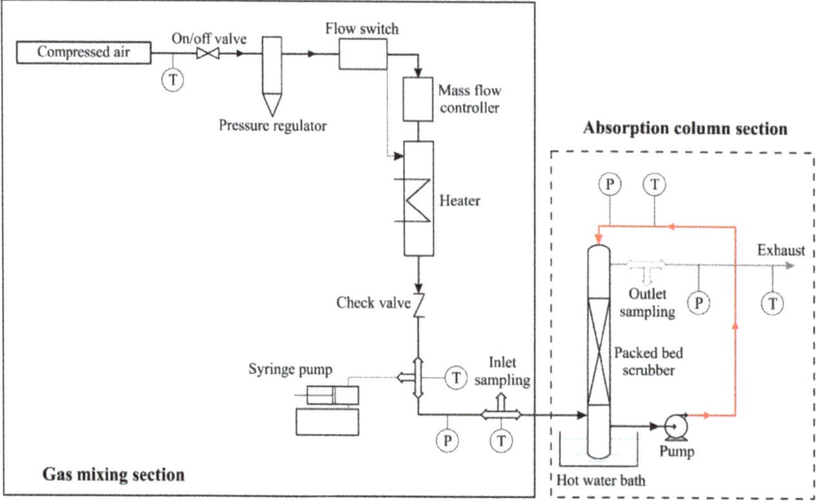

Figure 2. Schematic diagram of a bench-scale wet scrubbing set-up [27].

2.1. Mathematical Models of the Packed Column Used for Tar Absorption Processes

Modelling packed column aims to predict the overall performance of the packed column. The gas with a temperature $T_{V,in}$, flow rate $\dot{N}_V$ and mole fraction $y_{i,in}$ for the component i enters the bottom of the packed column, and it exit with temperature $T_{V,out}$, mass flow rate $\dot{N}_V$, and mole fraction $y_{i,\,out}$ for the component i. The liquid with temperature $T_{L,in}$, flow rate $\dot{N}_L$ mole fraction $x_{i,in}$ for the component i enters the top of the packed column, (countercurrent flow), and it exit with temperature $T_{L,out}$, flow rate $\dot{N}_L$, and mole fraction $x_{i,\,out}$ for the component i. Basically, two common models are used for calculating the absorption process parameters: the equilibrium model and the rate-based model [26].

2.2. Equilibrium Model

The equilibrium model based on the assumption of each plate in the absorption column is considered as a theoretical plate (equilibrium plate) which means that the vapor and the liquid leave any plate at thermodynamic equilibrium [19]. According to Seader et al. [28], the main assumptions of equilibrium model are as follows:

- Phase equilibrium is presented at each stage;
- There are no chemical reactions between the components of vapor and liquid;
- Entrainment of liquid drops in vapor and occlusion of vapor bubbles in the liquid is negligible.

In practice, this equilibrium takes place only at the interfaces between the vapor and liquid phases, so the efficiencies such as point and Murphree efficiencies are used in equilibrium model to account deviations from real equilibrium state [26]. For modelling the whole packed column, the packed bed is divided into stags (equilibrium stage). Figure 3 illustrates the typical entry and exit parameters of equilibrium plate stage. The equilibrium model of absorption process consists of well-known and accepted correlations called MESH equations that include the equations of component material balance, the equations of phase equilibrium, summation equations, and energy balance for each stage as following [21,28,29]:

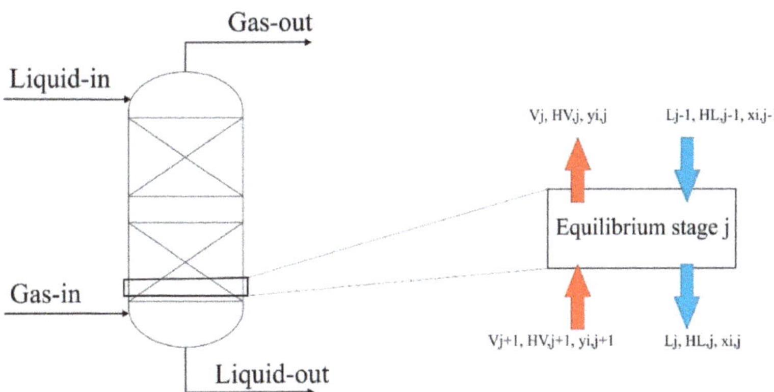

Figure 3. Equilibrium stage model [21].

Overall mass balance for stage j:

$$L_{j-1} - L_j - V_j + V_{j+1} = 0 \tag{1}$$

Component mass balance for stage j:

$$L_{j-1}x_{i,j-1} - L_j x_{i,j} - V_j y_{i,j} + V_{j+1}y_{i,j+1} = 0 \tag{2}$$

Energy balance equations for stage j:

$$L_{j-1}H^L_{j-1} - L_jH^L_j - V_jH^V_j + V_{j+1}H^V_{j+1} = 0 \tag{3}$$

Phase equilibrium

$$y_i - K_i x_i = 0 \tag{4}$$

Summation equations

$$\sum y_i = 1 \tag{5}$$

$$\sum x_i = 1 \tag{6}$$

where L, V are the molar flow rate of liquid and vapor respectively, K is equilibrium ratio, H is enthalpy, x_i, y_i are the mole fractions of component i in liquid and vapor phases respectively.

2.3. Rate-Based Model Description

The rate-based model consists of a set of well-accepted equations, which are modelled to calculate the mass and energy transfer across the interface using rate equation and mass transfer coefficients [26]. For the calculation of the mass transfer coefficient and interfacial area, the correlation by Billet and Schultes [30] can be applied.

2.4. Assumptions and Mathematical Model for Rate-Based Model

Assumptions for the rate-based approach according to Afkhamipour et al. [26] are summarized below:

- The reaction is quick and occurs in the liquid film;
- The absorption column is supposed to be adiabatic;
- The interfacial surface area is identical for both heat and mass transfer;
- The liquid-side heat transfer resistance is small compared to the gas phase, and the interface temperature is, therefore, identical as the bulk temperature;
- The type of flow is plug flow, and concentration and temperature change in the radial direction is negligible;
- Both liquid and gas phases are formally discussed as ideal mixtures

2.5. Material and Energy Balances

For easy and accurate calculations, the packed-bed column with a height of Z is divided into some stages. Figure 4 shows a stage j of the column, which represents a differential height of the column (j refers to the stage number, where i refers to the compounds). The material and energy balances around stage j are performed by using the MERSHQ equations (Equations of Material, Energy balances, Rate of mass and heat transfer, Summation of composition, the hydraulic equation of pressure drop, and equilibrium relation) presented by Taylor et al. [31] as the following:

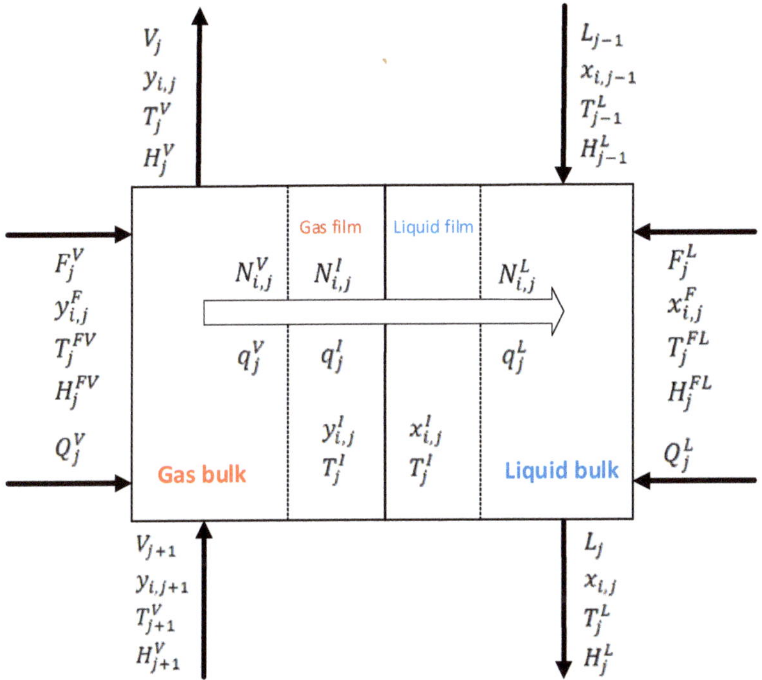

Figure 4. Rate-based stage model (Reproduced from ASPEN PLUS software manual).

Material balance for bulk liquid:

$$F_j^L x_{i,j}^F + L_{j-1} x_{i,j-1} + N_{i,j}^L + r_{i,j}^L - L_j x_{ij} = 0 \tag{7}$$

Material balance for bulk vapor:

$$F_j^V y_{i,j}^F + V_{j+1} y_{i,j+1} + N_{i,j}^V + r_{i,j}^V - V_j y_{i,j} = 0 \tag{8}$$

Material balance for liquid film:

$$N_{i,j}^I + r_{i,j}^{fL} - N_{i,j}^L = 0 \tag{9}$$

Material balance for vapor film:

$$N_{i,j}^V + r_{i,j}^{fV} - N_{i,j}^I = 0 \tag{10}$$

Energy balance for bulk liquid:

$$F_j^L H_j^{FL} + L_{j-1} H_{j-1}^L + Q_j^L + q_j^L - L_j H_j^L = 0 \tag{11}$$

Energy balance for bulk vapor:

$$F_j^V H_j^{FV} + V_{j+1} H_{j+1}^V + Q_j^V - q_j^V - V_j H_j^V = 0 \tag{12}$$

Energy balance for the liquid film

$$q_j^I - q_j^L = 0 \tag{13}$$

Energy balance for vapor film:

$$q_j^V - q_j^I = 0 \tag{14}$$

Phase equilibrium at the interface:

$$y_{i,j}^I - K_{i,j}x_{i,J}^I = 0 \tag{15}$$

Summations:

$$\sum_{i=1}^{n} x_{i,j} - 1 = 0 \tag{16}$$

$$\sum_{i=1}^{n} y_{i,j} - 1 = 0 \tag{17}$$

$$\sum_{i=1}^{n} x_{i,j}^I - 1 = 0 \tag{18}$$

$$\sum_{i=1}^{n} y_{i,j}^I - 1 = 0 \tag{19}$$

where F is the molar flow rate of feed. L, V are the molar flow rates of liquid and vapor respectively, N molar transfer rate, K is equilibrium ratio, r is reaction rate, H is enthalpy, Q is heat input to a stage, q is heat transfer rate, x_i , y_i are the mole fraction of component i in liquid and vapor phases respectively.

2.6. Mass Transfer through the Interfacial Area

The rate-based model is based on the two-film theory that describes the mass transfer between gas and liquid [26]. The film theory is based on the assumption that when two fluid phases are coming in contact with each other, a thin layer of stagnant fluid exists on each side of the interface [32]. From Figure 5, the partial pressure of component i drops from P_i at gas bulk to P_i^I at the interface [32,33]. This pressure difference creates a driving force $\left(P_i - P_i^I\right)$ for component i to transfer it from the gas bulk to gas film and then from the gas film to liquid film [33]. The accumulation of the component i at the liquid film creates a concentration difference between the liquid film and the liquid bulk. Similarly, this concentration difference creates a driving force $\left(C_i^I - C_i\right)$ for component i to transfer from the liquid film to the liquid bulk [32,33]. The molar flux N of component A from the bulk of one phase to the interface is written as below [33]:

$$N = k\Delta c \tag{20}$$

where N is the molar flux of the component (moles per unit area per unit time) and Δc is the driving force for mass transfer between the bulk and the interface. Consequently, Equation (11) can be written for phase and liquid phase [33] as:

$$N_{A,G} = K_G\left(P_i - P_i^I\right) \tag{21}$$

$$N_{A,L} = K_L\left(C_i^I - C_i\right) \tag{22}$$

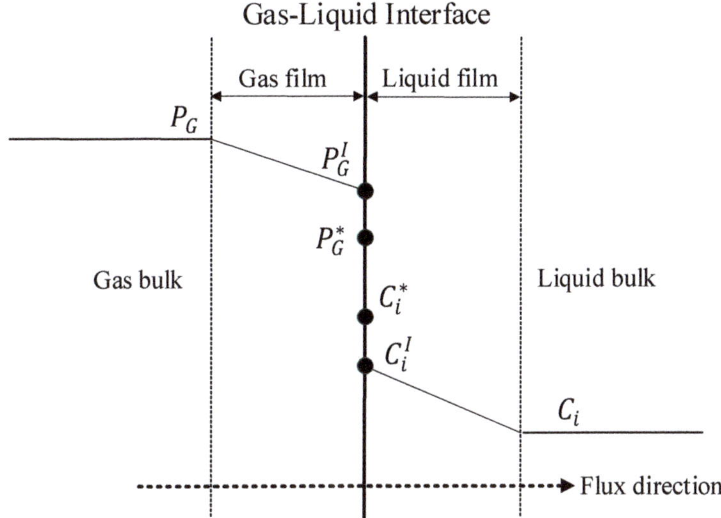

Figure 5. Concentration profiles based on the two-film model [34], * refers to the conditions at the equilibrium state.

At steady state, the flux of *i* from bulk gas to the interface must be equal to the flux of *i* from the interface to the bulk liquid [33]:

$$N = N_{L,i} = N_{G,i} \tag{23}$$

$$N = K_G\left(P_i - P_i^I\right) = K_L\left(C_i^I - C_i\right) \tag{24}$$

K_G and K_L are two different overall mass transfer coefficients (with different units). If equilibrium conditions exist at the interface, the overall mass transfer coefficient can be calculated as follows [35]:

$$\frac{1}{K_G} = \frac{1}{k_G} + \frac{H}{Ek_L} \tag{25}$$

where H is Henry's law constant, E is the enhancement factor. k_G and k_L are the mass transfer coefficients without reaction in the gas and liquid phase. The mass transfer coefficients and the wetted interfacial area for mass and heat transfer were calculated according to the correlations proposed by several studies [30,35,36] as following:

$$k_{L,i}a_{eff} = C_L\left(\frac{g}{v_L}\right)^{1/6}\left(\frac{D_{L,i}}{d_h}\right)^{1/2}a^{2/3}U_L^{1/3}\left(\frac{a_{eff}}{a}\right) \tag{26}$$

$$k_{G,i}a_{eff} = C_G(\varepsilon - h_L)^{-1/2}\left(\frac{a^3}{d_h}\right)^{1/2}D_{G,i}\left(\frac{U_G}{av_G}\right)^{3/4}\left(\frac{v_G}{D_{G,i}}\right)^{1/3}\left(\frac{a_{eff}}{a}\right) \tag{27}$$

$$\frac{a_{eff}}{a} = 1.5\,(ad_h)^{-0.5}\left(\frac{U_Ld_h}{v_L}\right)^{-0.2}\left(\frac{U_L^2\rho_Ld_h}{\sigma}\right)^{0.75}\left(\frac{U_L^2}{gd_h}\right)^{-0.45} \tag{28}$$

$$d_h = 4\frac{\varepsilon}{a} \tag{29}$$

where $k_{L,i}$ and $k_{G,i}$ and are the mass transfer coefficients of component *i* in the liquid and gas phase respectively, a_{eff} is the effective interfacial area per unit packed volume, a is the total surface area per unit packed volume, C_G and C_L are the gas and liquid and transfer coefficient parameter respectively; characteristic of the shape and structure of the packing, g is gravitational constant, v_L and v_G are kinematic viscosity of the liquid and gas phase respectively, $D_{L,i}$ and $D_{G,i}$ are diffusivity of component i in the liquid and gas phase respectively, d_h is hydraulic diameter of the dumped packing, ρ_L is density of liquid, U_L and U_G are velocity of liquid and gas phase respectively with reference to free column cross section, ε is void fraction of the packing, h_{La} is column liquid holdup, σ is liquid surface tension.

Theoretical liquid holdup correlation of Billet and Schultes [30] is given below:

$$h_L = \left(\frac{12}{g} v_L U_L a^2\right)^{1/3} \tag{30}$$

2.7. Heat Transfer through the Interfacial Area

In order to determine the heat transfer through the interfacial area, the Chilton-Colburn-Analogy is used [35]. The expressions for the analogy are taken from many studies [35,37]. These expressions lead to a heat transfer coefficient h as [35]:

$$h = k_G \left(\frac{\rho_G\left(c_p/M_{w,L}\right)\lambda^2}{D^2}\right)^{1/3} \tag{31}$$

where k_G is mass transfer coefficient, ρ_G is the density of the gas, c_p is specific molar heat capacity, $M_{w,L}$ is molecular weight of the liquid phase, λ is Thermal conductivity, D is the diffusion coefficient.

2.8. Thermodynamics Approaches for Prediction of Phase Behavior

The accuracy of equilibrium and rate-based models depend on the accurate prediction of the phase behavior properties of chemical species and their mixtures [19]. There are two conventional approaches for the prediction of phase behavior: approach (φ, φ) and approach (φ/γ) [38]. By approach (φ, φ), the fugacity coefficient φ. is applied for predicting the non-ideal behavior of both vapor and liquid phases [19]. The fugacity coefficient is calculated at the equilibrium condition according to Gebreyohannes et al. [38] as below:

$$f_i^V = f_i^L \tag{32}$$

$$f_i^V = \varphi_i^V y_i P \tag{33}$$

$$f_i^L = \varphi_i^L x_i P \tag{34}$$

$$ln\varphi_i^a = -\frac{1}{RT}\int_\infty^{Va}\left[\left(\frac{\partial p}{\partial n_i}\right)_{T,V,n_{i,j}} - \frac{RT}{V}\right]dV - lnZ_m^a \tag{35}$$

where f_i^L and f_i^V are Fugacity of component i in the liquid and gas phase respectively, φ_i^L and φ_i^V are Fugacity coefficient of the component i in liquid and gas phase respectively, P is pressure, x_i and y_i mole fraction of the f component i in liquid and gas phase respectively, φ_i^a is Fugacity coefficient of the component i where a refer to liquid or gas phase, V is total volume, R is Gas constant it has the value $0.08314\left[\text{L bar K}^{-1}\right]$, T is temperature, n_i is mole number of component i, Z_m^a Compression factor.

The approach (φ/γ) uses the fugacity coefficients (φ) and the activity coefficients (γ) to account the non-ideal behavior of vapor and liquid phase. The equations related to (φ/γ) approach are listed as below [38]:

$$f_i^V = f_i^L \tag{36}$$

$$f_i^V = \varphi_i^V y_i P \tag{37}$$

$$f_i^L = x_i \gamma_i f_i^{*,l} \tag{38}$$

where $f_i^{*,l}$ is liquid fugacity of pure component i at mixture temperature, γ_i is the liquid activity coefficient of component i. While φ_i^V is calculated according to Equation (26), the activity coefficients can be calculated from non-random two liquid model (NRTL) as below [39]:

$$ln\gamma_i = \frac{\sum_{j=1}^n x_j\tau_{ji}G_{ji}}{\sum_{k=1}^n x_kG_{ki}} + \sum_{j=1}^n \frac{x_jG_{ij}}{\sum_{k=1}^n x_kG_{kj}}\left(\tau_{ij} - \frac{\sum_{m=1}^n x_m\tau_{mj}G_{mj}}{\sum_{k=1}^n x_kG_{kj}}\right) \tag{39}$$

$$G_{ij} = exp\left(-a_{ij}\tau_{ij}\right) \tag{40}$$

$$\tau_{ij} = a_{ij} + \frac{b_{ij}}{T} + e_{ij}lnT + f_{ij}T \tag{41}$$

$$a_{ij} = c_{ij} + d_{ij}(T - 273.315K) \tag{42}$$

Here, a_{ij} is NRTL non-randomness constant for binary interaction, a_{ij}, b_{ij}, c_{ij}, d_{ij}, e_{ij}, and f_{ij} are binary parameters. The ASPEN PLUS physical property system has extensive property databanks for binary parameters for the model.

2.9. Model Specification

For simulating the packed bed absorber, the RateFrac model is adopted. The model flowsheet generated in ASPEN PLUS is shown in Figure 6. In the RateFrac model, the NRTL method is used. The flow model is countercurrent. The packed column height is divided into ten stages. The mass transfer coefficients and the wetted interfacial area for mass and heat transfer were calculated according to the empirical correlation of Billet and Schultes [30] with the constants CL = 2.4 and CV = 0.8 [36]. The heat transfer coefficient is estimated by Chilton–Colburn method [40]. Other relevant parameters are obtained by the default correlations of RateFrac [41] (see Tables 2–4). The soybean oil used as a solvent is a blend of acids (see Table 5). The tar was considered as a mixer of benzene, toluene, and ethylbenzene. The air participates in the process as tar holder—the water is used as the cooling liquid to adjust the temperature of the gas inlet to fit in the experimental data.

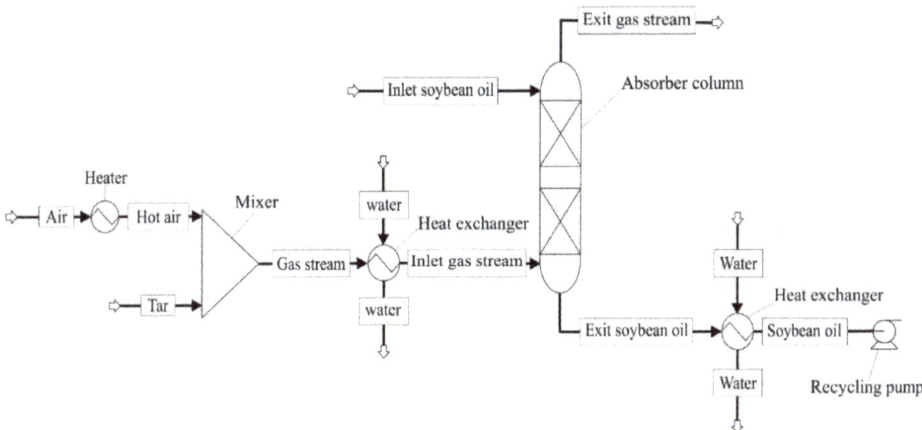

Figure 6. Flowsheet of process simulation in ASPEN PLUS.

Table 2. Binary diffusion coefficients (cm^2/s) used in Aspen PLUS at conditions of an experiment of 30 °C, flow rates 53 mL/min, and at a bed height of 0.5 m.

Component	N-HEXADECANOIC -ACID	STEARIC -ACID	OLEIC-ACID	LINOLEIC -ACID	LINOLENIC -ACID	BENZENE	TOLUENE	ETHYLBENZENE	AIR
N-HEXADECANOIC -ACID	0	4.31171×10^7	4.68659×10^7	4.64699×10^7	4.49859×10^7	1.06954×10^6	9.4904×10^7	8.58892×10^7	2.02164×10^6
STEARIC-ACID	4.31171×10^7	0	4.38352×10^7	4.38287×10^7	4.40934×10^7	1.10896×10^6	9.96534×10^7	9.036×10^7	2.07824×10^6
OLEIC-ACID	4.68659×10^7	4.38352×10^7	0	4.55134×10^7	4.54058×10^7	1.12982×10^6	9.97283×10^7	9.01854×10^7	2.14434×10^6
LINOLEIC-ACID	4.64699×10^7	4.38287×10^7	4.55134×10^7	0	4.53858×10^7	1.12457×10^6	9.93516×10^7	8.98562×10^7	2.13301×10^6
LINOLENIC-ACID	4.49859×10^7	4.40934×10^7	4.54058×10^7	4.53858×10^7	0	1.11272×10^6	9.88611×10^7	8.94878×10^7	2.10185×10^6
BENZENE	1.06954×10^6	1.10896×10^6	1.12982×10^6	1.12457×10^6	1.11272×10^6	0	5.44604×10^7	4.90094×10^7	6.55351×10^7
TOLUENE	9.4904×10^7	9.96534×10^7	9.97283×10^7	9.93516×10^7	9.88611×10^7	5.44604×10^7	0	5.42316×10^7	4.14606×10^7
ETHYLBENZENE	8.58892×10^7	9.036×10^7	9.01854×10^7	8.98562×10^7	8.94878×10^7	4.90094×10^7	5.42316×10^7	0	3.32784×10^7
AIR	2.02164×10^6	2.07824×10^6	2.14434×10^6	2.13301×10^6	2.10185×10^6	6.55351×10^7	4.14606×10^7	3.32784×10^7	0

Table 3. Overall binary mass transfer coefficients (kmol/s) used in Aspen PLUS at conditions of an experiment of 30 °C, flow rates 53 mL/min, and at a bed height of 0.5 m.

Component	N-HEXADECANOIC -ACID	STEARIC -ACID	OLEIC-ACID	LINOLEIC -ACID	LINOLENIC -ACID	BENZENE	TOLUENE	ETHYLBENZENE	AIR
N-HEXADECANOIC -ACID	0	4.9358×10^7	5.1459×10^7	5.12412×10^7	5.04163×10^7	7.77376×10^7	7.32277×10^7	6.9663×10^7	1.06877×10^6
STEARIC-ACID	4.9358×10^7	0	4.97674×10^7	4.97636×10^7	4.99137×10^7	7.91572×10^7	7.50376×10^7	7.14531×10^7	1.08363×10^6
OLEIC-ACID	5.1459×10^7	4.97674×10^7	0	5.0711×10^7	5.06511×10^7	7.98982×10^7	7.50658×10^7	7.13841×10^7	1.10073×10^6
LINOLEIC-ACID	5.12412×10^7	4.97636×10^7	5.0711×10^7	0	5.06399×10^7	7.97126×10^7	7.49239×10^7	7.12536×10^7	1.09782×10^6
LINOLENIC-ACID	5.04163×10^7	4.99137×10^7	5.06511×10^7	5.06399×10^7	0	7.92915×10^7	7.47387×10^7	7.11074×10^7	1.08977×10^6
BENZENE	7.77376×10^7	7.91572×10^7	7.98982×10^7	7.97126×10^7	7.92915×10^7	0	5.54719×10^7	5.26227×10^7	6.08513×10^7
TOLUENE	7.32277×10^7	7.50376×10^7	7.50658×10^7	7.49239×10^7	7.47387×10^7	5.54719×10^7	0	5.53553×10^7	4.84006×10^7
ETHYLBENZENE	6.9663×10^7	7.14531×10^7	7.13841×10^7	7.12536×10^7	7.11074×10^7	5.26227×10^7	5.53553×10^7	0	4.33625×10^7
AIR	1.06877×10^6	1.08363×10^6	1.10073×10^6	1.09782×10^6	1.08977×10^6	6.08513×10^7	4.84006×10^7	4.33625×10^7	0

Table 4. Heat transfer coefficients (Watt/m^2 K) used in Aspen PLUS at conditions of an experiment of 30 °C, flow rates 53 mL/min, and at a bed height of 0.5 m.

Heat Transfer Coefficients for Liquid (Watt/m^2 K)	Heat Transfer Coefficients for Vapor (Watt/m^2 K)
4512.399	48.685

Table 5. The composition of soybean oil [27].

Acids	Soybean Oil
Palmitic acid (16:0)	9%
Steric acid (18:0)	4.4%
Oleic acid (18:1)	26.4%
Linoleic acid (18:2)	51.6%
Linolenic acid (18:3)	6.8%

3. Results and Discussion

3.1. Model Validation

In this phase of simulation, the experimental data of the first minute reported by Bhoi [19] is used. The tar model used in this study was a mixture of compounds benzene, toluene, and ethylbenzene with mass fractions: 50% benzene, 30% toluene, and 20% ethylbenzene. The mass fraction values of these materials were selected because they are almost the mass fraction values of the tar compounds collected and measured from a fluidized bed gasifier [42]. The studied solvent temperatures for this experiment are 30, 40, and 50 °C. The studied solvent flow rates are 53, 63 and 73 mL/min. The studied heights of the packed bed are 0.5, 0.8, and 1.1 m. The pressure of both solvent and gas stream is 20 psig. The simulation results are presented in terms of removal efficiencies of tar compounds and reported as a function of operating parameters, i.e., the solvent temperature, the solvent mass flow rate, and the packed bed height.

Tar removal efficiency (η) was calculated using the following equation [19]:

$$\eta = \frac{C_{in} - C_{out}}{C_{in}} \tag{43}$$

where C_{in} is cona centration of tar compounds (benzene, toluene, and ethylbenzene) at the inlet of the column [*ppmv*], C_{out} is concentration of the tar compounds (benzene, toluene and ethylbenzene) at the outlet of the column [*ppmv*].

From the simulation, Figures 7–9 show a comparison between the experimental data and the results predicted by both rate-based and equilibrium-stage models. Profiles in these figures show how the removal efficiencies of tar components change by changing the critical target parameters: solvent temperature, the flow rate of solvent (soybean oil), and the packing bed height. As shown in the figures, the prediction results of the removal efficiencies of tar components for both rate-based model and the equilibrium model are high compared to the experimental data, but the results obtained from the rate-based model have a better prediction for experimental data in comparison with the equilibrium-stage model. For assessing the accuracy of the models, MAPE (the mean absolute percentage error) is calculated between the values calculated using the models and those obtained from empirical measurements. The values of MAPE are shown in the Tables 6–8.

Table 6. MAPE between the values calculated using the models (rate-based (RB) and equilibrium-stage (EQ)) and those obtained from empirical measurement at a bed height of 0.5 m.

MAPE at Flow Rates of 53 mL/min, and Bed Height of 0.5 m						
	Benzene		**Toluene**		**Ethylbenzene**	
Temperatures °C	**RB**	**EQ**	**RB**	**EQ**	**RB**	**EQ**
30 °C	1.0%	6.7%	0.1%	4.2%	1.9%	1.9%
40 °C	2.4%	8.7%	0.1%	5.5%	1.6%	3.0%
50 °C	6.2%	4.3%	1.0%	6.3%	1.6%	3.8%
MAPE at flow rates of 63 mL/min, and bed height of 0.5 m						
Temperatures °C	RB	EQ	RB	EQ	RB	EQ
30 °C	0.7%	4.7%	0.2%	3.0%	1.6%	1.3%
40 °C	1.3%	7.1%	0.1%	4.1%	1.5%	2.0%
50 °C	2.9%	14.6%	2.5%	8.3%	0.2%	4.6%
MAPE at flow rates of 73 mL/min, and bed height of 0.5m						
Temperatures °C	RB	EQ	RB	EQ	RB	EQ
30 °C	0.1%	4.2%	0.9%	3.4%	1.0%	1.3%
40 °C	2.0%	8.4%	1.0%	4.3%	1.1%	1.8%
50 °C	1.4%	11.1%	1.5 5	6.0%	0.0%	3.5%

Table 7. MAPE between the values calculated using the models (RB and EQ) and those obtained from empirical measurement at a bed height of 0.8 m.

MAPE at Flow Rates of 53 mL/min, and Bed Height of 0.8 m						
	Benzene		**Toluene**		**Ethylbenzene**	
Temperatures °C	**RB**	**EQ**	**RB**	**EQ**	**RB**	**EQ**
30 °C	0.9%	1.8%	2.9%	3.8%	1.2%	2.0%
40 °C	1.5%	3.9%	3.5%	5.0%	2.0%	3.1%
50 °C	6.3%	0.8%	3.5%	6.2%	2.8%	4.3%
MAPE at flow rates of 63 mL/min, and bed height of 0.8 m						
Temperatures °C	RB	EQ	RB	EQ	RB	EQ
30 °C	0.4%	1.1%	2.4%	3.0%	0.8%	1.3%
40 °C	1.0%	4.5%	3.1%	4.1%	2.1%	2.9%
50 °C	1.4%	4.4%	6.6%	8.3%	2.8%	3.8%
MAPE at flow rates of 73 mL/min, and bed height of 0.8 m						
Temperatures °C	RB	EQ	RB	EQ	RB	EQ
30 °C	0.9%	1.8%	3.4%	3.8%	1.7%	2.0%
40 °C	3.5%	5.8%	4.9%	5.6%	2.6%	3.1%
50 °C	1.3%	5.9%	4.5%	5.6%	3.1%	3.8%

Table 8. MAPE between the values calculated using the models (RB and EQ) and those obtained from empirical measurement at a bed height of 1.1 m.

MAPE at Flow Rates of 53 mL/min, and a 0.5 Bed Height of 1.1 m						
	Benzene		**Toluene**		**Ethylbenzene**	
Temperatures °C	**RB**	**EQ**	**RB**	**EQ**	**RB**	**EQ**
30 °C	0.0%	1.0%	3.0%	3.2%	1.3%	1.4%
40 °C	1.4%	1.3%	3.3%	3.7%	1.5%	1.7%
50 °C	5.5%	2.6%	5.1%	6.0%	4.2%	4.6%
MAPE at flow rates of 63 mL/min, and bed height of 1.1 m						
Temperatures °C	RB	EQ	RB	EQ	RB	EQ
30 °C	0.7%	1.1%	3.4%	3.5%	2.1%	2.1%
40 °C	1.8%	3.2%	4.7%	4.9%	2.7%	2.9%
50 °C	0.9%	2.2%	5.9%	6.4%	4.6%	4.8%
MAPE at flow rates of 73 mL/min, and bed height of 1.1 m						
Temperatures °C	RB	EQ	RB	EQ	RB	EQ
30 °C	1.4%	1.6%	4.0%	4.1%	2.1%	2.1%
40 °C	2.3%	3.1%	4.9%	5.0%	3.2%	3.3%
50 °C	0.5%	1.7%	3.8%	4.1%	2.2%	2.4%

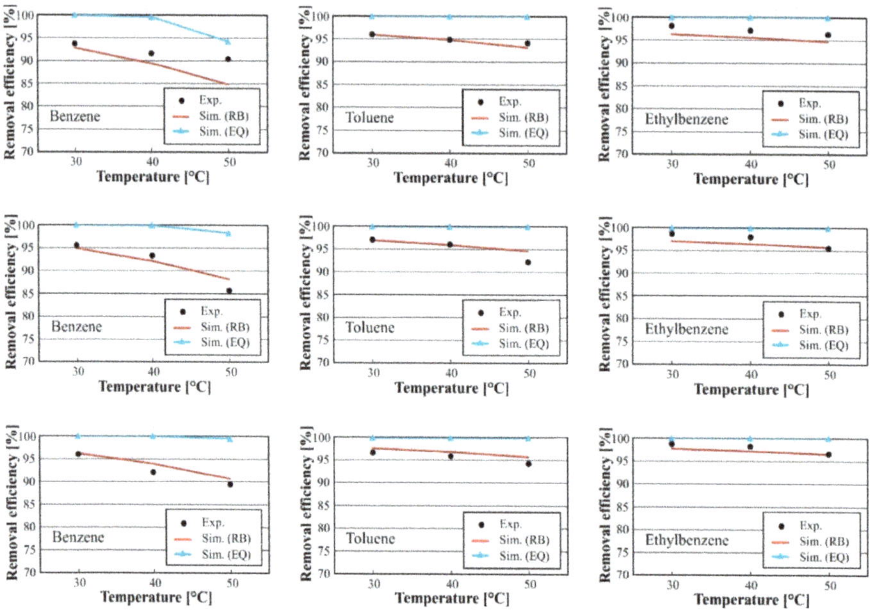

Figure 7. Effect of solvent temperature on the removal efficiency of tar components at a bed height of 0.5 m and different solvent volumetric flow rates of 53 mL/min (**above**), 63 mL/min (**middle**), 73 mL/min (**below**).

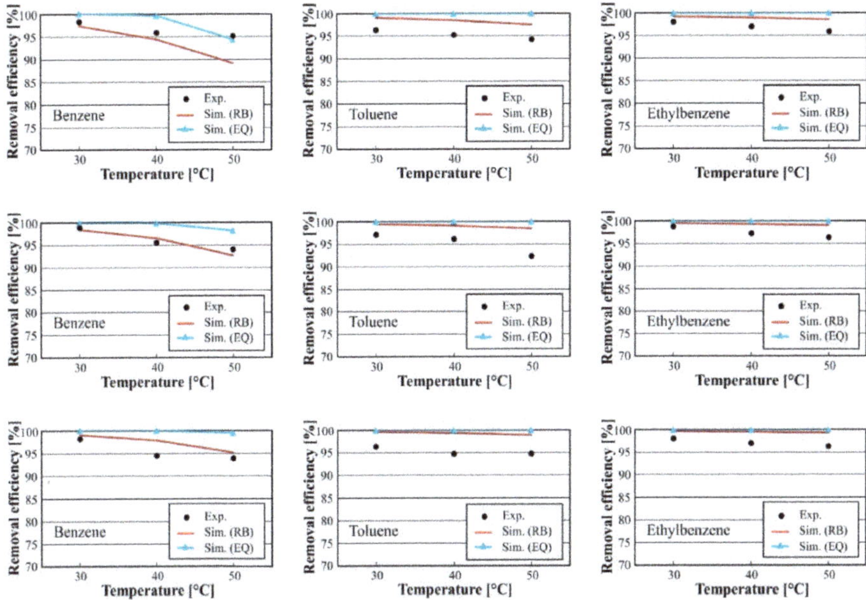

Figure 8. Effect of solvent temperature on the removal efficiency of tar components at a bed height of 0.8 m and different solvent volumetric flow rates 53 mL/min (**above**), 63 mL/min (**middle**), 73 mL/min (**below**).

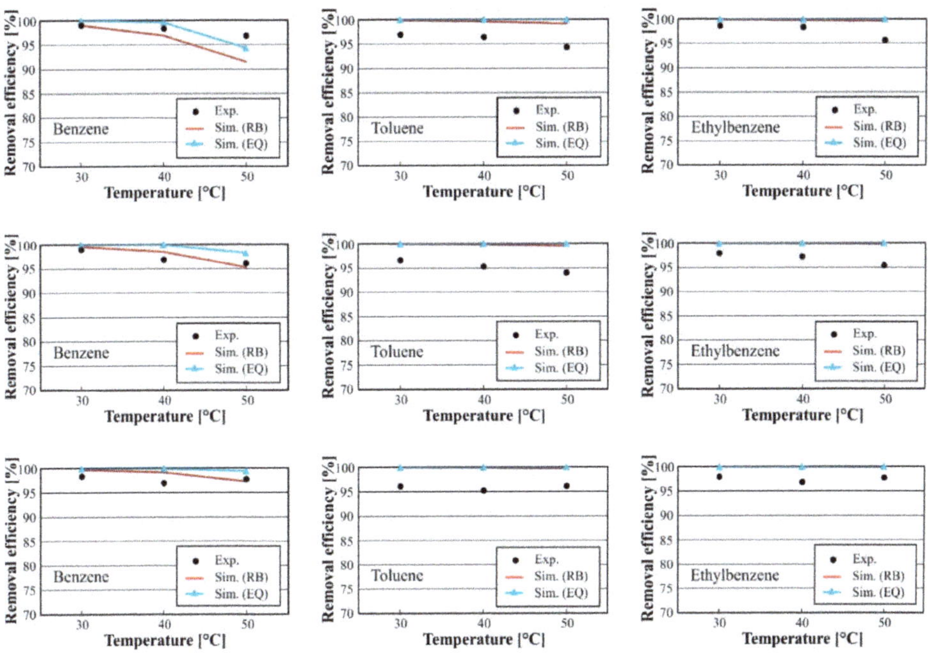

Figure 9. Effect of solvent temperature on the removal efficiency of tar components at a bed height of 1.1 m and different solvent volumetric flow rates of 53 mL/min (**above**), 63 mL/min (**middle**), 73 mL/min (**below**).

The deviation of results between the equilibrium model and the experimental data is because the equilibrium model assumes that the vapor and liquid left on the plates are in thermodynamic equilibrium [19]. In actual operation, the equilibrium between the gas and liquid phases is rare [29]. Although the rate-based model shows a good agreement with experimental data, there is a deviation. It may be explained because of the rate-based model based on the empirical correlations used for calculating the mass and heat transfer parameters [26]. It is clear that the deviation of results between the rate-based model and the experimental data increased by increasing the bed height. The explanation for this trend is that by increasing the packing bed height, the conditions will be close to the equilibrium state; this explains why the excellent agreement between the rate-based model and the equilibrium model increased by increasing the bed height. As a whole, the simulation results predicted by rate-based model are in the range of the experimental results.

3.2. Analysis of Tar Absorption Process

3.2.1. Effect of Solvent Temperature

Figures 7–9, show the effect of the temperature of soybean oil solvent on the removal efficiency of tar components at bed heights 0.5, 0.8, and 1.1 m, and solvent flow rate of 53 mL/min (above), 63 mL/min (middle), 73 mL/min (below). It is clear from the figures that the increase in solvent temperature has a significant effect on the removal efficiencies. The removal efficiency decreases with increasing the temperature from 30 to 50 °C. The principal reason for this effect is that by increasing the solvent temperature, the solubility of tar compounds decreased, hence increasing the equilibrium ratio (K-value) [19].

On the other hand, increasing the solvent temperature leads to increasing the wettability of the solvent due to the decreased viscosity, as a result, the mass transfer and removal efficiency increase [19]. The effect of decreasing the solubility is more significant than increasing the wettability on decreasing the removal efficiency, so the removal efficiency is reduced by increasing the solvent temperature. The trend of this effect is similar for all the tar components, but the change rates (decreasing rates) of removal efficiency are different from component to component. The change rate for benzene is higher than toluene and ethylbenzene, for example at operating conditions of the volumetric flow rate of 53mL/min and a bed height of 0.5 m. Here, the removal efficiency decreased for benzene by about 8% by increasing the temperature from 30 to 50 °C, i.e., the change rate of removal efficiency for benzene is 0.4%/°C. Whereas the change rate value is 0.14%/°C for toluene and 0.08%/°C for ethylbenzene. It is also observed that the effect of increasing the soybean oil solvent temperature on the change rates of removal efficiency is influenced by the increase of the solvent volumetric flow rate. Increasing the volumetric flow rate of the soybean oil leads to a change rate decrease of the removal efficiency. For benzene at bed height of 0.5 as an example, the change rate of the removal efficiency at volumetric flow rate of 53 mL/min is 0.4%/°C and decreases to value 0.33%/°C at volumetric flow rate of 63 mL/min and it continues decreasing to a value of 0.27%/°C at 73 mL/min.

Furthermore, increasing the bed height influences the change rate of the removal efficiency. By increasing the temperature, the change rate of the removal efficiency decreases by increasing the bed height. This trend is shown in toluene and ethylbenzene for example at a volumetric flow rate of 53 mL/min. Here, the change rate of removal efficiency for toluene at a bed height of 0.5 m is 0.14%/°C, and it decreases to value 0.077%/°C at a bed height of 0.8 m, and it continues decreasing to a value 0.034%/°C at a bed height of 1.1 m.

3.2.2. Effect of Bed Height and Solvent Volumetric Flow Rate

Figure 10 shows the effect of solvent volumetric flow on the removal tar efficiency. It is evident that an increase in the solvent volumetric flow rate has a significant effect on the removal efficiency. The removal efficiency is dramatically enhanced when the solvent volumetric flow rate is increased. The reason is that by increasing the solvent volumetric flow rate, the mass transfer rates of tar compounds increase, resulting in higher tar removal efficiencies [19]. The trend of this effect is similar for all the tar components, but the change rates (increasing rates) of removal efficiencies by increasing the solvent volumetric flow rate are different from component to component. The change rates for benzene are higher than toluene and ethylbenzene.

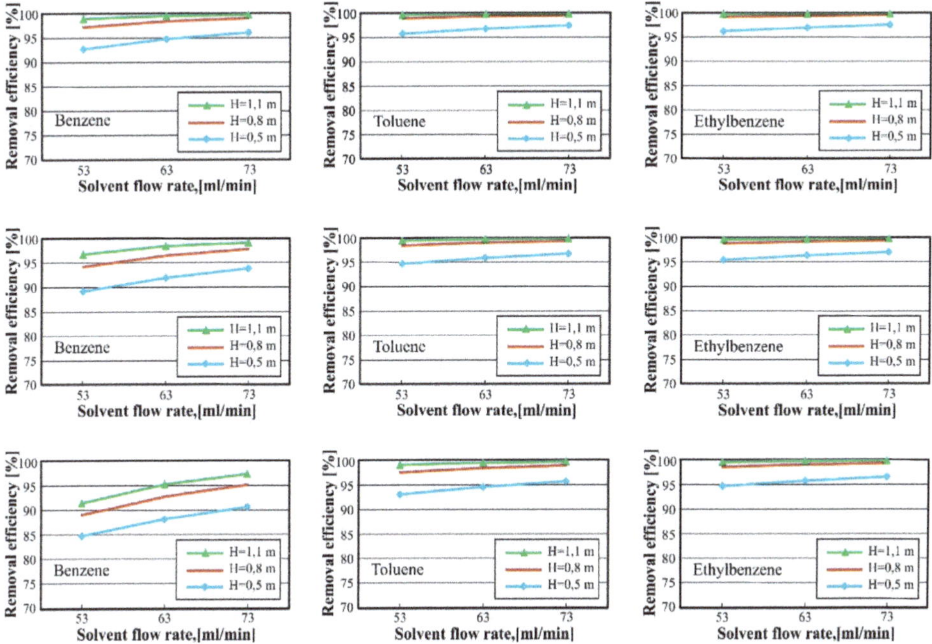

Figure 10. Effect of bed height on the removal efficiency of tar components at solvent volumetric flow rates of 53.63 and 73mL/min, the temperatures are 30 °C (**above**), 40 °C (**middle**), and 50 °C (**below**).

Furthermore, increasing the bed height enhances the removal efficiency because of the mass transfer increasing between gas and liquid [19]. This trend of effect is similar for all tar components, but the change rates (increasing rates) of removal efficiency are different from components to another. Here, the change rate for benzene is higher than toluene and ethylbenzene.

3.2.3. Optimum Operation Conditions

Selecting the optimum (most economical) operation conditions should consider the requirements of the process as well as the operation cost and annualized charges on equipment. The requirement for tar concertation depends on the intended end use of the produced gas (gas application) [3]. Several researchers reported that the tar concentration should be up to 50–100 mg/Nm^3 for ICE and less than 5 mg/Nm^3 for gas turbines [2]. According to Hlina et al. [43], the tar concentration should be less than 0.1 mg/Nm^3 for Fischer-Tropsch synthesis.

As previously mentioned, the removal efficiency is affected by three parameters: the temperature, the solvent volumetric flow rate, and the height of the bed. It is clear that increasing the bed height enhances the removal efficiency of tar components, but on the other hand, this leads to an increase in the amount of packing material needed to fill in a packed column which means more cost will be added to the total cost of the plant. Therefore, the bed height depends mainly on the requirement for tar concentration. It should meet these requirements considering the solvent volumetric flow rate to be set at a minimum value. Increasing the volumetric flow rate of the soybean oil solvent enhances the removal efficiency of tar components, but on the other hand, this leads to an increase in the operations cost. This operation cost results from the electrical energy consumed by the recycling pump. The optimum volumetric flow rate is linked to annualized charges on equipment. The total annualized cost should be calculated, which is equal to the sum of operation cost and annualized charges on equipment. The optimum flow rate of the solvent meets the minimum of the total annualized costs.

The temperature of the solvent has to be selected carefully. It has been observed that a decrease in the solvent temperature will increase the removal efficiency. However, the heat duty that should be added to the process to cool the solvent to the setup temperature (the inlet solvent temperature to absorber) should be considered. Therefore it is necessary to estimate the heat duty used to cool down the solvent temperature. The heat duty was calculated by ASPEN PLUS software. The results are illustrated in Figure 11; it clears that the heat duty to cool-down the solvent decreases by increasing the temperature of the solvent. From this curve, one can conclude that the solvent with high temperature is preferred for the deliberate process, but in a condition that this temperature achieves the requirement of tar concentration.

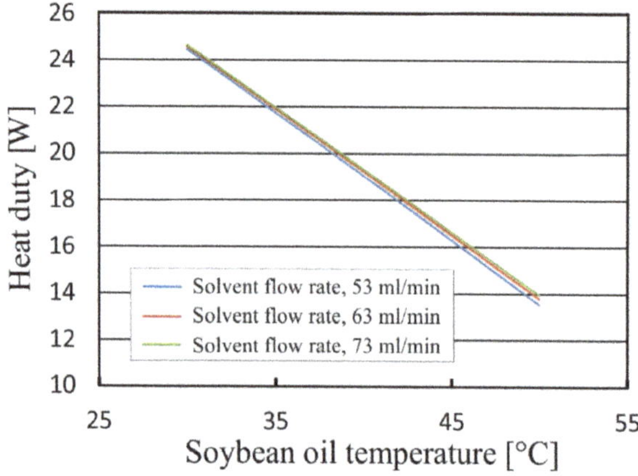

Figure 11. Heat duty of the heat exchanger.

4. Conclusions

A rate-based model and equilibrium model have been built by applied Aspen Plus software for simulation of tar absorption using soybean oil as a solvent. Both models has been validated against experimental data at different operation points. The experimental data published by Bhoi [19] was used for validation of the models. Comparison between the results predicted by two mathematical models (rate-based and equilibrium stage) and the experimental data shows that the rate-based model has a higher accuracy than the equilibrium model. The deviation of results between the rate-based model and the experimental data increases by increasing the bed height where the conditions will be close to the equilibrium state. An analytical study of tar absorption process by using soybean oil has been presented which reveals the following points:

The removal efficiencies are different between the tar compounds (benzene, toluene, and ethylbenzene). The ethylbenzene has the highest removal efficiency. The removal efficiencies η of these components can be ranked as follows: η-value of ethylbenzene > η-value of toluene > η-value of benzene. The difference of removal efficiencies can be explained because of the different K values for these components: the lower th K value, the higher the removal efficiency, i.e., K-value of ethylbenzene < K-value of toluene < K-value of benzene at specific pressure and temperature [19].

The slope curves of removal efficiency for benzene, toluene, and ethylbenzene between 30 and 50 °C were approximately 0.4, 0.14, and 0.08, respectively which means that decreasing the temperature by 1 °C will enhance the removal efficiency for benzene, toluene, and ethylbenzene approximately by 0.4%, 0.14%, and 0.08%, respectively.

Increasing the solvent volumetric flow rate enhances the removal efficiency. The slopes curve of removal efficiency for benzene, toluene, and ethylbenzene between 53 and 73 mL/min are approximately 0.17, 0.084, and 0.071, respectively. It is essential to consider the energy consumed by increasing the volumetric flow rate of the solvent.

Increasing the height of the packed bed has a significant effect in enhancing the removal efficiency.

A methodology for selecting the optimum (most economical) operation conditions has been presented which it is vital in case scaling lab-scale experiment for the pilot plant.

Finally, it should be mentioned that the built and validated model is an essential tool for studying tar absorption process because of its ability to predict the process performance with changing parameters and hence saving the cost and time.

Author Contributions: A.A. is responsible for preparing the original draft and developing the applied methodology. F.A. supported the writing process with his reviews and edits. The conceptualization was conducted by C.H. B.E. supervised the research progress and the presented work. All authors have read and agreed to the published version of the manuscript.

Funding: The authors received no specific funding for this work. The corresponding author would like to thank the Technical University of Darmstadt, enabling the open-access publication of this paper.

Nomenclatures

L	the molar flow rate of liquid [kmol/s]
V	the molar flow rate of vapor [kmol/s]
F	the molar flow rate of feed [kmol/s]
N	molar transfer rate [kmol/s]
$\dot{n}_i$	molar transfer rate per unit of square area $\left[\frac{kmol}{s.m^2}\right]$
c	Molar concentration $\left[\frac{kmol}{m^3}\right]$
K	Equilibrium ratio [−]
P	Gas pressure [atm]
r	Reaction rate [kmol/s]
H	Enthalpy [J/kmol]
Q	Heat input to a stage [J/s]
q	Heat transfer rate [J/s]
T	Temperature K
x	Liquid mole fraction [−]
y	Vapor mole fraction [−]
k_L	Binary mass transfer coefficient for the liquid [m/s]
k_G	Binary mass transfer coefficient for the gas [m/s]
C_L	Liquid mass transfer coefficient parameter, characteristic of the shape and structure of the packing [−]
C_V	Vapor mass transfer coefficient parameter, characteristic of the shape and structure of the packing [−]
g	Gravitational constant $\left[m/s^2\right]$
ρ_L, ρ_G	The density of the liquid, vapor $\left[kg/m^3\right]$
v_L, v_G	The viscosity of liquid, vapor $\left[m^2/s\right]$
D_L, D_G	Diffusivity of the liquid, vapor $\left[m^2/s\right]$
d_h	Hydraulic diameter m
U_L, U_G	velocity for the liquid, vapor [m/s]
ε	Void fraction of the packing [−]
a	Specific surface area $\left[m^2/m^3\right]$
a_{eff}	Effective surface area per unit volume of the column $\left[m^2/m^3\right]$
d_{eq}	Equivalent diameter [m]
σ	Liquid surface tension [N/m]
h^L	Heat transfer coefficient for liquid $\left[W/m^2K\right]$

h^V	Heat transfer coefficient for vapor $\left[\mathrm{W/m^2K}\right]$
nc	The number of components $[-]$
C_p	Specific molar heat capacity $[\mathrm{J/kmol\ K}]$
λ	Thermal conductivity $[\mathrm{W/m\ K}]$
h_L	Volumetric liquid holdup $\left[\mathrm{m^3}\right]$
f_i^V	Fugacity of component i in the vapor phase [bar]
f_i^L	Fugacity of component i in the liquid phase [bar]
V	Total volume [L]
φ_i^a	Fugacity coefficient of component i $[-]$
P	Pressure [bar]
R	Gas constant $=0.08314$ $\left[\mathrm{L\ bar\ K^{-1}}\right]$
T	Temperature [K]
n_i	Mole number of component i $[-]$
Z_m^a	Compression factor $[-]$

Subscripts

F	Feed
f	Film
I	Interface
L	Liquid
V	Vapor
a	Liquid or Vapor
G	Gas
i	Component
n	Number of components
j	Stage number

Abbreviations

ASPEN PLUS	Simulation software program
GC	Gas chromatography
NRTL	Non-random two-liquid model
ICE	Internal combustion engine
MESH	Equations of material, energy balances, summation of composition, and equilibrium relation
MERSHQ	Equations of material, energy balances, rate of mass and heat transfer, summation of composition, hydraulic equation of pressure drop, and equilibrium
RB	Rate-based model
EQ	Equilibrium-stage model
MAPE	The mean absolute percentage error

References

1. Van Paasen, S.; Kiel, J. Tar formation in fluidised-bed gasification-impact of gasifier operating conditions. *Acknowledgement/Preface* **2004**, *14*, 26.
2. Anis, S.; Zainal, Z. Tar reduction in biomass producer gas via mechanical, catalytic and thermal methods: A review. *Renew. Sustain. Energy Rev.* **2011**, *15*, 2355–2377. [CrossRef]
3. Milne, T.A.; Evans, R.J.; Abatzaglou, N. *Biomass Gasifier "Tars": Their Nature, Formation, and Conversion*; National Renewable Energy Laboratory: Golden, CO, USA, 1998.
4. Devi, L.; Ptasinski, K.J.; Janssen, F.J.; Van Paasen, S.V.; Bergman, P.C.; Kiel, J.H. Catalytic decomposition of biomass tars: Use of dolomite and untreated olivine. *Renew. Energy* **2005**, *30*, 565–587. [CrossRef]
5. Boerrigter, H.; Van Paasen, S.; Bergman, P.; Könemann, J.; Emmen, R.; Wijnands, A. OLGA tar removal technology. *Energy Res. Centre Neth. ECN-C–05-009* **2005**.
6. Unger, C.; Ising, M. Mechanismen und Bedeutung der Teerbildung und Teerbeseitigung bei der termochemischen Umwandlung fester Kohlenstoffträger. *DGMK-Tagungsbericht* **2002**, *2*, 131–142.

7. Li, C.; Suzuki, K. Tar property, analysis, reforming mechanism and model for biomass gasification—An overview. *Renew. Sustain. Energy Rev.* **2009**, *13*, 594–604. [CrossRef]

8. Neeft, J.; Knoef, H.; Onaji, P. *Behaviour of Tar in Biomass Gasification Systems: Tar Related Problems and Their Solutions*; Novem: Sittard, The Netherlands, 1999.

9. Kübel, M. *Teerbildung und Teerkonversion bei der Biomasseevergasung-Anwendung der Nasschemischen Teerbestimmung nach CEN-Standard*; Cuvillier: Göttingen, Germany, 2007.

10. Wolfesberger, U.; Aigner, I.; Hofbauer, H. Tar content and composition in producer gas of fluidized bed gasification of wood-Influence of temperature and pressure. *Environ. Prog. Sustain. Energy* **2009**, *28*, 372–379. [CrossRef]

11. Devi, L.; Ptasinski, K.J.; Janssen, F.J. A review of the primary measures for tar elimination in biomass gasification processes. *Biomass Bioenergy* **2003**, *24*, 125–140. [CrossRef]

12. Nakamura, S.; Kitano, S.; Yoshikawa, K. Biomass gasification process with the tar removal technologies utilizing bio-oil scrubber and char bed. *Appl. Energy* **2016**, *170*, 186–192. [CrossRef]

13. Dyment, J.; Watanasiri, S. *Acid Gas Cleaning Using DEPG Physical Solvents: Validation with Experimental and Plant Data*; Aspen Technology Inc.: Bedford, MA, USA, 2015.

14. Pilling, M.; Holden, B.S. *Choosing Trays and Packings for Distillation*; American Institute of Chemical Engineers CEP: New York, NY, USA, 2009; pp. 44–50.

15. Mudhasakul, S.; Ku, H.-M.; Douglas, P.L. A simulation model of a CO_2 absorption process with methyldiethanolamine solvent and piperazine as an activator. *Int. J. Greenh. Gas Control.* **2013**, *15*, 134–141. [CrossRef]

16. Phuphuakrat, T.; Namioka, T.; Yoshikawa, K. Absorptive removal of biomass tar using water and oily materials. *Bioresour. Technol.* **2011**, *102*, 543–549. [CrossRef] [PubMed]

17. Bergman, P.C.; van Paasen, S.V.; Boerrigter, H. The novel "OLGA" technology for complete tar removal from biomass producer gas. In *Pyrolysis and Gasification of Biomass and Waste, Expert Meeting*; osti.gov: Strasbourg, France, 2002.

18. Paethanom, A.; Nakahara, S.; Kobayashi, M.; Prawisudha, P.; Yoshikawa, K. Performance of tar removal by absorption and adsorption for biomass gasification. *Fuel Process. Technol.* **2012**, *104*, 144–154. [CrossRef]

19. Bhoi, P.R. *Wet Scrubbing of Biomass Producer Gas Tars Using Vegetable Oil*; Oklahoma State University: Stillwater, OK, USA, 2014.

20. Ozturk, B.; Yilmaz, D. Absorptive Removal of Volatile Organic Compounds from Flue Gas Streams. *Process. Saf. Environ. Prot.* **2006**, *84*, 391–398. [CrossRef]

21. Mofarahi, M.; Khojasteh, Y.; Khaledi, H.; Farahnak, A. Design of CO_2 absorption plant for recovery of CO_2 from flue gases of gas turbine. *Energy* **2008**, *33*, 1311–1319. [CrossRef]

22. Pandya, J. Adiabatic gas absorption and stripping with chemical reaction in packed towers. *Chem. Eng. Commun.* **1983**, *19*, 343–361. [CrossRef]

23. Tontiwachwuthikul, P.; Meisen, A.; Lim, C. CO_2 absorption by NaOH, monoethanolamine and 2-amino-2-methyl-1-propanol solutions in a packed column. *Chem. Eng. Sci.* **1992**, *47*, 381–390. [CrossRef]

24. Tobiesen, F.A.; Svendsen, H.F.; Juliussen, O. Experimental validation of a rigorous absorber model for CO_2 postcombustion capture. *AIChE J.* **2007**, *53*, 846–865. [CrossRef]

25. Mores, P.L.; Scenna, N.; Mussati, S. A rate based model of a packed column for CO_2 absorption using aqueous monoethanolamine solution. *Int. J. Greenh. Gas Control.* **2012**, *6*, 21–36. [CrossRef]

26. Afkhamipour, M.; Mofarahi, M. Comparison of rate-based and equilibrium-stage models of a packed column for post-combustion CO_2 capture using 2-amino-2-methyl-1-propanol (AMP) solution. *Int. J. Greenh. Gas Control.* **2013**, *15*, 186–199. [CrossRef]

27. Bhoi, P.; Huhnke, R.L.; Kumar, A.; Payton, M.E.; Patil, K.N.; Whiteley, J.R. Vegetable oil as a solvent for removing producer gas tar compounds. *Fuel Process. Technol.* **2015**, *133*, 97–104. [CrossRef]

28. Seader, J.; Henley, E.J.; Roper, D.K. *Separation Process Principles: Chemical and Biochemical Operations*; John Wiley and Sons, Inc.: Hoboken, NJ, USA, 2011.

29. Ramesh, K.; Aziz, N.; Shukor, S.A.; Ramasamy, M. Dynamic rate-based and equilibrium model approaches for continuous tray distillation column. *J. Appl. Sci. Res.* **2007**, *3*, 2030–2041.

30. Billet, R.; Schultes, M. Prediction of Mass Transfer Columns with Dumped and Arranged Packings. *Chem. Eng. Res. Des.* **1999**, *77*, 498–504. [CrossRef]

31. Taylor, R.; Krishna, R. *Multicomponent Mass Transfer*; John Wiley & Sons: Hoboken, NJ, USA, 1993.

32. Whitman, W.G. The two film theory of gas absorption. *Int. J. Heat Mass Transf.* **1962**, *5*, 429–433. [CrossRef]

33. Ngo, T.H. Gas Absorption into Emulsions (Doctoral Dissertation Published by Universitätsbibliothek Braunschweig. Available online: https://nbn-resolving.org/urn:nbn:de:gbv:084-13043009095 (accessed on 5 March 2013).

34. Khan, F.; Krishnamoorthi, V.; Mahmud, T. Modelling reactive absorption of CO_2 in packed columns for post-combustion carbon capture applications. *Chem. Eng. Res. Des.* **2011**, *89*, 1600–1608. [CrossRef]

35. Simon, L.L.; Elias, Y.; Puxty, G.; Artanto, Y.; Hungerbuhler, K. Rate based modeling and validation of a carbon-dioxide pilot plant absorbtion column operating on monoethanolamine. *Chem. Eng. Res. Des.* **2011**, *89*, 1684–1692. [CrossRef]

36. Bhoi, P.; Huhnke, R.L.; Kumar, A.; Patil, K.N.; Whiteley, J.R. Design and development of a bench scale vegetable oil based wet packed bed scrubbing system for removing producer gas tar compounds. *Fuel Process. Technol.* **2015**, *134*, 243–250. [CrossRef]

37. Bird, R.B.; Stewart, W.E.; Lightfoot, E.N.; Meredith, R.E. Reviewer Transport Phenomena. *J. Electrochem. Soc.* **1961**, *108*, 78C. [CrossRef]

38. Gebreyohannes, S.; Yerramsetty, K.; Neely, B.J.; Gasem, K.A.M. Improved QSPR generalized interaction parameters for the nonrandom two-liquid activity coefficient model. *Fluid Phase Equilibria* **2013**, *339*, 20–30. [CrossRef]

39. Renon, H.; Prausnitz, J.M. Local compositions in thermodynamic excess functions for liquid mixtures. *AIChE J.* **1968**, *14*, 135–144. [CrossRef]

40. Chilton, T.H.; Colburn, A.P. Mass Transfer (Absorption) Coefficients Prediction from Data on Heat Transfer and Fluid Friction. *Ind. Eng. Chem.* **1934**, *26*, 1183–1187. [CrossRef]

41. Plus, A. *Aspen Plus Documentation Version V7. 3*; Aspen Tech: Cambridge, MA, USA, 2011.

42. Cateni, B.G. *Effects of Feed Composition and Gasification Parameters on Product Gas from a Pilot Scale Fluidized Bed Gasifier*; Oklahoma State University: Stillwater, OK, USA, 2007.

43. Hlina, M.; Hrabovsky, M.; Kavka, T.; Konrad, M. Tar measurement in synthetic gas produced by plasma gasification by solid phase microextraction (SPME). In Proceedings of the ISPC Conference, Philadelphia, PA, USA, 24–29 July 2011; pp. 24–29.

 *applied sciences*

Article

Steam Gasification of Lignite in a Bench-Scale Fluidized-Bed Gasifier Using Olivine as Bed Material

Elisa Savuto [1,*], Jan May [2], Andrea Di Carlo [3], Katia Gallucci [3], Andrea Di Giuliano [3] and Sergio Rapagnà [1]

[1] Faculty of Bioscience and Agro-food and environmental technology, University of Teramo, Via R. Balzarini 1, 64100 Teramo, Italy; srapagna@unite.it
[2] Institute for Energy Systems and Technology, Technische Universität Darmstadt, Otto-Berndt-Straße 2, 64287 Darmstadt, Germany; jan.may@est.tu-darmstadt.de
[3] Industrial Engineering Department, University of L'Aquila, Piazzale E. Pontieri 1, Monteluco di Roio, 67100 L'Aquila, Italy; andrea.dicarlo1@univaq.it (A.D.C.); katia.gallucci@univaq.it (K.G.); andrea.digiuliano@univaq.it (A.D.G.)
* Correspondence: esavuto@unite.it

Received: 19 February 2020; Accepted: 20 April 2020; Published: 23 April 2020

Abstract: The gasification of lignite could be a promising sustainable alternative to combustion, because it causes reduced emissions and allows the production of syngas, which is a versatile gaseous fuel that can be used for cogeneration, Fischer-Tropsch synthesis, or the synthesis of other bio-fuels, such as methanol. For the safe and smooth exploitation of syngas, it is fundamental to have a high quality gas, with a high content of H_2 and CO and minimum content of pollutants, such as particulate and tars. In this work, experimental tests on lignite gasification are carried out in a bench-scale fluidized-bed reactor with olivine as bed material, chosen for its catalytic properties that can enhance tar reduction. Some operating parameters were changed throughout the tests, in order to study their influence on the quality of the syngas produced, and pressure fluctuation signals were acquired to evaluate the fluidization quality and diagnose correlated sintering or the agglomeration of bed particles. The effect of temperature and small air injections in the freeboard were investigated and evaluated in terms of the conversion efficiencies, gas composition, and tar produced.

Keywords: lignite; lignite gasification; fluidized-bed gasifier; olivine

---

1. Introduction

The environmental issues associated with global warming have resulted in a strong tendency towards the reduction of greenhouse gas emissions in the field of energy production. This trend has thus drawn attention to clean-coal technologies, such as gasification, which, compared to conventional combustion, can reduce CO_2 emissions by up to 90% [1,2]. The gasification process consists of partial oxidation of an organic feedstock that takes place at high temperatures (around 700–1200 °C) in the presence of a gasification agent (air, steam, oxygen, or a combination of these). Gasification consists of a combination of chemical processes, such as drying, pyrolysis, combustion, and partial oxidation. The main product is syngas, which is a mixture of gases with a high calorific value, typically composed of H_2, CO, CO_2, and CH_4. Some undesired products are also generated during the process, such as particulate, acid gases, and tars (condensable heavy hydrocarbons) [3], which have to be removed in order to make the gas usable for energy purposes or the production of base chemicals.

Gasification can be a promising option for exploiting abundant carbonaceous solid resources, such as lignite [4], for the production of a versatile fuel gas with a wide range of possible final uses. The product gas can in fact be used for the production of electricity in integrated gasification combined cycle systems (IGCC) or fuel cells [5], or it can be exploited for the synthesis of liquid fuels [6] and

Fischer-Tropsch synthesis [7,8], thus reducing imports of oil and natural gas from outside Europe and reducing emissions from the transport sector. Lignite, with its high heating value and low volatile content compared to biomass, could be a very suitable feedstock for the gasification process [9,10].

Olivine, a mineral mainly composed of magnesium, iron oxide, and silica, is used as an inventory of the fluidized-bed reactor, because of its advantageous qualities. As stated in the literature, olivine is recommended as bed material because of its reported activity in tar reduction, which is comparable to that of calcined dolomite. Furthermore, olivine has a stronger attrition resistance, compared to dolomite, which makes it more suitable as bed material [11,12].

Previous works have already reported lignite gasification in fluidized-bed reactors. Bayarsaikhan et al. studied lignite steam gasification (particles of 500–1000 μm) in a fluidized-bed reactor with a bed of silica and alumina in the range of temperatures of 850–950 °C [13]. Additionally, Kern et al. investigated lignite gasification (particles of 2–6 mm) in the dual fluidized-bed gasifier developed by Vienna University with olivine as bed material, at an operating temperature of 850 °C, with steam/fuel ranging from 0.9 to 1.4 [14]. Furthermore, Karimipour et al. performed a statistical analysis based on the experimental results of lignite gasification (particles of 70–500 μm) in a fluidized-bed reactor of silica sand particles with steam and oxygen as oxidizing agents [15].

The novelty of the present work is the assessment of the fluidized-bed technology for the steam gasification of small particle size lignite (around 50 μm) pre-treated by the WTA process (fluidized-bed drying process with internal waste heat utilization, in german: Wirbelschichttrocknung mit interner Abwärmenutzung), with a bed of olivine particles in a temperature range of 750–850 °C and a steam/fuel ratio equal to 0.65. Furthermore, in this work, the operating temperature was changed in order to study its effect on the gas quality, and in some of the experimental tests, air injections were added in the freeboard, in order to reproduce, on a smaller scale, the oxygen added in the post gasification zone of the High Temperature Winkler (HTW) gasifier [2,16], and thus to assess its effectiveness in tar reduction.

This work was carried out within the European project LIG2LIQ [17], whose aim is to develop an economically efficient concept for the production of liquid fuels, such as Fischer-Tropsch fuels or methanol, from lignite and solid recovered fuel from municipal waste by means of HTW gasification technology. Therefore, in the first step, the concept of HTW gasification, which was optimized with respect to the cold gas efficiency and carbon conversion efficiency, using lignite/waste mixtures as feedstocks for the production of syngas, was studied on a small scale. This paper describes the results of gasification experiments using only lignite as fuel in a laboratory fluidized bed. In the second step, experiments with a mixture of municipal waste and lignite were carried out.

Lignite gasification tests were carried out in a bench-scale fluidized-bed reactor, with the aim of investigating the best conditions to produce a high-quality syngas, with a low tar content and high H_2 and CO fraction, foreseeing downstream processes for Fischer-Tropsch or methanol synthesis. Tests were carried out, as mentioned above, changing the operating temperature and carrying out small air injections in the freeboard of the gasifier, in order to study the effect of a temperature increase in the upper part of the reactor aimed at enhancing tar conversion. The results were evaluated in terms of the syngas composition, tar content, gas yield, and conversion rates. Additionally, pressure fluctuation signals were acquired in the reactor freeboard during experimental tests, in order to evaluate the fluidization quality at the explored process conditions and to detect possible sintering or particle agglomeration within the bed inventory; in fact, these phenomena can increase the particles' average diameter, and then negatively affect the fluidization properties [18].

Therefore, the aim of the work was to evaluate the quality of the product gas obtained from the gasification of fine lignite in a fluidized bed of olivine particles at different temperatures, and to study the effect of air injections in the enhancement of tar reduction. In addition, the analysis of the materials before and after tests and the pressure fluctuation analysis helped to assess the eventuality of undesired phenomena, such as ash melting and the aggregation of bed particles, which could lead to defluidization of the bed.

2. Materials and Methods

2.1. Experimental Test Rig

The bench-scale gasification reactor represented in Figure 1 was used to carry out the experimental tests. The gasifier consisted of a cylindrical stainless steel reactor (internal diameter of 100 mm and height of 850 mm) externally heated with a 6 kW electric furnace. Steam was used as a gasification agent and a flow of nitrogen was added in order to fluidize the bed; they were sent to a wind-box to be mixed and pre-heated, and then fed from the bottom of the gasifier through a porous ceramic distribution plate. In order to investigate the effect of an increase of temperature in the freeboard, in some tests, a small stream of air was injected in the freeboard through a steel tube of a 6 mm diameter. The bed material used was sintered, calcined olivine particles provided by Magnolithe GmbH [19] with a $d_{3,2}$ diameter of 317 μm and density of 3000 kg/m^3, with the following composition by weight: MgO, 48%–50%; SiO$_2$, 39%–42%; and Fe$_2$O$_3$, 8%–10%. As reported in the literature, calcination at high temperatures (around 1100 °C or higher) allows the iron oxides contained in the olivine particles to emerge at the surface, and thus to be available for catalytic reactions [20–22]. The height of the bed in the reactor was approximately 200 mm. The feedstock consisted of WTA lignite, and Rhenish lignite from a process of fluidized-bed drying with internal waste heat utilization [23], supplied by RWE Power AG. Lignite was continuously fed to the bed of the reactor by means of a screw feeder and a feeding probe that delivered the material directly to the fluidized bed. The feeding probe was purged with a small N$_2$ flow, in order to help the fall of the feedstock and to avoid the material from clogging the tube. The reactor was designed to host a ceramic filter candle in the freeboard above the bed, through which the product gas was forced to pass to exit the gasifier; in this way, the solid particulate remained on the external surface of the candle and the dust-free gas could exit the reactor.

The product gas downstream of the gasifier was sent to heat exchangers for gas cooling and condensation of the residual steam. Circulation of the gas was granted by a vacuum pump. The flow of the dry product gas was measured with a mass flow controller, and its composition was analysed by online gas analyzers; a slipstream of the gas produced, about 1 Nl/min, was sent to the tar sampling unit, carried out following the specification of the standard CEN/TS 15439. The gas passed through five impinger bottles containing 2-propanol and kept in a cold bath at −20°C, to help the condensation of tars. The gas stream was moved by a vacuum pump and its flow rate was controlled by a mass flow controller. Finally, the main gas stream and the slipstream used for tar sampling were both sent to the vent.

Temperatures were measured by means of three K-type thermocouples: one was positioned in the reactor bed (T1); one was located in the freeboard (T2); and the other was situated at the exit of the candle, just at the outlet of the filter (T3). The operating temperature was considered the average of the values measured from T1, T2, and T3. Differential pressures were measured by means of pressure probes through the candle (ΔP1) and through the reactor (ΔP2). The pressure probes were connected to U-tube manometers, and were able to measure in the range of 0.5–90 mbar.

The fluidized-bed bench-scale reactor was equipped with a vertical probe in its freeboard, for acquisitions of pressure fluctuation signals. The probe was connected to a piezoelectric pressure transducer, in turn transmitting its signal to a charge amplifier KISTLER 5019A, and the operation parameters were tuned so as to obtain the highest amplification, without overloading. The resulting amplified voltage signal was then digitally converted and stored on a PC, provided with a tailored LABVIEW® routine. The data acquisition frequency was 100 Hz, which is much higher than the values typically observed in gas-fluidized beds under study (less than 10 Hz). The duration of each acquisition was 2–3 min, to ensure their repeatability and significance [24]. Stored signals were processed by means of a MATLAB® script, which calculated their standard deviations, as well as the power spectral density function (PSDF), by fast Fourier transform [25]. Standard deviations are directly related to the size of bubbles erupting at the upper bed surface (the higher the standard deviation, the bigger the bubbles), while PSDF allows dominant frequencies of pressure fluctuations to be identified, related to the number of erupting bubbles [18].

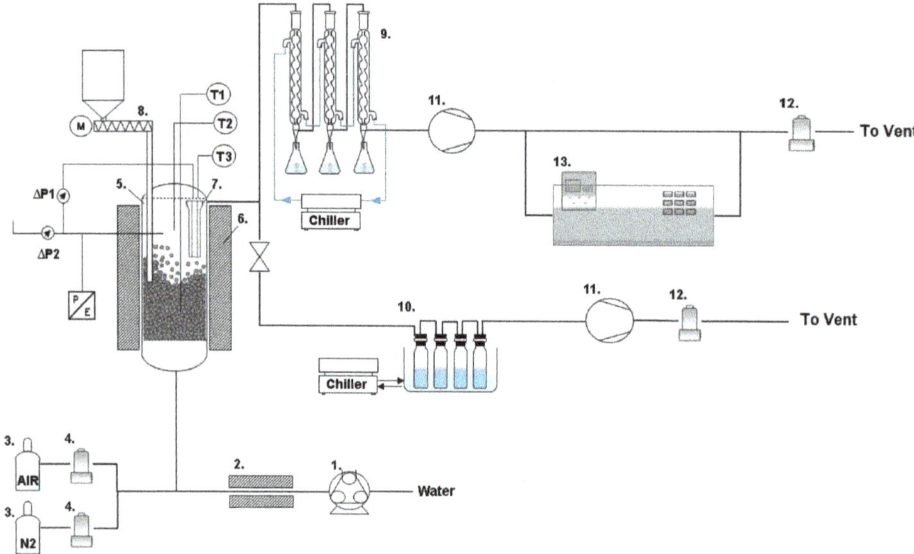

Figure 1. Scheme of the bench-scale gasification test rig. 1—water pump; 2—steam generator; 3—air and N_2 gas tanks; 4—air and N_2 mass flow controllers; 5—bubbling fluidized-bed gasifier; 6—electric furnace; 7—ceramic filter candle (OD of 60 mm, length of 440 mm); 8—screw conveyor for feeding fuel; 9—heat exchangers for steam condensation; 10—tar sampling unit; 11—vacuum pumps; 12—syngas mass flow controllers; 13—gas analyzers.

The lignite gasification tests were carried out using olivine as bed material, with a constant feeding rate of lignite and steam. For each test, a new batch of calcined olivine was inserted as bed material in the reactor, in order to compare tests with equivalent initial conditions avoiding the accumulation of ash in the bed material, which could have beneficial effects, such as the enhancement of tar conversion [26,27]. The operating temperature was changed in the tests between 750 and 850 °C and air injections were added in three of the six tests. Tests #1, #2, and #3 were carried out at operating temperatures of approximately 750, 800, and 850 °C, respectively. Tests #4, #5, and #6 had the same input conditions adopted in the first three tests, but with an additional air stream of 8 Nl/min injected in the freeboard of the reactor.

The operating conditions used are summarized in Table 1.

Table 1. Operating conditions of the test campaign.

Test	#1	#2	#3	#4	#5	#6
Feedstock (g/min)			12.75			
Operating Temperature (°C)	750	800	850	750	800	850
Air Injections (Nl/min)	-	-	-	8	8	8
Steam flow (g/min)			8.25			
N_2 flow (Nl/min)	8.83	17.20	7.39	7.32	7.44	7.44
Steam/Fuel (g/g)			0.65			

2.2. Analysis of Products

Downstream of the steam condensers and the vacuum pump, a slipstream of the product gas was sent to online gas analyzers for an evaluation of the composition. Online analyzers allowed H_2, CO, CO_2, CH_4, H_2S, and NH_3 to be detected (ABB URAS, LIMAS, CALDOS, and ULTRAMAT 6 Siemens).

The water content in the product gas was calculated from the quantity of water collected in the flasks connected to the steam condensers. The water conversion η_{wc} (%) was thus calculated as

$$\eta_{wc} = \frac{\dot{m}_{water,\,in} - \dot{m}_{water,\,out}}{\dot{m}_{water,\,in}} \times 100, \tag{1}$$

where $\dot{m}_{water,\,in}$ and $\dot{m}_{water,\,out}$ are the mass flows of the water input and output, respectively.

The quantity of dry product gas, measured by means of a mass flow controller, allowed the gas yield Y_{gas} $\left(Nm^3/kg_{feedstock}\right)$ to be calculated:

$$Y_{gas} = \frac{F_{gas,\,out}}{F_{feedstock,\,in}}, \tag{2}$$

where $F_{gas,\,out}$ is the total dry N_2-free volume of gas flow produced, and $F_{feedstock,\,in}$ is the mass flow of the input feedstock.

From the analysis of the gas composition and the carbon content in the feedstock, it was possible to calculate the carbon conversion X_C (%):

$$X_C = \frac{n_{CO} + n_{CO_2} + n_{CH_4}}{n_{C_{in}}} \times 100, \tag{3}$$

where n_i represents the moles of the i carbonaceous species in the product gas (CO, CO_2, and CH_4), and $n_{C_{in}}$ represents the total moles of C in the feedstock input.

Post-combustion was carried out after gasification: an air stream was fed to the reactor and the gaseous products (CO and CO_2) were analysed and quantified, in order to evaluate the amount of the residual un-reacted char in the reactor. From this result, it was possible to calculate the char yield Y_{char} (%):

$$Y_{char} = \frac{m_{char}}{m_{feedstock}} \times 100, \tag{4}$$

where m_{char} is the mass of the residual char estimated by the post-combustion in the gasifier, and $m_{feedstock}$ is the mass of the total feedstock fed.

The cold gas efficiency η_{CG} (%) was calculated as

$$\eta_{CG} = \frac{LHV_{gas}F_{gas,\,out}}{LHV_{lignite}F_{feedstock,\,in}}, \tag{5}$$

where LHV_{gas} and $LHV_{lignite}$ are the lower heating values of the gas produced and lignite, expressed in MJ/Nm^3 and MJ/kg, respectively.

The liquid tar samples collected in the impinger bottles were analysed offline by means of HPLC (Hitachi "Elite LaChrom" L-2130) for the detection and quantification of heavy hydrocarbons in the product gas. The tar compounds chosen as representative of a typical tar composition [28] were as follows: phenol, toluene, styrene, indene, naphthalene, acenaphthylene, fluorene, phenanthrene, anthracene, fluoranthene, and pyrene.

After each test, the bed material was extracted from the reactor and sieved, in order to separate the fly-ashes accumulated in the bed during the experimental run from the olivine particles; samples of ashes were thus collected for analysis after tests. Afterwards, the ashes were analysed by means of SEM/EDS and XRD analyses, in order to study their morphology and the present elements and compounds.

3. Results and Discussion

3.1. Analysis of Materials Pre-Test

3.1.1. Lignite

The characterization of lignite, supplied by Technische Universität Darmstadt (TUDA), and its particle size distribution, are reported in Table 2 and Figure 2, respectively. The particle size analysis was performed with a Malvern Mastersizer 2000.

Table 2. Proximate and ultimate analysis of lignite.

	Weight %
Total Moisture content (%)	16.09
Ash content (dry basis) (%)	13.86
Volatile matter (dry basis) (%)	38.57
C (dry basis) (%)	67.51
H (dry basis) (%)	4.88
N (dry basis) (%)	0.76
S (dry basis) (%)	1.11

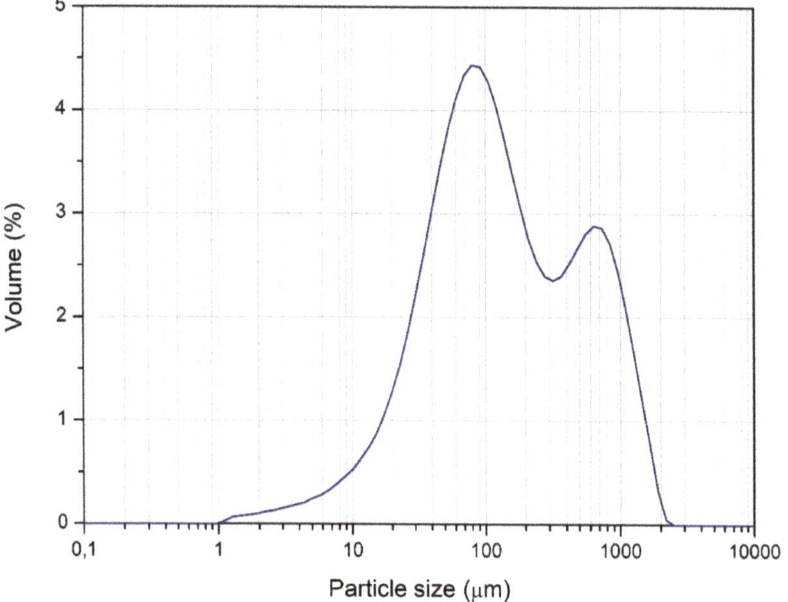

Figure 2. Particle size distribution of lignite.

Figure 2 shows that lignite has a bimodal particle size distribution, meaning that there are two main particle diameters: a high content of particles with a smaller diameter (around 70 µm) and another relevant fraction of particles with a larger diameter (around 600 µm). The $d_{3,2}$ diameter of lignite given by the particle size analysis is 44.08 µm.

As the average diameter of the feedstock particles is very small, working with the filter candle in the freeboard of the gasifier is recommended, in order to avoid the entrainment of lignite particles outside of the reactor with the gas flow.

3.1.2. Olivine before Tests

The particle size analysis of olivine before the gasification tests is reported in Figure 3.

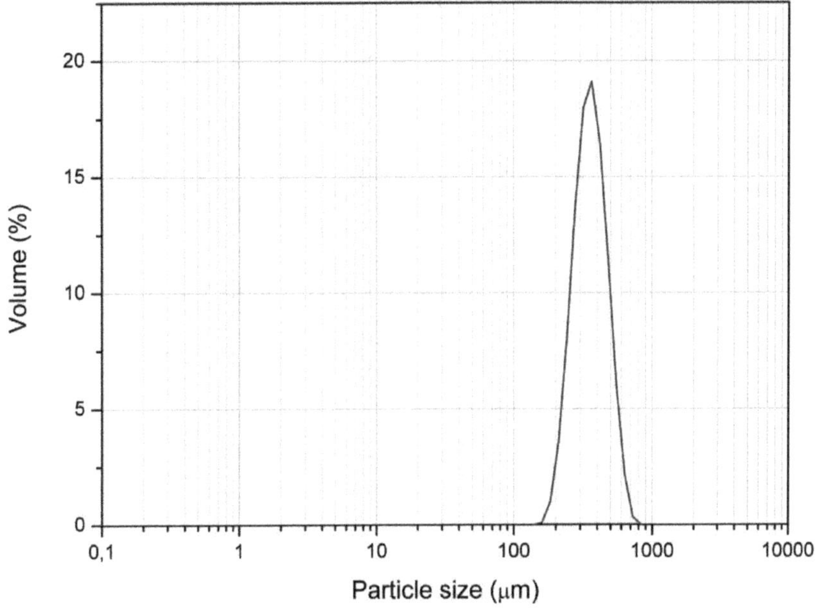

Figure 3. Particle size distribution of olivine.

The $d_{3,2}$ diameter of olivine given by the particle size analysis is 316.75 µm.

3.2. Gasification Results

The results obtained are summarized in Table 3.

Table 3. Results of lignite gasification tests.

Test	#1	#2	#3	#4	#5	#6
Bed Temperature (°C)	752	813	841	757	800	850
Avg Temperature (°C)	757	825	842	741	786	818
Air Injections (Nl/min)	-	-	-	8	8	8
Steam/Fuel	0.61	0.64	0.65	0.65	0.65	0.66
Length test (min)	120	120	120	121	66	50
H_2O conversion (%)	38.91	57.86	56.07	28.14	32.92	35.04
C conversion (%)	66.12	68.52	75.41	62.50	67.51	90.17
Gas yield (Nm^3/kg)	1.35	1.41	1.37	1.03	1.15	1.46
H_2 (%) dry N_2-free	55.11	54.28	53.16	49.78	49.38	48.48
CO (%)dry N_2-free	22.75	27.91	31.03	19.93	23.01	30.69
CO_2 (%) dry N_2-free	18.56	13.53	12.87	26.92	24.73	17.76
CH_4 (%) dry N_2-free	3.58	4.28	2.95	3.37	2.89	3.08
NH_3 (ppm) dry N_2-free	1094	1355	761	902	957	886
H_2 (Nl/min)	10.12	9.65	10.07	7.63	7.80	10.06
Char yield (%)	17.06	8.93	8.00	14.91	7.86	
η_{CG} (%)	57.11	61.27	58.21	44.58	42.17	58.94
Tar content (mg/Nm^3)	2461	938	481	2259	N.A.	305
Mass balance (err %)	1.25	6.11	4.95	3.25	5.62	

The length of tests #5 and #6 was shorter compared to the previous tests; however, the gas composition analysed online during the tests reached a steady state after around 10 min, giving stable values for the entire length of the test; a 60 min period was therefore considered sufficient for an evaluation of the results. The amounts of char and ash produced during gasification were evaluated in relation to the total amount of lignite fed during the experimental run. Consequently, the evaluation of their content was not affected by the length of the test.

From the comparison of the results obtained in the first three tests, it is possible to observe that, in general, at higher temperatures, the product gas has a higher quality in terms of the H_2O and C conversions and gas yield. Moreover, the gas composition in the tests carried out at 850 °C displays an increase in CO and decrease in CO_2, probably caused by the higher extent of steam gasification reactions that takes place for higher temperatures, and the lower extent of the WGS reaction, enhanced at lower temperatures (~600 °C). Starting from the syngas compositions obtained in the six experimental runs, the equilibrium contents of CO and CO_2 were calculated with the software Aspen Plus, in order to compare the experimental and equilibrium compositions. The equilibrium compositions calculated were approximately equal to the gas compositions obtained in the experimental tests, meaning that the WGS reaction had reached equilibrium. For higher temperatures, the char yield is also lower, which is proof of the higher conversion of carbon and thus lower amount of solid un-reacted residual char. Furthermore, the amount of tar produced at higher temperatures is lower compared to the cases with lower temperatures. In the tests conducted at 750 °C, the tar content is around 2.5 g/Nm^3, while at 850 °C, it reduces to 0.5 g/Nm^3. Regarding the tar content in general, it was observed that in tests with lignite, the amount of tar produced is lower compared to in similar tests carried out with biomass in the same experimental reactor with olivine as bed material [27,29,30]. Biomass gasification tests in similar conditions carried out at 800 °C produced tar contents ≥3300 mg/Nm^3, while lignite gasification at the same temperature (test #2) produced a tar content <950 mg/Nm^3. This phenomenon could be related to the lower volatile content of lignite compared to biomass (~50% versus ~70%, respectively [27]). In fact, volatile matter has been reported to make organic feedstocks more susceptible to tar formation [31].

The tests with and without air injections in the freeboard were evaluated in a comparison. As expected, in the tests with air injections, there is a higher content of CO_2 in the product gas. Furthermore, in the tests with air injections, the H_2O conversion was lower, probably because, being a product of combustion, it increases during the reaction. The H_2 content and its production in terms of Nl/min are lower compared to the case without air injection. It is possible that some of the produced H_2 was consumed in the combustion reactions with the injected O_2. The difference between the H_2 produced in test #2 (without air injections) and test #5 (with air injections), and their corresponding difference in H_2O content in the syngas are both in the order of 0.1 mol/min. This supports the hypothesis that the missing H_2 in the tests with air injections has been combusted and converted into additional H_2O, as confirmed by the consistency of the reported values of molar flows. In all of the tests, the H_2S content was approximately 400 ppm on a dry N_2-free basis. The NH_3 content, favored by the presence of steam as a gasification agent [32], was around 1000 ppm, with lower values at 850 °C. The higher NH_3 content at 800 °C, noticed both in the tests with and without air injections, was also observed by Xie at al. in gasification experiments on coal macerals [33], and could be related to the combination of two effects: the increase of NH_3 production from the N-containing structures in coal enhanced in steam gasification with a higher temperature [32,34,35], and the thermal decomposition of NH_3 occurring for increasing temperatures, as found in the literature [36]. For tests #4, #5, and #6, in which air was injected in the freeboard, the presence of O_2 increases the possibility of NH_3 combustion and consequently decreases its content in the product gas, showing the same trend as a function of temperature.

As mentioned above, in tests #4, #5, and #6, combustion of part of the gas took place, as expected, as a consequence of the air injections, especially those performed in order to increase the temperature in the freeboard and enhance the reactions of tar decomposition. In spite of the combustion reactions, it was observed that the tar content was not really affected by the air injections. The explanation for this could be that in the bench-scale gasifier, the temperature of the reactor is controlled by the electric furnace, so the temperature increase caused by the combustion did not have a relevant effect for the promotion of the tar conversion reactions. In addition, due to the reduced dimensions of the bench-scale reactor, the air injections in the freeboard are close to the exit of the gasifier, and consequently, tars could have had too little residence time to decompose. Moreover, the high content of inert N_2 in the air injected, which dilutes the O_2, could be the cause of the attenuation of the temperature increase due to the combustion, thus reducing the beneficial effect of the air injections.

A global mass balance was carried out, taking into account the mass flows of lignite and steam as inputs for the duration of the test. The outputs were calculated as the sum of the masses of the gases produced, the liquid water condensed downstream of the reactor, the tar contents in the samples (reported as the total gas flow), the ash content separated and collected from the bed material after the tests, and the char and residual carbon in the reactor (including the carbon particles deposited on the surface of the filter candle), which were evaluated from the post-combustion carried out after each test.

3.3. Analysis of Materials after the Test

Lignite Ash

The ashes produced during the tests were collected and analysed with SEM/EDS. Figures 4 and 5 show that in some spots of the analysed areas, Si and Mg are found together, probably corresponding to olivine particles still present in the ashes after their separation from the bed material. Furthermore, it was noticed that Ca, which is largely present in lignite ashes [37,38], was often found together with S in some spots analysed, as highlighted by the colored maps showing the presence of single elements in the particles.

The ashes generated during the tests were also analysed with XRF and XRD. The elements detected and quantified with the XRF analysis are reported in Table 4. The value of loss of ignition was 43.28%. The result of the XRD analysis is reported in Figure 6.

The results of the XRF analysis show that the elements present in major quantities are Ca, S, Fe, Si, Mg, and Na, as observed in the results obtained from the SEM/EDS analysis.

Figure 6 shows the diffraction spectrum of the ashes collected in the reactor after gasification and post-combustion. The broad halo visible at low Bragg angles indicates the presence of amorphous phases typical of ashes [39]. The peaks identified show the presence of K and Ca oxides present in the ashes, and of Si, Mg, and Fe compounds, present in both the ashes and the olivine particles [40]. Furthermore, the phase $CaSO_4$ was identified, as proof of the affinity between Ca and S already observed in the results of the SEM/EDS analysis. The observation of Ca and S together, in the characterization analysis of the ashes, could be proof of the capacity of Ca to react with S, as reported in the literature [41,42], and thus to retain the sulphur compounds in the ashes.

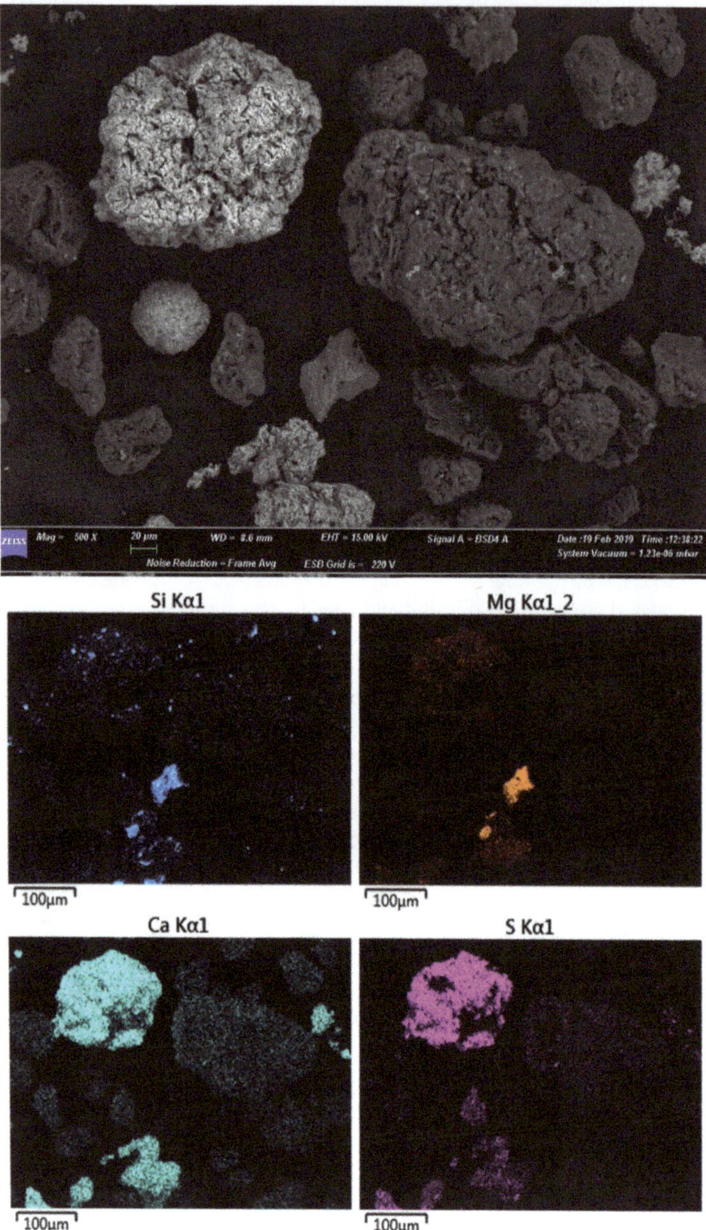

Figure 4. SEM/EDS of ash from lignite (image 1).

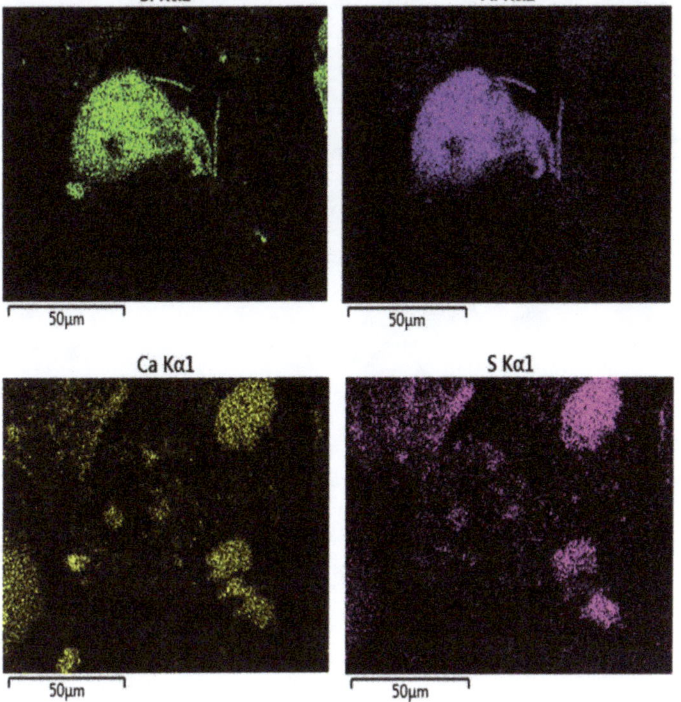

Figure 5. SEM/EDS of ash from lignite (image 2).

Table 4. XRF analysis of lignite ash produced during the gasification test.

Element	Concentration (%)	Absolute Error (%)
Na	0.705	0.037
Mg	2.399	0.016
Al	0.149	0.0046
Si	3.690	0.006
S	6.577	0.004
Cl	0.113	3×10^{-4}
K	0.094	0.0011
Ca	12.460	0.01
Ti	0.108	7×10^{-4}
Cr	0.072	2.5e-04
Mn	0.176	4×10^{-4}
Fe	5.964	0.005
Ni	0.067	3×10^{-4}
Cu	0.0143	1.4×10^{-4}
Sr	0.145	1×10^{-4}
Ba	0.169	0.0011

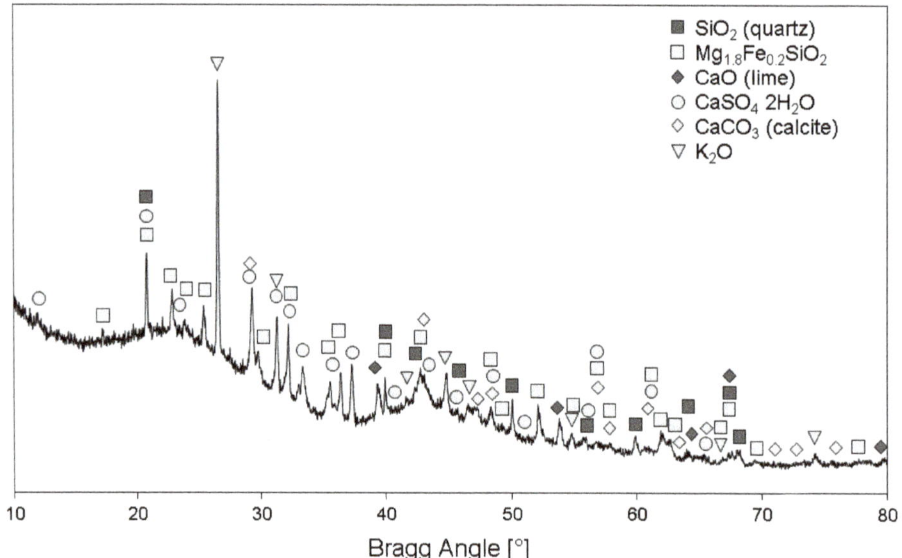

Figure 6. XRD analysis of lignite ash.

3.4. Pressure Fluctuation Analysis

More than 150 acquisitions of pressure fluctuation signals were performed during the six tests discussed above, always depicting the situation exemplified by Figure 7 for test #1. During the preliminary heating of the reactor, under an N_2 flowrate high enough to fluidize the bed, PSDF resulted in dominant frequencies of around 3–4 Hz (Figure 7a), which were compatible with the desired bubbling fluidization regime (usually less than 10 Hz) [24] and were assumed to be characteristic of the fresh olivine bed inventory. As soon as the gasification session started, a series of low-frequency phenomena (<1 Hz) took action with a high power spectral density, partially disguising those related to bed bubbles in the PSDF; the latter were still detectable, maintaining their dominant characteristic frequencies at 3–4 Hz (Figure 7b). The low-frequency phenomena were associated with the peristaltic pump feeding water and the instantaneous devolatilization of lignite particles. As further confirmation of this last observation, pressure fluctuation signals were also acquired during post-combustion, when water and lignite were no longer fed. In related PSDF, dominant frequencies clearly reappeared within the range of 3–4 Hz, without any high power spectral density disturbance at less than 1 Hz (Figure 7c).

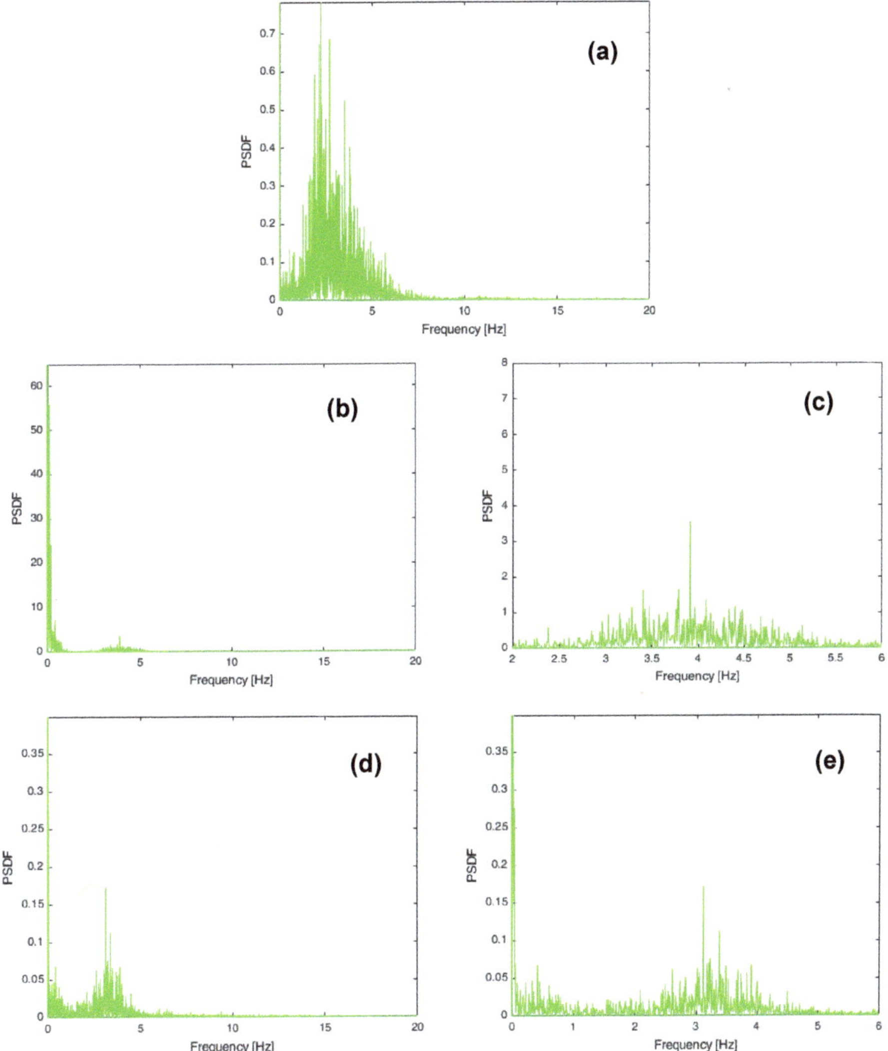

Figure 7. Power spectral density function (PSDF) of pressure fluctuations signals from test #1: pre-heating under N$_2$, T = 767 °C (**a**); gasification (**b**) with magnification of a 2–6 Hz range (**c**); and post-combustion (**d**) with magnification of a 0–6 Hz range (**e**).

For the case shown in Figure 7, standard deviations of pressure fluctuation signals were 0.98 mbar for preliminary heating (Figure 7a), 3.14 mbar during gasification (Figure 7b), and 0.39 mbar for post-combustion (Figure 7c), with trends and orders of magnitude representative of all tests. The fresh olivine beds during preliminary heating, approaching temperatures of the gasification, had pressure fluctuations with a standard deviation of around 1 mbar. It increased several times during gasification (because of the powerful low-frequency phenomena mentioned above), and then returned to the order of 1 mbar in the post-combustion phase.

All these observations allowed us to conclude that, for the investigated process conditions, the olivine bed inventory did not undergo modifications able to modify its fluidization quality, as sintering between olivine particles or due to lignite and ashes. This would have caused an increase of the average particle diameter and then a drop in the fluidization quality, detectable by PSDF modifications. SEM analyses confirmed the absence of particle sintering or agglomeration. Figure 8 shows an SEM image of olivine after the test, in which it is possible to see that the dimensions of the particles are approximately between 200 and 400 μm, similar to the particle size of olivine before the test. No sign of agglomeration or particle sintering is observed from the SEM images. In the tests carried out in this work, the absence of agglomeration issues was probably proof of the suitability of the operating temperatures chosen, which were lower than the ash melting point. Furthermore, fluidized-bed technology has the well-known advantage of guaranteeing a good mixing of the materials and thus a uniform distribution of temperatures across the reactor volume, avoiding the presence of hot spots that could be responsible for ash melting.

The evaluation of the agglomerate formation for longer operational times, and therefore with an increased ash content due to accumulation, was not taken into account, because the tests carried out in this work aimed to reproduce the operation of the HTW gasifier, in which the ash produced during the process is discharged from the bottom of the reactor [43], and consequently, the accumulation of high contents of ash does not take place.

Figure 8. SEM analysis of olivine after the test.

4. Conclusions

Steam gasification tests with lignite were carried out in a bench-scale fluidized-bed gasifier, in order to study the quality of the gas produced at different operating conditions. The correct operation of the gasification process with small lignite particles as feedstock (~44 μm) was possible thanks to the ceramic filter candle integrated in the freeboard of the gasifier, which prevented the entrainment of particles outside the reactor. The bed material used was olivine, and the Steam/ Fuel ratio (S/F) was kept at approximately 0.65. The effect of the temperature and air injections in the freeboard was investigated in terms of the gas composition and tar produced.

The results obtained showed that the increase of the operating temperature caused an improvement of the gas quality, in particular, higher conversion rates and gas yields, and a lower amount of tar produced. Tests carried out with air injections in the freeboard did not show the desired effect of tar reduction, probably because the combustion of part of the syngas did not cause the increase of temperature expected in the externally heated bench-scale gasifier used in this work. However, the amount of tar produced was smaller compared to other tests carried out with biomass at similar operating conditions in the same bench-scale gasifier. Lignite could be less prone to tar production because of its lower volatile content compared to lignocellulosic biomass.

The ashes produced during the gasification test were analysed with XRD and SEM/EDS analysis, and an affinity between Ca and S was noticed, probably indicating the capacity of Ca to retain S in the ashes.

Pressure fluctuations were acquired in the freeboard of the fluidized bed during each test, in order to diagnose possible alterations in the fluidization quality, related to sintering phenomena involving bed particles. The results from the signal analyses, in terms of dominant frequencies in the power spectral density functions and standard deviations, did not show any worsening of the fluidization quality for all investigated gasification conditions. This was confirmed by the SEM analysis, which did not exhibit clusters of particles.

Author Contributions: Conceptualization, S.R., K.G. and A.D.C.; methodology, S.R., K.G. and A.D.C.; software, A.D.G.; validation, K.G., A.D.C. and S.R.; formal analysis, E.S., J.M. and A.D.G.; investigation, E.S., J.M. and S.R.; resources, K.G., S.R. and A.D.C.; data curation, E.S., J.M. and A.D.G.; writing—original draft preparation, E.S.; writing—review and editing, E.S., K.G., A.D.C., A.D.G., J.M.; visualization, E.S. and A.D.G.; supervision, K.G. and A.D.C.; project administration, K.G.; funding acquisition, K.G. All authors have read and agree to the published version of the manuscript.

Funding: This research was funded by the European Commission managed Research Fund for Coal and Steel (RFCS), grant number 796585. The APC was funded by the University of L'Aquila with the budget of the EU Project LIG2LIQ for dissemination activity.

Acknowledgments: The authors kindly acknowledge the financial support of the European Project LIG2LIQ (RFCS-01-2017 n°796585) co-funded by the European Commission managed Research Fund for Coal and Steel (RFCS). In particular, the authors would like to thank the ICHPW and TUDA, partners of the LIG2LIQ Project, for their characterizations, as well as Fabiola Ferrante (University of L'Aquila), for the XRF analysis. Moreover, the authors would like to thank RWE Power AG for the lignite feedstock supplied. We also give thanks for the financial support provided by the DFG in the framework of the Excellence Initiative, Darmstadt Graduate School of Excellence Energy Science and Engineering (GSC 1070).

Abbreviations

IGCC	Integrated Gasification Combined Cycle
HTW	High Temperature Winkler
OD	Outer Diameter
WTA	Fluidized-bed drying process with internal waste heat utilization (German: Wirbelschichttrocknung mit interner Abwärmenutzung)
HPLC	High Performance Liquid Chromatography
Avg	Average
PSDF	Power Spectral Density Function
S/F	Steam to Fuel
N.A.	Not Available
XRD	X-ray Diffraction
XRF	X-ray fluorescence
SEM	Scanning Electron Microscope
EDS	Energy-dispersive X-ray Spectroscopy

Symbols

η_{wc}	Water conversion
$\dot{m}_{water,\,in}$	Mass flow of water input
$\dot{m}_{water,\,out}$	Mass flow of water output
Y_{gas}	Gas yield
$F_{gas,\,out}$	Total dry N_2-free volume gas flow produced
$F_{fuel,\,in}$	Mass flow of the input fuel in the gasifier
n_i	Moles of the carbonaceous species in the product gas (CO, CO_2, CH_4),
$n_{C_{in}}$	Total moles of C in the feedstock input
Y_{char}	Char yield
LHV_{gas}	Lower heating value of the product gas
$LHV_{lignite}$	Lower heating value of lignite
m_{char}	Mass of the residual char estimated by the post-combustion in the gasifier
m_{fuel}	Mass of the total fuel fed to the gasifier
X_C	Carbon conversion (%)
η_{CG}	Cold gas efficiency (%)

References

1. Stiegel, G.J.; Maxwell, R.C. Gasification technologies: The path to clean, affordable energy in the 21st century. *Fuel Process. Technol.* **2001**, *71*, 79–97. [CrossRef]
2. Krause, D.; Herdel, P.; Ströhle, J.; Epple, B. HTW^{TM}-gasification of high volatile bituminous coal in a 500 kWth pilot plant. *Fuel* **2019**, *250*, 306–314. [CrossRef]
3. Milne, T.A.; Evans, R.J.; Abatzaglou, N. *Biomass Gasifier "'Tars'": Their Nature, Formation, and Conversion*; National Renewable Energy Laboratory: Golden, CO, USA, 1998.
4. Stec, M.; Czaplicki, A.; Tomaszewicz, G.; Słowik, K. Effect of CO2 addition on lignite gasification in a CFB reactor: A pilot-scale study. *Korean J. Chem. Eng.* **2018**, *35*, 129–136. [CrossRef]
5. Aravind, P.V.; De Jong, W. Evaluation of high temperature gas cleaning options for biomass gasification product gas for Solid Oxide Fuel Cells. *Prog. Energy Combust. Sci.* **2012**, *38*, 737–764. [CrossRef]
6. Sikarwar, V.S.; Zhao, M.; Fennell, P.S.; Shah, N.; Anthony, E.J. Progress in biofuel production from gasification. *Prog. Energy Combust. Sci.* **2017**, *61*, 189–248. [CrossRef]
7. Venvik, H.J.; Yang, J. Catalysis in microstructured reactors: Short review on small-scale syngas production and further conversion into methanol, DME and Fischer-Tropsch products. *Catal. Today* **2017**, *285*, 135–146. [CrossRef]
8. Wang, H.; Pei, Y.; Qiao, M.; Zong, B. Design of bifunctional solid catalysts for conversion of biomass-derived syngas into biofuels. In *Production of Biofuels and Chemicals with Bifunctional Catalysts*; Springer Nature: Singapore, 2017; pp. 137–158, ISBN 9789811051371.
9. Matsuoka, K.; Hosokai, S.; Kato, Y.; Kuramoto, K.; Suzuki, Y.; Norinaga, K.; Hayashi, J.I. Promoting gas production by controlling the interaction of volatiles with char during coal gasification in a circulating fluidized bed gasification reactor. *Fuel Process. Technol.* **2013**, *116*, 308–316. [CrossRef]
10. Benedikt, F.; Fuchs, J.; Schmid, J.C.; Müller, S.; Hofbauer, H. Advanced dual fluidized bed steam gasification of wood and lignite with calcite as bed material. *Korean J. Chem. Eng.* **2017**, *34*, 2548–2558. [CrossRef]
11. Rapagnà, S.; Gallucci, K.; Foscolo, P.U. Olivine, dolomite and ceramic filters in one vessel to produce clean gas from biomass. *Waste Manag.* **2017**. [CrossRef]
12. Devi, L.; Ptasinski, K.J.; Janssen, F.J.J.G. A review of the primary measures for tar elimination in biomass gasification processes. *Biomass Bioenergy* **2002**, *24*, 125–140. [CrossRef]
13. Bayarsaikhan, B.; Sonoyama, N.; Hosokai, S.; Shimada, T.; Hayashi, J.I.; Li, C.Z.; Chiba, T. Inhibition of steam gasification of char by volatiles in a fluidized bed under continuous feeding of a brown coal. *Fuel* **2006**, *85*, 340–349. [CrossRef]
14. Kern, S.; Pfeifer, C.; Hofbauer, H. Gasification of lignite in a dual fluidized bed gasifier - Influence of bed material particle size and the amount of steam. *Fuel Process. Technol.* **2013**, *111*, 1–13. [CrossRef]
15. Karimipour, S.; Gerspacher, R.; Gupta, R.; Spiteri, R.J. Study of factors affecting syngas quality and their interactions in fluidized bed gasification of lignite coal. *Fuel* **2013**, *103*, 308–320. [CrossRef]

16. Herdel, P.; Krause, D.; Peters, J.; Kolmorgen, B.; Ströhle, J.; Epple, B. Experimental investigations in a demonstration plant for fluidized bed gasification of multiple feedstock's in 0.5 MW th scale. *Fuel* **2017**, *205*, 286–296. [CrossRef]

17. LIG2LIQ Eu Project. Available online: https://www.lig2liq.eu/ (accessed on 17 February 2020).

18. Gibilaro, L.G. *Fluidization Dynamics The Formulation and Applications of a Predictive Theory for the Fluidized State*; Butterworth-Heinemann: Oxford, UK, 2001.

19. Magnolithe Steel Industry Products. Available online: http://www.magnolithe.at/pages/en/firma/fa_rohstoffe.htm (accessed on 17 February 2020).

20. Rauch, R.; Pfeifer, C.; Bosch, K.; Hofbauer, H.; Swierczynski, D.; Courson, C.; Kiennemann, A. Comparison of different olivines for biomass steam gasification. *Proc. Conf. Sci. Therm. Chem. Biomass Convers.* **2004**, 799–809.

21. Christodoulou, C.; Grimekis, D.; Panopoulos, K.D.; Pachatouridou, E.P.; Iliopoulou, E.F.; Kakaras, E. Comparing calcined and un-treated olivine as bed materials for tar reduction in fluidized bed gasification. *Fuel Process. Technol.* **2014**, *124*, 275–285. [CrossRef]

22. Michel, R.; Ammar, M.R.; Poirier, J.; Simon, P. Phase transformation characterization of olivine subjected to high temperature in air. *Ceram. Int.* **2013**, *39*, 5287–5294. [CrossRef]

23. Klutz, H.J.; Moser, C.; Block, D.R.P. WTA Fine Grain Drying: Module for Lignite-fired Power Plants of the Future-Development and Operating Results of the Fine Grain Drying Plant. *VGB Power Tech.* **2006**, *86*, 11.

24. Gallucci, K.; Jand, N.; Foscolo, P.U.; Santini, M. Cold model characterisation of a fluidised bed catalytic reactor by means of instantaneous pressure measurements. *Chem. Eng. J.* **2002**, *87*, 61–71. [CrossRef]

25. Gallucci, K.; Gibilaro, L.G. Dimensional cold-modeling criteria for fluidization quality. *Ind. Eng. Chem. Res.* **2005**, *44*, 5152–5158. [CrossRef]

26. Rapagna, S.; D'orazio, A.; Gallucci, K.; Foscolo, P.U.; Nacken, M.; Heidenreich, S. Hydrogen rich gas from catalytic steam gasification of biomass in a fluidized bed containing catalytic filters. *Chem. Eng. Trans.* **2014**, *37*, 157–162.

27. D'Orazio, A.; Rapagnà, S.; Foscolo, P.U.; Gallucci, K.; Nacken, M.; Heidenreich, S.; Di Carlo, A.; Dell'Era, A. Gas conditioning in H2 rich syngas production by biomass steam gasification: Experimental comparison between three innovative ceramic filter candles. *Int. J. Hydrogen Energy* **2015**, *40*, 7282–7290. [CrossRef]

28. Di Marcello, M.; Gallucci, K.; Rapagnà, S.; Gruber, R.; Matt, M. HPTLC and UV spectroscopy as innovative methods for biomass gasification tars analysis. *Fuel* **2014**, *116*, 94–102. [CrossRef]

29. Rapagnà, S.; Gallucci, K.; Di Marcello, M.; Foscolo, P.U.; Nacken, M.; Heidenreich, S.; Matt, M. First Al2O3based catalytic filter candles operating in the fluidized bed gasifier freeboard. *Fuel* **2012**, *97*, 718–724. [CrossRef]

30. Savuto, E.; Di Carlo, A.; Steele, A.; Heidenreich, S.; Gallucci, K.; Rapagnà, S. Syngas conditioning by ceramic filter candles filled with catalyst pellets and placed inside the freeboard of a fluidized bed steam gasifier. *Fuel Process. Technol.* **2019**, *191*, 44–53. [CrossRef]

31. Font Palma, C. Modelling of tar formation and evolution for biomass gasification: A review. *Appl. Energy* **2013**, *111*, 129–141. [CrossRef]

32. Tian, F.J.; Yu, J.; McKenzie, L.J.; Hayashi, J.I.; Li, C.Z. Conversion of fuel-N into HCN and NH3 during the pyrolysis and gasification in steam: A comparative study of coal and biomass. *Energy Fuels* **2007**, *21*, 517–521. [CrossRef]

33. Xie, K.C.; Lin, J.Y.; Li, W.Y.; Chang, L.P.; Feng, J.; Zhao, W. Formation of HCN and NH33 during coal macerals pyrolysis and gasification with CO_2. *Fuel* **2005**, *84*, 271–277. [CrossRef]

34. Chang, L.P.; Xie, Z.L.; Xie, K.C. Study on the formation of NH_3 and HCN during the gasification of brown coal in steam. *Process Saf. Environ. Prot.* **2006**, *84*, 446–452. [CrossRef]

35. Li, C.Z.; Tan, L.L. Formation of NOx and SOx precursors during the pyrolysis of coal and biomass. Part III. Further discussion on the formation of HCN and NH3 during pyrolysis. *Fuel* **2000**, *79*, 1899–1906.

36. Zhou, J.; Masutani, S.M.; Ishimura, D.M.; Turn, S.Q.; Kinoshita, C.M. Release of fuel-bound nitrogen during biomass gasification. *Ind. Eng. Chem. Res.* **2000**, *39*, 626–634.

37. Demirbaş, A.; Aslan, A. Evaluation of lignite combustion residues as cement additives. *Cem. Concr. Res.* **1999**, *29*, 983–987.

38. Ilic, M.; Cheeseman, C.; Sollars, C.; Knight, J. Mineralogy and microstructure of sintered lignite coal fly ash. *Fuel* **2003**, *82*, 331–336.

39. Sathonsaowaphak, A.; Chindaprasirt, P.; Pimraksa, K. Workability and strength of lignite bottom ash geopolymer mortar. *J. Hazard. Mater.* **2009**, *168*, 44–50.

40. Vamvuka, D.; Pitharoulis, M.; Alevizos, G.; Repouskou, E.; Pentari, D. Ash effects during combustion of lignite/biomass blends in fluidized bed. *Renew. Energy* **2009**, *34*, 2662–2671.

41. Galloway, B.D.; Sasmaz, E.; Padak, B. Binding of SO3 to fly ash components: CaO, MgO, Na$_2$O and K$_2$O. *Fuel* **2015**, *145*, 79–83.

42. Hu, Y.; Watanabe, M.; Aida, C.; Horio, M. Capture of H$_2$S by limestone under calcination conditions in a high-pressure fluidized-bed reactor. *Chem. Eng. Sci.* **2006**, *61*, 1854–1863.

43. Toporov, D.; Abraham, R. Gasification of low-rank coal in the High-Temperature Winkler (HTW) process. *J. South. Afr. Inst. Min. Metall.* **2015**, *115*, 589–597.

Article

A New Design of an Integrated Solar Absorption Cooling System Driven by an Evacuated Tube Collector: A Case Study for Baghdad, Iraq

Adil Al-Falahi *, Falah Alobaid and Bernd Epple

Institut Energiesysteme und Energietechnik, Technische Universität Darmstadt, Otto-Berndt-Straße 2, 64287 Darmstadt, Germany; falah.alobaid@est.tu-darmstadt.de (F.A.); bernd.epple@est.tu-darmstadt.de (B.E.)
* Correspondence: adil.al-falahi@est.tu-darmstadt.de; Tel.: +49-6151-16-20724; Fax: +49-6151-16-22690

Received: 12 April 2020; Accepted: 21 May 2020; Published: 23 May 2020

Abstract: The electrical power consumption of refrigeration equipment leads to a significant influence on the supply network, especially on the hottest days during the cooling season (and this is besides the conventional electricity problem in Iraq). The aim of this work is to investigate the energy performance of a solar-driven air-conditioning system utilizing absorption technology under climate in Baghdad, Iraq. The solar fraction and the thermal performance of the solar air-conditioning system were analyzed for various months in the cooling season. It was found that the system operating in August shows the best monthly average solar fraction (of 59.4%) and coefficient of performance (COP) (of 0.52) due to the high solar potential in this month. Moreover, the seasonal integrated collector efficiency was 54%, providing a seasonal solar fraction of 58%, and the COP of the absorption chiller was 0.44, which was in limit, as reported in the literature for similar systems. A detailed parametric analysis was carried out to evaluate the thermal performance of the system and analyses, and the effect of design variables on the solar fraction of the system during the cooling season.

Keywords: solar cooling; solar cooling system; TRNSYS; absorption chiller; performance and analysis; solar energy

---

1. Introduction

There is growing demand for air conditioning in hot climate countries (due to increase in internal loads in buildings), and greater demand for thermal comfort by its users; thus, it is becoming one of the most important types of energy consumption [1]. Accordingly, the consumption of electrical power by refrigeration equipment begins to cause problems in the supply network on the hottest summer days.

Most buildings are provided with electrically driven vapor compression chillers. Currently, the energy for air conditioning is expected to increase tenfold by 2050 [2]. In Iraq, the demand for cooling and air conditioning is more than 50%–60% of total electricity demand (48% in the residential sector) [3]; thus, it contributes to increased CO_2 emissions, which could increase by 60% by 2030, compared to the beginning of the century (even though we urgently need to reduce) [4]. On the other hand, mechanical compression chillers utilize various types of halogenated organic refrigerants, such as HCFCs (hydrochlorofluorocarbons), which still contribute to the depletion of the ozone layer; this is why many of these refrigerants have been banned or are in the process of being banned.

To enhance a building's energy efficiency, solar-driven cooling systems seem to be an attractive alternative to conventional electrical driven compression units, as they achieve primary energy savings and reduce greenhouse gas emissions for solar fractions higher than about 50% [5]. They use refrigerants that do not harm the ozone layer and demand little external electric power supply.

The simulations of lithium bromide (LiBr)/water (H_2O) absorption cooling systems have a long history, but a general model for all circumstances is still elusive. Bani Younes et al. [6] presented a simulation of a LiBr–H_2O absorption chiller of 10 kW capacity for a small area of 100 m^2 under three different zones in Australia. They concluded that the best system configuration consists of a 50 m^2 flat plate collector and a hot water storage tank of 1.8 m^3. In Tunisia, a feasibility and sensitivity analysis of the solar absorption cooling system was conducted by Barghouti et al. [7] using TRNSYS (University of Wisconsin-Madison, Madison, WI, USA, 1994) and EES software. They concluded that a house of 150 m^2 required 11 kW of absorption chiller, with 30 m^2 of flat plate solar collectors and a 0.8 m^3 storage tank to cover the cooling load.

For their part, Martínez et al. [8] compared the simulation of a solar cooling system using TRNSYS software, with real data from a system installed in Alicante, Spain. The air-conditioning system was composed of a LiBr–H_2O absorption chiller with 17.6 kW capacity and 1 m^3 hot storage tank. The results show an approximation between the measured and simulated data, where the coefficient of performance (COP) of the absorption chiller from the experimental data was 0.691 while the COP of the simulated system reached a value of 0.73.

Burckhartyotros [9] described a 250 m^2 field of vacuum tube solar thermal collectors, which provided hot water at temperatures of about 90 °C, to drive lithium bromide/water absorption chiller with a capacity of 95 kW, utilized to cover the thermal loads for a building of 4000 m^2, which included offices, laboratories, and a public area.

Ketjoy et al. [10] evaluated the performance of a LiBr–H_2O absorption chiller with 35 kW cooling capacity, integrated with 72 m^2 of evacuated tube collectors (ETC) and an auxiliary boiler. They found that the solar absorption system had high performance with a ratio of 2.63 m^2 of collector area for each kW of air-conditioning.

A solar parabolic trough collector has been used beside a single effect LiBr–H_2O absorption chiller [11]. Peter Jenkins [12] studied the principles of the operation of the solar absorption cooling system. The total solar area was 1450 m^2. Wang [13] investigated the effect of large temperature gradients and serious nanoparticles, photothermal conversion efficiency on direct absorption solar collectors.

Hamza [14] studied the development of a dynamic model of a 3TR (Ton of Refrigeration) single-effect absorption cooling cycle that employs LiBr–water as an absorbent/refrigerant pair, coupled with an evacuated tube solar collector and a hot storage unit.

Rasool Elahi [15] studied the effect of using solar plasma for the enhancement operation of solar assisted absorption cycles. Behi [16] presented an applied experimental and numerical evaluation of a triple-state sorption solar cooling module. The performance of a LiCl–H_2O based sorption module for cooling/heating systems with the integration of external energy storage has been evaluated. Special design for solar collectors was investigated by Behi [16]. Related to thermo-economics, Salehi [17] studied the feasibility of solar assisted absorption heat pumps for space heating. In this study, single-effect LiBr–H_2O and NH_3-H_2O absorption, and absorption compression-assisted heat pumps were analyzed for heating loads of 2MW (Mega Watt). Using the geothermal hot springs as heat sources for refrigerant evaporation, the problem of freezing was prevented. The COP ranged between 1.4 and 1.6. Buonomano et al. [18] studied the feasibility of a solar assisted absorption cooling system based on a new generation flat plate ETC integrated with a double-effect LiBr–H_2O absorption chiller. The results of the experiment show that maximum collector efficiency is above 60% and average daily efficiency is about 40%, and they show that systems coupled with flat-plate ETC achieve a higher solar fraction (77%), in comparison with 66.3% for PTC (Parabolic Through Collector) collectors.

Mateus and Oliveira [19] performed energy and detailed economic analysis of the application of solar air conditioning for different buildings and weather conditions. According to their analysis, they consider that the use of vacuum tube collectors reduces the solar collector surface area of about 15% and 50% in comparison with flat solar collectors. According to the final report of the European Solar Combi+ project, the use of evacuated tube collectors allows for greater energy savings (between 15% and 30%) but the investment increases significantly [20].

Shirazi et al. [21] simulated four configurations of solar-driven LiBr–H_2O air-conditioning systems for heating and cooling purpose. Their simulation results revealed that the solar fraction of 71.8%

and primary energy conservation of 54.51% could be achieved by the configuration that includes an absorption chiller with a vapor compression cycle as an assistance cooling system.

Vasta et al. [22] analyzed the performance of an adsorption cycle under different climate zones in Italy. It was concluded that the performance parameters were influenced significantly by the design variables. They found that with the dry and wet cooler, the solar fraction could archive values of 81% and 50% at lower solar collector areas. In addition, it was found that the COP could reach 57% and 35% in the same collector arrangement.

In recent years, research projects on solar refrigeration have been carried out to develop new equipment, reducing costs and stimulating integration into the building air conditioning market.

Calise et al. [23] carried out a transitional simulation model using the TRNSYS software. The building was 1600 m^2 building; the system included the vacuum tube collectors of 300 m^2 and a LiBr–H$_2$O absorption chiller. It was found that a higher coefficient of performance (COP) was 0.80; optimum storage volume of 75 L/m^2 was determined when the chiller cooling capacity was 157.5 kW.

Djelloul et al. [24] simulated a solar air conditioning system for a domestic house using TRNSYS software. They indicate that to cover the cooling load of a house of 120 m^2 the best air-conditioning system configuration consisted of a single-effect Yazaki absorption chiller of 10 kW, 28 m^2 flat plate collectors with 35° inclination, and a hot storage tank of 0.8 m^3. They concluded that the ratio of the collector area per kW cooling was 2.80 m^2/kW.

In Iraq, the conventional electricity grid is not working well as the country struggles to recuperate from years of war [25]. However, Iraq is blessed with an abundance of solar energy, which is evident from the average daily solar irradiance, ranging from 6.5–7 kWh/m^2 (which is one of the highest in the world). This corresponds to total annual sunshine duration ranging between 2800–3000 h [26]. Accordingly, solar cooling technology promotion in Iraq appears to be of high importance, concerning development, and is part of the government's new strategy for promoting renewable energy projects.

It is clear from the literature that solar energy has a great influence on refrigeration and/or air conditioning processes. Different types and configurations of solar collectors have been applied for such purposes. The most used type was the evacuated tube collector (ETC). Moreover, it was noticed that LiBr–H$_2$O have been used for most of the research activities in this regard [27,28].

The aim of this work is to provide (1) a valuable roadmap related to solar-driven cooling systems operating under the Iraq climate to allow for sustained greenhouse gas emission reductions in the residential air conditioning sector, and (2) energetic performance analysis of solar driven cooling systems to investigate the best system design parameters.

2. Design Aspects

2.1. Thermal Solar Cooling System Description

Solar cooling technology uses the solar hot water system as an energy resource for the sorption cycle. The solar absorption cooling system (SACS) under investigation contains two main parts (see Figure 1): the heat medium production and cold medium production. The heat medium production includes solar thermal collectors, a solar tank, auxiliary boiler, two pumps, and a distribution cycle. The cold medium production integrates an absorption chiller, a cooling tower, and two circulating pumps connected, respectively, to the absorber and evaporator. The energy harvested from the incident solar radiation heats the water in a field of the evacuated tube collector (ETC). Then, the hot water flows into a solar tank and is subsequently transported to the absorption chiller through the auxiliary boiler to produce chilled water, which circulates through a conventional distribution system of individual fan coils to deliver cold air to the building. An auxiliary heater is activated if the hot water temperature is not sufficient to drive the chiller. The cooling water dissipates the heat of the absorber and condenser of the chiller through the cooling tower. Figure 1 shows all of the elements that will be taken into account in the simulation and is described, in detail, in the following subsections.

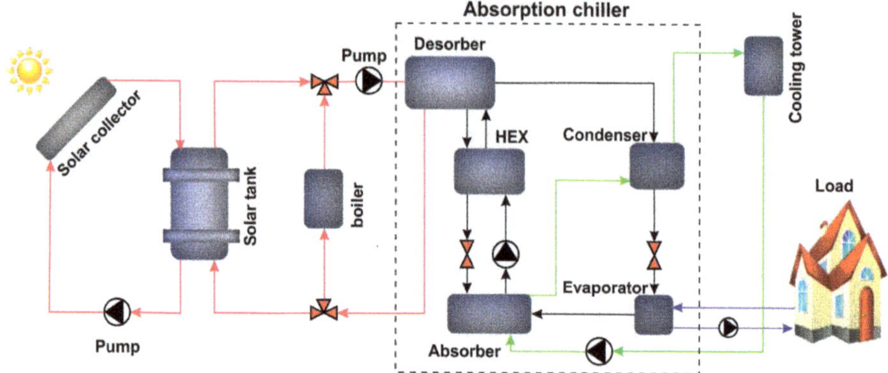

Figure 1. Solar absorption cooling system components.

2.1.1. Solar Collector

The evacuated tube collector (ETC) is the most popular solar collector in the world and excels in cloudy and cold conditions. The Apricus ETC-30 solar collector has been selected in this study [29]. The ETC is made up of two concentric glass tubes; the interior acts as a collector and the exterior as a cover. The elimination of air between the tubes reduces energy loss. An advantage of flat absorber vacuum tubes, from architectural integration, is that they can be installed on a horizontal or vertical surface, and the tubes can be rotated so that the absorber is at the appropriate inclination [30]. The collector thermal efficiency η_c is given in Equation (1):

$$\eta_c = \eta + a_1 \frac{\Delta T}{I_T} - a_2 \frac{(\Delta T)^2}{I_T} \tag{1}$$

where: η is the optical efficiency, a_1 and a_2 present, respectively, loss coefficient, ΔT refers to the difference between the average water temperature through solar collector T_m and the ambient temperature T_a and I_T is the total radiation incident on the absorber surface, for modeling the evacuated tube collectors (ETCs) in TRNSYS, needs an external file of incidence angle modifier (IAM) both longitudinal and transversal, which can be gained from manufacturer catalog. The performance specifications of the ETC are listed in Table 1 [29].

Table 1. Technical specifications of the Apricus evacuated tube collector (ETC)-30 solar collector [29].

Variable	Units	Value
Absorber area	m^2	2.4
Optical performance (η)	-	0.845
Loss coefficient (a_1)	W/(m²·K)	1.47
Loss coefficient (a_2)	W/(m²·K)	0.01

2.1.2. Solar Tank

The capacity of the hot storage tank is a decisive step in the solar system design and depends on the type of installation of three factors: the installed area of collectors, the operating temperature, and the time difference between the capture and storage. In installations for solar cooling, some authors have used values of 25 to 100 L/m² of the collector area [21]. For the calculation of the solar tank, we will assume that the hot water is stratified. The stratified storage tank comprises N nodes, the i node energy balance is given in Equation (2) [31]:

$$M_i C_P \frac{dT_i}{dT} = \dot{m}_s C_P (T_{i-1} - T_i) - \dot{m}_L C_P (T_{i+1} - T_i) - UA_i (T_i - T_a) \tag{2}$$

where: M_i is the fluid mass at the node i, C_P is the fluid specific heat, $\dot{m}_s$ is the mass flow rate from the heat source side, $\dot{m}_L$ is the mass flow rate of the load side, U is the overall losses from the solar tank to the environment, A_i is the surface transfer area, T_i is the node temperature, and T_a is the ambient temperature. An overall heat transfer coefficient for heat loss between the storage tank and the environment of 1.5 kJ/(h·m²·K) will be assumed, close to that used by Barghouti et al. [31].

2.1.3. Auxiliary Boiler

To operate the absorption chiller when the captured radiation is insufficient and the solar tank is depleted, an auxiliary system is employed to maintain the thermal energy at the desired level to drive the thermal chiller, the thermal energy $\dot{Q}_{aux}$ supplied by the auxiliary boiler can be calculated by Equation (3):

$$\dot{Q}_{aux} = \frac{\dot{m} \, C_P (T_{set} - T_{in}) + UA_{aux} (T_{aux} - T_o)}{\eta_{aux}} \tag{3}$$

where: T_{in} is the fluid inlet temperature, T_{set} is the thermostat set temperature, UA_{aux} refers to the overall coefficient of loss to the environment, η_{aux} is the auxiliary heater efficiency, and T_{aux} is the average temperature can be calculated by Equation (4):

$$T_{aux} = \frac{(T_{set} - T_{in})}{2} \tag{4}$$

In this work, a boiler with a nominal power of 60 kW (Q_e/COP = 35/0.70 = 50 KW) has been selected, and a performance of 90%, which will be assumed constant. For the auxiliary boiler, the parallel arrangement is preferred to prevent its operation from contributing to the heating of the water in the storage tank.

2.1.4. Heat Rejection System: Cooling Tower

A heat rejection system is attached to the thermal absorption chiller in order to evacuate the heat from the absorber and the condenser of the chiller and eject it to the ambient air. In this paper, the counterflow mechanical wet cooling tower was selected with the Baltimore Aircoil Company (BAC, Madrid, Spain) [32], The selected tower is the FXT-26 model, which has the nominal operating conditions listed in Table 2 and it is capable of dissipating all the heat evacuated by the absorption chiller under any environmental conditions in the location of installation (in our case, Baghdad, Iraq). The counterflow, the forced draft-cooling tower can be modeled in TRNSYS, based on the number of transfer units (NTU) [33]:

$$NTU = c \left[\frac{\dot{m}_a}{\dot{m}_w} \right]^{(n+1)} \tag{5}$$

where: $\dot{m}_w$ and $\dot{m}_a$ are the mass flow rates of water and air, respectively, c and n are the coefficients of mass transfer constant and exponent; their values are given by the manufacturer's curves in this paper, the selected values of c and n are 0.5 and −0.856 respectively.

Table 2. Technical characteristics of the FXT-26 cooling tower [32].

Characteristics	Value	Units
Cooling tower capacity	105	kW
Wet temperature	25	°C
Cooling water temperature	35-30	°C
Airflow rate	16	m^3/h
Electrical power	0.75	kW

2.1.5. Cooling Cycle: Absorption

The proposed chiller simulated here is the single-effect LiBr–H$_2$O absorption chiller YAZAKI WFC-SC10 (Yazaki Energy Systems Inc., Plano, TX, USA) with a nominal coefficient of performance (COP) of 0.70 and nominal cooling capacity $\dot{Q}_e$ of 35 kW. The technical specifications of the chiller are listed in Table 3 [34]. For the analysis of facilities, we will always assume a maximum demand capable of being satisfied by this chiller to cover the cooling load (in our case the maximum demand will be 25 kW). The simulation program required data from the chiller catalog that describes the chiller's operating map in order to determine the operating variables. The absorption machines are usually characterized by two basic parameters:

→ COP nominal. COP$_{nom}$. (0.7 for Yazaki WFC-10, Yazaki Energy Systems Inc., Plano, TX, USA)

→ $\dot{Q}_e$ Nominal evaporator power $\dot{Q}_{e,nom}$. (35 kW for Yazaki WFC-10, Yazaki Energy Systems Inc., Plano, TX, USA)

Table 3. Specifications of the YAZAKI WFC-SC10 absorption chiller [34].

Characteristic	Unit	Value
Cooling capacity	kW	35
Chilled water outlet /inlet temp.	°C	7/12.5
Cooling water outlet /inlet temp.	°C	35/31
Heating water outlet /inlet temp.	°C	88/83
Chilled water flowrate	m^3/h	11
Cooling water flow rate	m^3/h	36.7
Heating water flow rate	m^3/h	17.3
Electric power consumption	kW	0.21

From these, the nominal generator power $\dot{Q}_{g,nom}$ is immediately available by simply dividing the nominal cooling power $\dot{Q}_{e,nom}$ by the nominal coefficient of performance COP$_{nom}$. The TRNSYS model also requires entering the target temperature to be obtained at the outlet of the evaporator $T_{e,set}$, as well as the temperatures and flows entering the three external circuits: evaporator T_{ei}, condenser T_{ci} and generator T_{gi}. In this way, the model can determine the load regime in which the chiller works. Under these conditions, two situations can occur: if there is sufficient output power available on the evaporator, the set temperature will be reached. If not, the lowest possible value will be reached with the available power.

The instant heat $\dot{Q}_{remove}$ that should be removed from the incoming flow of the child as well as the load fraction f_{Load} are determined by Equations (6) and (7).

$$\dot{Q}_{remove} = \dot{m}_e C_{p,e} (T_{ei} - T_{e,set})$$
(6)

$$f_{Load} = \frac{\dot{Q}_{remove}}{\dot{Q}_{e,nom}} \tag{7}$$

With the load fraction and the temperatures indicated above (set, evaporator, condenser, and generator) it is possible to access the configuration file, whose structure will be commented on later, and which has been made from the chiller operation curves offered by the manufacturer for a set of operation points, establishing two basic parameters:

→ Fraction capacity $f_{capacity}$: is the ratio of the evaporator's output power to the nominal power of the chiller. With the manufacturer's data for each of the established operating points, the quotient between the output power it has in each of these conditions and the nominal power of the evaporator is evaluated.

$$f_{capacity} = \frac{\dot{Q}_e}{\dot{Q}_{e,nom}} \tag{8}$$

where $\dot{Q}_e$ is the output power under the particular conditions;

→ Energy input fraction $f_{Energyinput}$: is the ratio of the generator power to the nominal generator power necessary to satisfy the evaporator power. Similarly, it is obtained from the operation curves as:

$$f_{Energyinput} = \frac{\dot{Q}_g}{\dot{Q}_{g,nom}} = \frac{\dot{Q}_e}{\dot{Q}_{e,nom}} \cdot \frac{COP_{nom}}{COP} \tag{9}$$

where $\dot{Q}_g$ and COP are the values for the particular evaluation conditions obtained from the manufacture's curves. The maximum output power $\dot{Q}_{e,max}$ that the chiller will be able to offer on the evaporator for each of the conditions evaluated is calculated from Equation (10).

$$\dot{Q}_{e,max} = f_{capacity} \cdot f_{Energyinput} \cdot \dot{Q}_{e,nom} \tag{10}$$

On the other hand, the output power of the evaporator will be the minimum between the maximum power it is capable of offering in each of the conditions, and the demand is given by Equation (11).

$$\dot{Q}_e = Min \ of \left(\dot{Q}_e, \dot{Q}_{remove} \right) \tag{11}$$

With this evaporator power value, the flow rate, and the inlet temperature, the outlet temperature of the evaporator T_{eo} can be determined. Logically, at partial loads.

$$T_{eo} = T_{ei} - \frac{\dot{Q}_e}{\dot{m}_e \cdot C_{pe}} \tag{12}$$

The generator demand is taken from the energy input fraction $f_{Energyinput}$ (whose value has been given by the operating curve file for the operating conditions), multiplied by the standardized generator power.

$$\dot{Q}_g = f_{Energyinput} \cdot \dot{Q}_{g,nom} = f_{Energyinput} \cdot \frac{\dot{Q}_{e,nom}}{COP_{nom}} \tag{13}$$

The output temperature is an immediate value, the input temperature, and the generator flow rate are known as shown in Equation (14):

$$T_{go} = T_{gi} - \frac{\dot{Q}_g}{\dot{m}_g \cdot C_{pg}} \tag{14}$$

If it is assumed that the machine is adiabatic and, therefore, has no heat loss or gain; the power in the condenser is equal to the sum of the generator plus the evaporator:

$$\dot{Q}_c = \dot{Q}_e + \dot{Q}_g \tag{15}$$

The output temperature of the condenser T_{co} is calculated in the same way as for the evaporator and the generator:

$$T_{co} = T_{ci} - \frac{\dot{Q}_c}{\dot{m}_c \cdot C_{pc}} \tag{16}$$

Finally, The COP of the chiller is determined by Equation (17) [35]:

$$COP = \frac{\dot{Q}_e}{\dot{Q}_g + \dot{Q}_{aux}} \tag{17}$$

2.2. Meteorological Data

This section will highlight the analysis of potential solar data in Baghdad, Iraq. The main reason for this is to discover the potential power of renewable energy available at the location of operation. The meteorological conditions of Iraq correspond to a warm and dry climate during the summer season. Iraq has abundant solar energy capability with a significant amount of sunlight throughout the year as it is located in the Global Sunbelt. Solar energy can be widely deployed throughout two-thirds of Iraq. In the western and southern areas, daily average radiation ranges between 2800 and 3000 h, with relatively high average daily solar radiation of 6.5–7 kWh/m^2. The direct and global solar irradiation is given in Figure 2, [26]. Thence, the study location has great potential for solar energy, allowing sufficient use of solar thermal power as a main prime mover for the absorption cooling system.

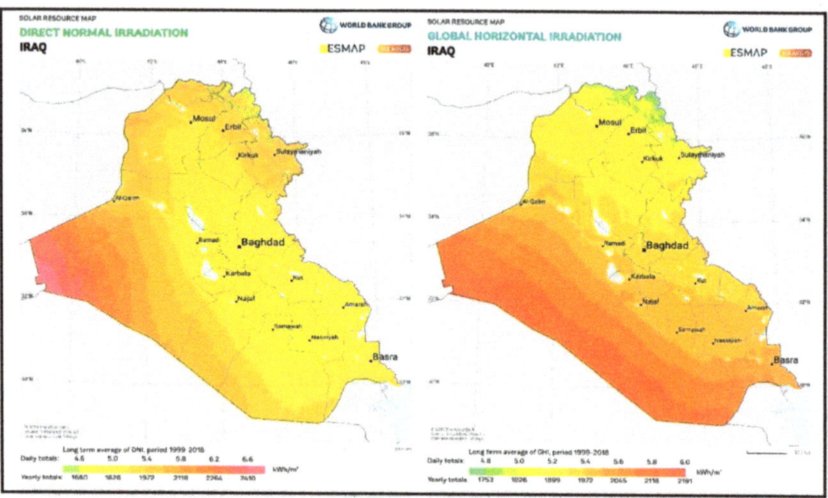

Figure 2. Iraq solar annual direct normal and global horizontal Irradiation map © 2019 The World Bank, Source: Global Solar Atlas 2.0, Solar resource data: Solargis [26].

The solar radiation data and environmental conditions used correspond to the TMY2 (typical meteorological year) format for Baghdad, the capital of Iraq (latitude is 33.3 N, longitude is 44.6 E, and Altitude 3.8 m). These data are provided by TRNSYS and have been obtained with Version 5 of the Meteonorm program.

Solar insolation varies according to the time of year. The daily highest solar irradiation of the globe is almost 8 kWh/m^2 and the daily highest temperature reaches over 45 °C (sometimes in summer season, the temperatures exceed 50 °C). A cooling effect is needed for seven months (April–October). During these months, the sunshine lasts for almost 10 hours per day, with an average total daylight of 13 hours per day [36].

2.3. House Profile and Cooling Loads

The proposed methodology was applied to a residential house located in Baghdad, Iraq. The house layout, wall layer details, and various construction components are given in Appendix A. As for the design of the house: the windows are on the north, east, and west walls; overhangs have a projection factor (overhang depth/window height) of 0.6. Two doors are on the north and east sides. Windows and doors are not specified on the south wall (to minimize heat gain through radiation). The window-to-gross-wall area is kept at 29%. The zone temperature is specified as 25 °C; the new design envelope specifications are as follows:

- The window-to-gross-wall area should not be greater than 35%.
- Overhangs should be placed on the east, west, and south windows of the building with a projection factor (overhang depth/window height) of greater than 0.5.
- Lighting devices should have an efficiency of 60 lumens/W.
- Specific lighting, 15 W/m^2.
- Specific gain (equipment and people), 15 W/m^2.
- Occupation rate 0.05 occupants/m^2.

Concerning the house under study, the monthly cooling demand of the house is variable during the summer season. An enormous portion of that variable is involved in the cooling load configuration due to the (transient) storage nature inherent in the cooling load. The sum of the components of the cooling load gives the total load of the house building. The calculation of the cooling demand was carried out using CARRIER software, version 4.04 (Carrier Software Systems, Syracuse, NY, USA, 2015), based on weather data for Baghdad. The inside conditions: temperature 25 °C and relative humidity 50%, ambient summer design dry-bulb temperature 48 °C, coincident wet-bulb 26 °C and 18.9 °C (daily range). Solar insolation varies according to the time of year, especially from April to October. The house's peak load occurs in August at 4 p.m., where the maximum outside temperature is about 49 °C. This value should be adopted for design purposes. Figure 3 illustrates the design temperature profiles for August; the maximum total cooling load according to CARRIER software is 25 kW was in August. Figure 4 shows the percentage of the various peak cooling load components.

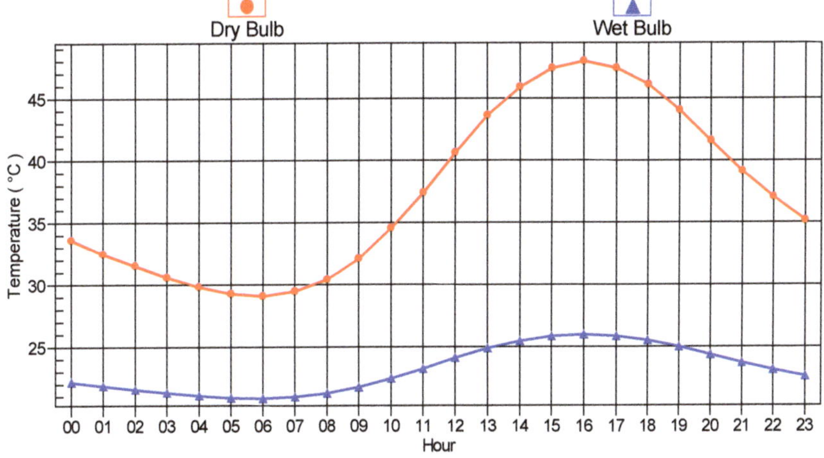

Figure 3. Design temperature profiles for August.

2.4. System Modeling

The TRNSYS library includes modules (TYPES) that represent the equipment commonly used in energy systems, modules for processing meteorological data, and modules for processing simulation results. The modular structure of TRNSYS gives great flexibility to analyze different types of energy systems. The representation of SACS in TRNSYS, described in the previous section, is illustrated in Figure 5. Table 4 provides information about the most important components of the system, the type of module that represents them, and provides some parameters of the basic design. Figure 5 and Table 4 do not include other components and flows of less importance. The period of simulation in TRNSYS was seven months (cooling season) from 1April 1 (2160 h) until the end of October (7296 h), with a step time simulation of one minute. Baghdad meteorological data was gained from TMY2.

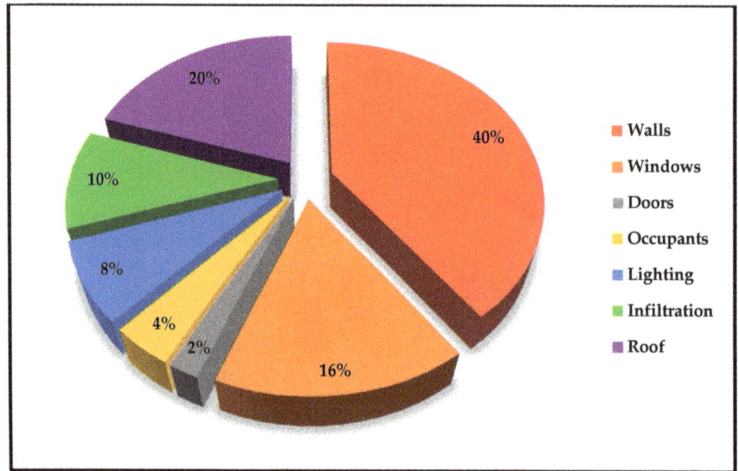

Figure 4. Contribution of the various cooling load component.

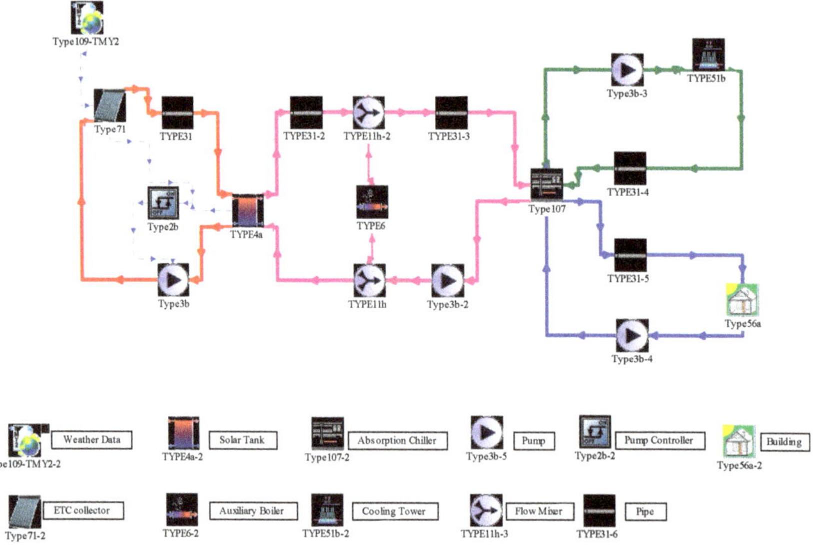

Figure 5. Diagram of the TRNSYS model of the solar absorption cooling system (SACS).

Some of the simplifying assumptions used in the calculation are indicated below:

- The electrical energy consumed by the pumps is neglected.
- Pumps are not supposed to transmit thermal energy to the fluid.
- When the pumps are running the mass, flows remain constant.
- The limit capacity of the chiller is assumed to correspond to a cooling water temperature of 27 °C.

In developing the model, the recommendations by some authors who have simulated the behavior of solar absorption cooling systems with TRNSYS, or other applications, have been taken into account [19,31].

Table 4. List of the most important components of the TRNSYS model.

Component	Type TRNSYS	Parameters (Base Design Values)
Solar Collector	TYPE 71a	Apricus ETC-30 (Table 3) Number of collectors (12) Inclination (30°)
Hot water tank	TYPE 4a	Volume (50 L/m^2 collector)
Auxiliary boiler	TYPE 6	Efficiency (90%)
Absorption chiller	TYPE 107	YAZAKI WFC SC10 (Table 5)
Cooling tower	TYPE 51b	B.A.C. FXT-26 (Table 4)
Weather data	TYPE 109—TMY2	Location: Baghdad, Iraq
Collector pump	TYPE 3b	Flow rate 50 (L/h)/m^2 of collector
Collector pump control	TYPE 2b	Maximum accumulator temperature (90 °C) Minimum collector gain (5 °C)
Pipe	TYPE 31	
Flow mixer	TYPE 11h	
Building	TYPE 56a	

3. Performance Analysis

3.1. Solar Fraction

The SACS performance can be evaluated using solar fraction (solar coverage). This factor demonstrates the solar energy contribution in chilled water production [37]; the following equation enables the calculation of the solar fraction.

$$SF = \frac{\dot{Q}_s}{\dot{Q}_S + \dot{Q}_{aux}} \tag{18}$$

where $\dot{Q}_s$ solar gained energy and $\dot{Q}_{aux}$ is energy from the auxiliary heater. Q_s can be calculated by:

$$\dot{Q}_s = \dot{Q}_c - \sum \dot{Q}_{loss} \tag{19}$$

where $\dot{Q}_c$ is useful collectors' energy and $\dot{Q}_{loss}$ is the system losses energy.

3.2. Primary Energy Saving

The primary energy (PE) savings is the saved primary energy, electric, and fossil. These values are mathematically described below, in order to evaluate the primary energy consumption of a solar system and a conventional one:

$$PE_{save} = \Delta PE_{fossil} + \Delta PE_{eletricity} \tag{20}$$

$$\Delta PE_{fossil} = \left(\frac{Q_{heat\,fossil,ref} - Q_{aux,total}}{\eta_{boiler} \cdot C_{con,fossil}} \right) \tag{21}$$

$$\Delta PE_{ele} = \left(\frac{P_{el,ref,tot} - P_{el,sc,tot}}{C_{con,elec}} \right) \tag{22}$$

$$Relative\ PE_{save} = \frac{PE_{save}}{PE_{ref}} \tag{23}$$

$$PE_{ref} = \frac{Q_{heat\,fossil,ref}}{\eta_{boiler} \cdot C_{con,fossil}} + \frac{P_{el,ref,tot}}{C_{con,ele}} \tag{24}$$

where:

η_{boiler} is the efficiency of auxiliary boiler 0.9;

$Q_{heat\,fossil,ref}$ is required heat for both space heating and DHW (Domestic Hot Water) in the conventional system (kWh).

$Q_{aux,total}$ is the produced energy by auxiliary heater (kWh).

$C_{con,fossil}$, $C_{con,ele}$ are the primary energy conversion factors for heat and electricity from fossil fuel, 0.95 $kWh_{heat,fossil}/kWh_{PE}$ and 0.5 $kWh_{elec,fossil}/kWh_{PE}$.

3.3. Electric Efficiency of the Total System

The electric efficiency is the relationship of the total heating and cooling energy generation to the required electricity for this production. The total system electrical efficiency $\eta_{ele,tot}$ is given by:

$$\eta_{ele,tot} = \frac{(Q_{cold})}{\left(P_c + P_{cw} + P_{el,chiller} + P_{el,CT} + P_{el,PS} + P_{el,boiler} \right)} \tag{25}$$

where:

P_c is the consumed electricity by a pump that feeds the chiller (kWh).

P_{cw} is the consumed electricity by cooling water loop pump (kWh).

$P_{el,chiller}$ is the consumed electricity by the chiller (kWh).

$P_{el,CT}$ is the electrical power of fan cooling tower (kWh).

$P_{el,PS}$ is the consumed electricity by solar loops pumps (kWh).

$P_{el,boiler}$ is the consumed electricity by boiler (kWh).

4. Results and Discussion

4.1. House Energy Balance Analysis

In this section, the thermal energy balance of SACS for the house under study was established through the evaluation of harvested solar energy, the delivered energy from a hot solar tank, the energy from the auxiliary boiler, and the necessary energy to satisfy the load. Table 5 shows the important and vital information efficiency parameters (result of SACS) for the summer season.

Table 5. Most relevant data and results for the cooling season operation (kWh).

Month	Incident Energy	Collected Energy	Solar Tank Energy	Aux. Boiler Energy	Load Energy	Collector Efficiency (%)	COP	Solar Fraction (%)
April	17,329	9489	2350	1760	4110	54.75	0.39	57.17
May	19,624	11,346	3642	1954	5596	57.81	0.41	65.08
June	30,632	16,358	6243	4720	10,963	53.40	0.45	56.94
July	33,685	18,509	7496	5153	12,649	54.94	0.51	59.26
August	35,173	19,245	7889	5391	13,280	54.71	0.52	59.40
September	29,627	15,296	4948	4430	9378	51.62	0.40	52.76
Oct.	11,953	5830	2162	2617	4779	48.77	0.40	45.23
Total	178,023	96,073	34,730	26,025	60,755	53.96	0.44	57.16

Figures 6 and 7 show, respectively, the energy contribution of the integrated gas boiler and solar field during the cooling season. The analyses results show that the useful energy of the solar field was 34,730 kWh and the energy delivered by the boiler was 26,025 kWh, indicating that the total season solar fraction (also called solar coverage) to the load was about 58% (see Figure 7). It is clearly seen that the system operating in May had the highest average solar fraction (a value of about 65%) due to the higher value of captured energy and the lower cooling load. Contrarily, the system presenting the lowest average value of solar fraction (45%) operated under October weather conditions. This is because of the lower energy supplied by the storage tank and lower harvested energy by ETC collectors (see Table 5). This outcome reflects the effect of solar irradiation on the energy performance of SACS.

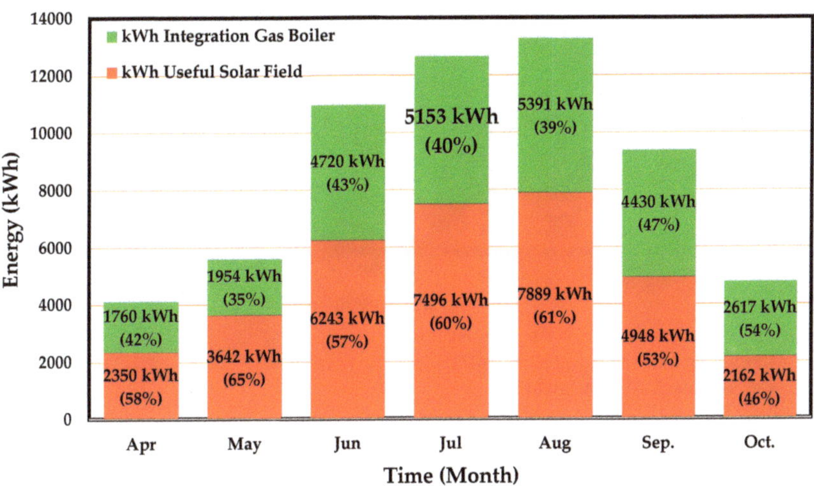

Figure 6. Energy contribution of integration gas boiler and solar field.

It is also clear that there is a significant impact on the solar fraction from the weather data each month, particularly, solar irradiation that has a direct influence on the energy generated by the ETC field. It is possible to observe this by referring to Equation (18).

Figure 8 shows the energy contribution of solar irradiation, energy from solar collectors, and solar tank. The average monthly values of incident solar radiation energy on the solar collectors was 178,023 kWh, while the total captured solar energy was 96,073 kWh, and the energy from the solar tank was 34,730 kWh, which implies that the efficiency of ETC collectors during the cooling season was about 54% (see Table 5).

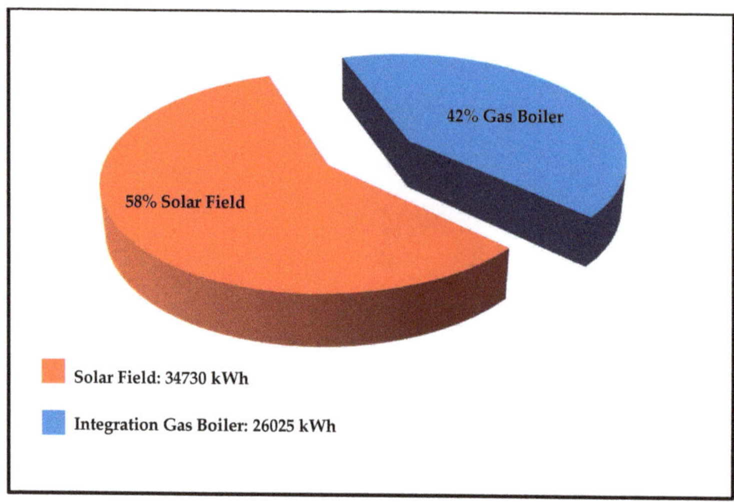

Figure 7. Solar coverage during cooling season.

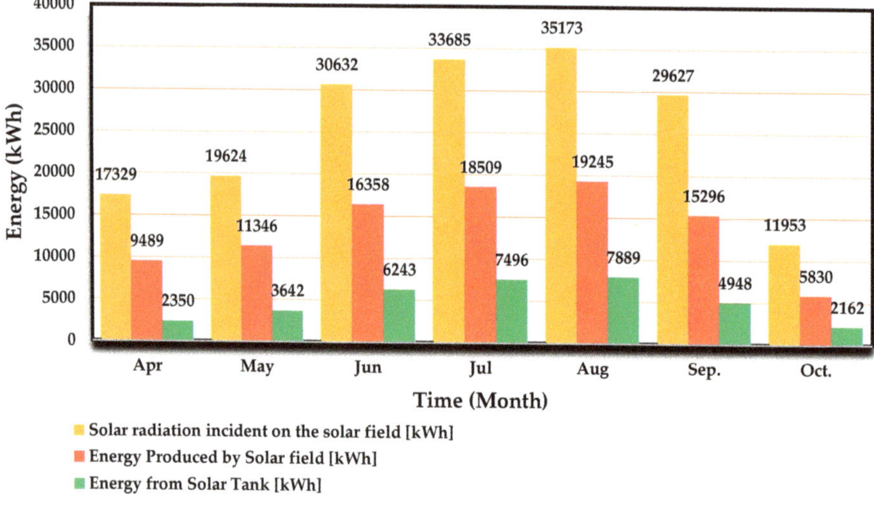

Figure 8. Energy contribution of solar field and solar tank.

Table 5 outlines the evaluation of the COP over six months, indicating that the average COP of SACS ranges between 0.39 and 0.52. It was also found that the system, operating in July and August, has the best average COP (a value of 0.51 and 0.52), respectively, due to a large amount of captured energy by ETC and a large cooling load that is led to higher solar coverage. Moreover, the lowest average COP (0.39) was recorded in April. Based on Equation (17), we can conclude that the variation in COP (see Table 5) through the six months is directly reported to the thermal energy at the input and output of the generator, and evaporator of the absorption chiller. The COP strongly depends on the flows of energy in these two parts. In general, the generator is influenced by the solar radiation of each month, while the evaporator is affected by building a cooling load, which depends on the ambient outdoor temperature of each month.

4.2. Primary Energy Analysis

The target of this analysis is to find the configuration that optimizes the system performance. Sensitivity analysis is presented under a different number of collectors (areas) and solar tank sizing. The number of collectors, and storage size, in the base case is 12 (30 m^2) and 50 L/m^2 (1500 L), respectively, compared with the base case. The sensitivity analysis includes changing the surface collector area from 25 m^2 to 35 m^2; the solar tank volume varied from 1000 l to 2000 l. The results are displayed in Table 6.

Table 6. Primary energy performance for various collector areas and solar tank volume.

		Number of Collectors			Solar Tank Volume L		
		10 (25 m^2)	12 (30 m^2)	14 (35 m^2)	1000	1500	2000
	Solar Fraction%	53.1	62.3	70.2	61.5	62.3	63.9
	$\eta_{ele,total}$	11.2	11.5	11.9	11.6	11.9	12.1
	PEsave kWh$_{PE}$	1361	3759	5661	4669	5761	6342
PE save	PEref (Primary Energy References) kWh$_{PE}$	15,469	15,545	15,666	15,628	15,666	15,306
	Relative %	8.8	24.8	36.8	29.9	36.8	40.6

It is shown that, with a greater collector area, the best results were obtained. A 16.6% increase in collector surface area is followed by an increase in the solar fraction and relative PE saved, 12.6% and 48.3%, respectively. Regarding solar tank volume, it is clearly seen that variation of tank volume does not present a significant influence on solar fraction, electrical efficiency, and PE relative; a solar tank volume increasing of 33.3% reflects on increasing in a solar fraction of about 2.5%, and PE relative around 10.3%. From the previous results, it can be recommended to use a collector area of 35 m^2 and a storage tank volume of 2000 L in order to achieve better performance than that reached in the base case.

4.3. Parametric Analysis

In this section, a parametric analysis has been carried out, taking into consideration the main important design parameters: the collector slope, water flow rate through the collector, number of collectors, and the solar tank size. In all analyses carried out below, the value of a single parameter is modified keeping the rest in the value corresponding to the base design.

4.3.1. Effect of Collector Slope

The inclination angle of the collector has a significant impact on the overall SACS performance. Figure 9 shows the variation in solar coverage with the collector field inclination in Baghdad. The evaluation based on a change in the tilt angle from 5° to 50° by a step of 5° was carried out in order to compute the optimum angle of the solar field that provides the highest solar fraction. The change in this variable shows that the tilt angles (15°, 20°, 25°, 30°) give a higher solar fraction, contrarily to the last three angles where solar fraction decreases. The reason for this difference is solar radiation perpendicularity that provides optimal results during summer with low tilt angle values, which help capture more solar radiation, as reported by Shariah and Elminir [38,39]. The optimum tilt values giving higher solar fraction are between 15° and 25°; therefore, operating at optimum value for tilt angles can readily expand the amount of solar energy incident and, thus, enhance both the thermal and economic efficiency of the SACS.

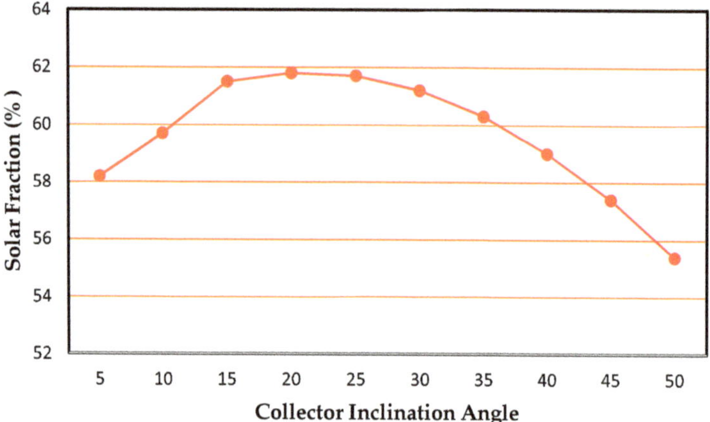

Figure 9. Solar coverage variation with the solar collector tilt.

4.3.2. Effect of Water Flow Rate

In the literature, the hot water flow rate values through solar collectors range from 20 to 80 L/h per m^2 of collector area are recommended for panels connected in parallel, as in our case. The variation of the solar fraction, with the hot water flow rate through the solar collector array, is indicated in Figure 10. The water flow rate varied from 20 to 55 (L/h)/m^2 of collector area. A change of water flow from 20 to 40 (L/h)/m^2 causes only a 0.9% increase in solar fraction; increasing the flow rate over an optimum value (40 (L/h)/m^2) will lead to drops in a solar fraction of about 0.2%. It is evident that the results obtained depict small changes in solar fraction and allow us to affirm that this parameter does not present a significant impact on solar coverage; it is in alignment with the results obtained by Beckman [40,41].

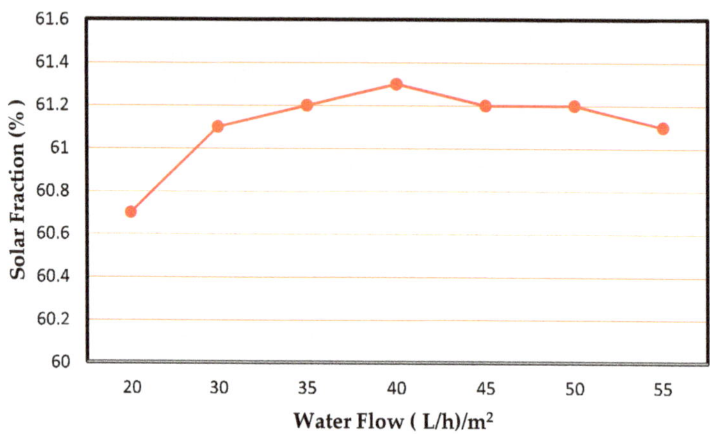

Figure 10. Variation of the solar fraction with the flow of water circulating through the collector.

4.3.3. Effect of Solar Field Area

The area and the number of solar collectors play an important role in determining the optimal configuration of the capture solar system. The collector surface has a decisive effect on the efficiency and feasibility of SACS. The simulation was carried out to establish the influence of this parameter on the overall performance of SACS under study, based on the collector's tilt angle 20°. The area of each collector was 2.5 m^2, the water flow rate was 40 L/h per m^2, the solar tank volume was 30 L/m^2;

the lower and upper solar tank temperatures were $T_{lower} = 75\,°C$, $T_{upper} = 90\,°C$. Figure 11 depicts the variation of the solar fraction with the number of collectors installed. The evaluation involves changing the number of a collectors from 4 to 24 ($10\,m^2$ to $60\,m^{2)}$ by a step of 2 ($2\,m^2$). It is clear that an increase in the collector surface area tends to enhance solar coverage due to the proportion between the captured energy from the ETC field and solar fraction, according to simulation results displayed in Figure 11. It is predicted that the solar coverage stays constant, especially at the higher solar surface field ($>55\,m^2$). As an example, an evacuated tube collector operating in Baghdad, inclination angle 30 degrees, presents a solar coverage of about 88.1% for 22 collectors ($55\,m^2$) and 88.3% for 24 collectors ($60\,m^2$). The stability in solar fraction SF (Solar Fraction), which was also achieved in published works Bahria and Assilzadeh [42,43], indicates that the system achieves its optimum level, and any additional increase in the surface field leads to overproduction of thermal energy, which can cause technological problems and significantly increase the initial investment. Therefore, with equal investment costs, the best design will be the one that offers the greatest coverage.

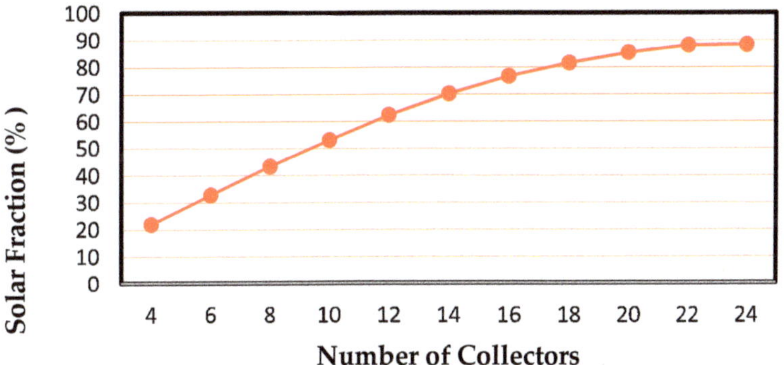

Figure 11. Solar coverage as a function of the number of collectors.

4.3.4. Effect of Solar Tank Capacity

This section examines the influence of the solar tank capacity on the solar fraction. The literature recommends values of storage solar tank capacity from 20 to $100\,L/m^2$ of collector area for installations where the time delay between collection and consumption does not exceed 24 h. The solar fraction is not significantly affected by the change in storage tank capacity, as shown in Figure 12. It is clear that increasing the solar tank capacity has a slight effect on the solar fraction. A change in solar accumulator capacity from 10 to $55\,L/m^2$ of the collector area obtains an increase in solar coverage of only 60.6% to 61.1%, respectively, with this difference (0.5%) observed—that the effect of the solar tank size on solar coverage is not significantly high. The optimum capacity of the solar tank, $30\,L/m^2$, gives solar coverage of 61.6%. Figure 12 depicts that the oversized solar tank will cause a decrease in solar fraction due to increases in thermal losses. The result in Figure 12 is in alignment with that of Beckman [40,41]. Therefore, it is not surprising that the optimal accumulator capacity is at the lower values of the recommended range.

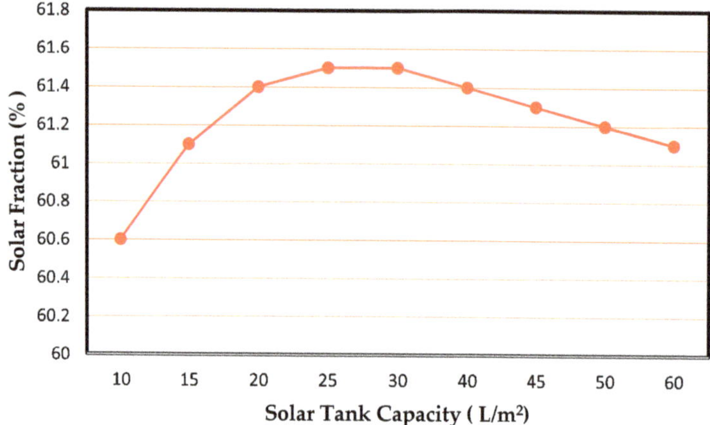

Figure 12. Solar coverage as a function of storage tank volume.

4.3.5. Effect of Solar Tank Temperature

Tables 7 and 8 present the results obtained for the variation of the solar fraction with the lower and upper temperatures of the solar tank that defines the operating of the absorption chiller with solar heat. The absorption chiller will operate with water from the solar accumulator tank when the upper temperature of the storage tank is between these limits and it is possible to completely cover the cold demand. Concerning the upper temperature, the value of 90 °C used in the basic design seems reasonable; a higher value of top solar tank temperature would improve the solar coverage somewhat, but it should be taken into account that the limit of 95 °C, imposed by the absorption chiller, cannot be exceeded. In fact, tank temperature affects, as well, the inlet temperature of the generator, since the hot water directly supplies the chiller generator. As for the lower temperature, the advantage of using values as small as possible is clear. This is because more solar heat can be used and the efficiency of the collector can be improved.

Table 7. Variation of the solar coverage with the lower temperature of the solar tank.

Temperature °C	70	72.5	75	77.5	80
Solar Fraction%	63.1	62.2	61.2	59.9	57.9

Table 8. Variation of solar coverage with an upper temperature of the solar tank.

Temperature °C	85	87.5	90	92.5	95
Solar Fraction%	60.7	61.0	61.2	61.3	60.6

5. Conclusions

This article presented a detailed analysis of the performance of a solar driven-absorption cooling system as alternative technology for air conditioning of a house, under hot and dry climate in Baghdad, Iraq. Various parameters influencing the solar fraction and the solar cooling performance of the proposed system have been discussed.

Based on the results of the simulation performed in this work, it was revealed that weather conditions have a significant effect on the performance of the solar absorption air conditioning system, with peak loads during the summer months. August presented the highest performance. The relevant average COP achieved a value of 0.52 while the solar fraction was 59.4%.

The results (of energy analysis contribution) during the summer season showed that the amount of energy incident was 178,023 kWh, while the total energy harvested was 96,073 kWh, which implies that the efficiency of ETC collectors during the cooling season is about 54%. It was also found that the solar energy supplied by the solar tank was 34,730 kWh and the energy delivered by the boiler was 26,025 kWh, indicating that the total seasonal solar coverage was about 58%.

The results of the primary energy analysis evidenced the use of a collector area of 35 m^2; a storage tank volume of 2000 L presented better performance than that reached in the base case.

Parametric analysis results showed the best configuration for the design of SACS. It was found that the solar collector tilt angle is significantly affected by the incident solar irradiation. An optimal value (between 15° and 25°) of the inclination presented a higher solar fraction. It was also found that increasing the water flow rate through the collectors does not indicate a significant effect on solar coverage.

The surface collector analysis revealed that, in general, the increase of the installed surface of the collector field leads to improve the solar fraction values. It was also found that the solar fraction remained the same for an area larger than 55 m^2. Therefore, it is concluded that the appropriate collector surface selection should be carried out together with economic and technical feasibility (mainly cost analysis and land availability) in order to achieve the best profitability of the system.

Moreover, the change in the size of the solar tank has no significant impact on the solar fraction (60.6% and 61.1% for 10 L/m^2 and 55 L/m^2 of collector area, respectively). Increasing the upper solar tank temperature improves the solar coverage, but it should be <95 °C.

Finally, this work provides a roadmap for designers, in particular, to ensure that all of the operating and design variable effects are taken into consideration when developing a solar air-conditioning cycle under the Iraq climate. Additionally, the model can be employed to carry out thermo-economic comparisons of the system using various types of collectors.

Author Contributions: Conceptualization, A.A.-F.; methodology, A.A.-F.; validation, A.A.-F.; formal analysis, A.A.-F.; investigation, A.A.-F.; data curation, A.A.-F.; writing—original draft preparation, A.A.-F.; writing—review and editing, F.A.; supervision, B.E. All authors have read and agreed to the published version of the manuscript.

Funding: The authors received no specific funding for this work. The corresponding author would like to thank the Technical University of Darmstadt, enabling the open-access publication of this paper.

Conflicts of Interest: The authors declare no conflict of interest.

Nomenclatures, Subscripts and Abbreviations

Nomenclatures

A	area (m^2)
a	loss coefficient
η	efficiency
I	radiation incident (W/m^2)
$\dot{m}$	mass flow rate (kg/s)
C_P	specific heat (kPa)
$\dot{Q}$	heat transfer rate (kW)
T	temperature (°C)
U	overall heat transfer coefficient (kW/m^2·K)
$\dot{W}$	power (kW)

Subscripts

a	Ambient
aux	Auxiliary
c	Condense
e	Evaporator
f	Fraction
g	Generator
i	Node
L	Load
nom	Nominal

min	Minimum
max	Maximum
o	Outlet
s	Source
set	Set
w	Water

Abbreviations

COP	Coefficient of performance
DHW	Domestic hot water
EES	Engineering Equation Solver
ETC	Evacuated tube collector
HVAC	Heating, ventilation, and air conditioning
IEA	International Energy Agency
IAM	Incidence angle modifier
NTU	Number of transfer unit
PE	Primary energy
SACS	Solar absorption cooling system
SF	Solar fraction
TMY	Typical meteorological year
TRNSYS	Transient System Simulation Program

Appendix A

The House under Study

The house building is double dwelling, connected by internal stairs, as shown in Figure A1 [37]. Each floor has an area of 97 m^2 and a height of 3 m. The ground floor includes an entrance, living room, bedroom, kitchen, and bathroom. The first floor contains the master room, bedrooms, and bathroom. The wall layer details and various construction components of the house are given in Tables A1 and A2.

The maximum total calculated cooling load, according to CARRIER software, is about 25 kW. Figure 4 shows the percentage of the various peak cooling load components estimated by CARRIER software. The load through the walls was 40% of the total cooling load due to the higher temperature difference between outdoor and indoor temperature; the heat envelope transmission aggravates cooling demand. The transmission losses could be reduced significantly with envelope higher thickness insulation. Additionally, ETC collectors mounted on the first floor roof of the house could reduce the cooling load due to shading, though not all are eliminated. The load on the roof is 20% of the overall cooling load. The actual total estimated cooling load would be about 21, which is in alignment with the data obtained in [37].

Table A1. Wall layers details.

Wall Details Outside Surface Color Dark Absorptivity 0.900 Overall U-Value 0.415 W/(m²·K) Wall Layers Details (Inside to Outside)					
Layers	**Thickness mm**	**Density kg/m³**	**Specific Ht·kJ/(kg·K)**	**R-Value (m²·K)/W**	**Weight kg/m²**
Inside surface resistance	0.000	0.0	0.00	0.00200	0.0
Cement bounded	12.000	1600.0	1.34	0.04200	19.2
Insulation	50.000	32.0	0.90	1.66600	1.6
Hollow block	200.000	1922.0	0.84	0.40000	384.4
13 mm gypsum board	12.700	800.9	1.09	0.07890	10.2
Outside surface resistance	0.000	0.0	0.00	0.00200	0.0
Air space	0.000	0.0	0.00	0.16026	0.0
Outside surface resistance	0.000	0.0	0.00	0.05864	0.0
Totals	274.700	-	-	2.40980	415.4

Table A2. Construction components of the house.

Component	Exterior Roof	Exterior Glass	Exterior Wooden Door	Exterior Steel Door
U value (W/m^2 °C)	1.670388	5.888993	2.087049	6.07040

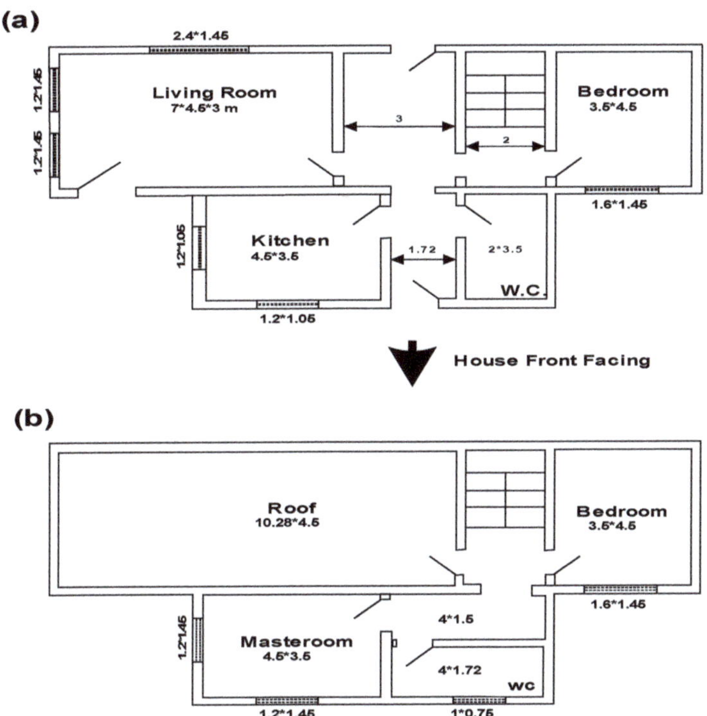

Figure A1. Layout diagram of the modern domestic solar house (a) ground floor (b) first floor.

References

1. Allouhi, A.; El Fouih, Y.; Kousksou, T.; Jamil, A.; Zeraouli, Y.; Mourad, Y. Energy consumption and efficiency in buildings: Current status and future trends. *J. Clean. Prod.* **2015**, *109*, 118–130. [CrossRef]
2. Dahl, R. Cooling concepts: Alternatives to air conditioning for a warm world. *Environ. Health Perspect.* **2013**, *121*, a18–a25. [CrossRef] [PubMed]
3. Rashid, S.; Peters, I.; Wickel, M.; Magazowski, C. Electricity Problem in Iraq. In *Economics and Planning of Technical Urban Infrastructure Systems*; HafenCity Universität Hamburg: Hamburg, Germany, 2012.
4. Allouhi, A.; Kousksou, T.; Jamil, A.; El Rhafiki, T.; Mourad, Y.; Zeraouli, Y. Economic and environmental assessment of solar air-conditioning systems in Morocco. *Renew. Sustain. Energy Rev.* **2015**, *50*, 770–781. [CrossRef]
5. Petela, K.; Szlęk, A. Assesment of Passive Cooling in Residential Application under Moderate Climate Conditions. *Ecol. Chem. Eng. A* **2016**, *23*, 211–225.
6. Baniyounes, A.M.; Rasul, M.G.; Khan, M.M.K. Assessment of solar assisted air conditioning in Central Queensland's subtropical climate, Australia. *Renew. Energy* **2013**, *50*, 334–341. [CrossRef]
7. Balghouthi, M.; Chahbani, M.H.; Guizani, A. Feasibility of solar absorption air conditioning in Tunisia. *Build. Environ.* **2008**, *43*, 1459–1470. [CrossRef]
8. Martínez, P.J.; Martínez, J.C.; Lucas, M. Design and test results of a low-capacity solar cooling system in Alicante (Spain). *Sol. Energy* **2012**, *86*, 2950–2960. [CrossRef]

9. Burckhart, H.J.; Audinet, F.; Gabassi, M.-L.; Martel, C. Application of a novel, vacuum-insulated solar collector for heating and cooling. *Energy Procedia* **2014**, *48*, 790–795. [CrossRef]

10. Ketjoy, N.; Rawipa, Y.; Mansiri, K. Performance evaluation of 35 kW LiBr H_2O solar absorption cooling system in Thailand. *Energy Procedia* **2013**, *34*, 198–210. [CrossRef]

11. Leiva-Illanes, R.; Escobar, R.; Cardemil, J.M.; Alarcón-Padilla, D.-C. Comparison of the levelized cost and thermoeconomic methodologies—Cost allocation in a solar polygeneration plant to produce power, desalted water, cooling and process heat. *Energy Convers. Manag.* **2018**, *168*, 215–229. [CrossRef]

12. Jenkins, P.; Elmnifi, M.; Younis, A.; Emhamed, A.; Alshilmany, M. Design of a Solar Absorption Cooling System: Case Study. *J. Power Energy Eng.* **2020**, *8*, 1–15. [CrossRef]

13. Wang, K.; He, Y.; Kan, A.; Yu, W.; Wang, D.; Zhang, L.; Zhu, G.; Xie, H.; She, X. Significant photothermal conversion enhancement of nanofluids induced by Rayleigh-Bénard convection for direct absorption solar collectors. *Appl. Energy* **2019**, *254*, 113706. [CrossRef]

14. Mukhtar, H.K.; Said, S.A.; El-Sharaawi, M.I. Dynamic performance of solar powered vapor absorption cooling system in dhahran—Saudi Arabia. In Proceedings of the 2018 5th International Conference on Renewable Energy: Generation and Applications (ICREGA), Al Ain, United Arab Emirates, 26–28 February 018.

15. Elahi, R.; Nagassou, D.; Trelles, J.; Mohsenian, S.; Trelles, J. Enhanced solar absorption by CO_2 in thermodynamic noneq. *Sol. Energy* **2019**, *195*, 369–381. [CrossRef]

16. Behi, M.; Mirmohammadi, S.A.; Ghanbarpour, M.; Behi, H.; Palm, B. Evaluation of a novel solar driven sorption cooling/heating system integrated with PCM storage compartment. *Energy* **2018**, *164*, 449–464. [CrossRef]

17. Salehi, S.; Yari, M.; Rosen, M.A. Exergoeconomic comparison of solar-assisted absorption heat pumps, solar heaters and gas boiler systems for district heating in Sarein Town, Iran. *Appl. Therm. Eng.* **2019**, *153*, 409–425. [CrossRef]

18. Buonomano, A.; Calise, F.; Dentice D'accadia, M.; Ferruzzi, G.; Frascogna, S.; Palombo, A.; Russo, R.; Scarpellino, M. Experimental analysis and dynamic simulation of a novel high-temperature solar cooling system. *Energy Convers. Manag.* **2016**, *109*, 19–39. [CrossRef]

19. Mateus, T.; Oliveira, A.C. Energy and economic analysis of an integrated solar absorption cooling and heating system in different building types and climates. *Appl. Energy* **2009**, *86*, 949–957. [CrossRef]

20. Fedrizzi, R.; Franchini, G.; Mugnier, D.; Melograno, P.N.; Theofilidi, M.; Thuer, A.; Nienborg, B.; Koch, L.; Fernandez, R.; Troi, A.; et al. Assessment of Standard Small-Scale Solar Cooling Configurations within the SolarCombi+ Project. In Proceedings of the 3rd International Conference Solar Air-Conditioning, Palermo, Italy, 30 September–2 October 2009.

21. Shirazi, A.; Taylor, R.A.; White, S.D.; Morrison, G.L. Transient simulation and parametric study of solar-assisted heating and cooling absorption systems: An energetic, economic and environmental (3E) assessment. *Renew. Energy* **2016**, *86*, 955–971. [CrossRef]

22. Vasta, S.; Palomba, V.; Frazzica, A.; Costa, F.; Freni, A. Dynamic simulation and performance analysis of solar cooling systems in Italy. *Energy Procedia* **2015**, *81*, 1171–1183. [CrossRef]

23. Calise, F.; d'Accadia, M.D.; Palombo, A. Transient analysis and energy optimization of solar heating and cooling systems in various configurations. *Sol. Energy* **2010**, *84*, 432–449. [CrossRef]

24. Djelloul, A.; Draoui, B.; Moummi, N. Simulation of a solar driven air conditioning system for a house in dry and hot climate of algeria. *Courr. Savoir* **2013**, *15*, 31–39.

25. Hassan, L.; Moghavvemi, M.; Mohamed, H.A.F. Impact of UPFC-based damping controller on dynamic stability of Iraqi power network. *Sci. Res. Essays* **2011**, *6*, 136–145.

26. The World Bank. Available online: https://solargis.com/maps-and-gis-data/download/iraq (accessed on 5 December 2019).

27. Nafey, A.S.; Sharaf, M.A.; García-Rodríguez, L. A new visual library for design and simulation of solar desalination systems (SDS). *Desalination* **2010**, *259*, 197–207. [CrossRef]

28. Sharaf Eldean, M.A.; Soliman, A.M. A new visual library for modeling and simulation of renewable energy desalination systems (REDS). *Desalin. Water Treat.* **2013**, *51*, 6905–6920. [CrossRef]

29. Available online: http://www.apricus.com/en/america/products/solar-collectors/ap-30/ (accessed on 6 December 2019).

30. Peuser, F.; Remmers, K.; Schnauss, M. *Solar Thermal Systems: Successful Planning and Construction*; Routledge: Abingdon, UK, 2013.

31. Balghouthi, M.; Chahbani, M.H.; Guizani, A. Solar powered air conditioning as a solution to reduce environmental pollution in Tunisia. *Desalination* **2005**, *185*, 105–110. [CrossRef]

32. Available online: http://www.BaltimoreAircoil.com (accessed on 16 December 2019).

33. ASHRAE. *Heating, Ventilating, and Air-Conditioning Systems and Equipment*, SI ed.; American Society of Heating, Refrigerating and Air-Conditioning Engineers, Inc.: Atlanta, GA, USA, 2012.

34. Available online: http://www.yazakienergy.com (accessed on 31 January 2020).

35. Rodríguez-Hidalgo, M.D.C.; Rodríguez-Aumente, P.A.; Lecuona, A.; Legrand, M.; Ventas, R. Domestic hot water consumption vs. solar thermal energy storage: The optimum size of the storage tank. *Appl. Energy* **2012**, *97*, 897–906.

36. Allouhi, A.; Kousksou, T.; Jamil, A.; Bruel, P.; Mourad, Y.; Zeraouli, Y. Solar driven cooling systems: An updated review. *Renew. Sustain. Energy Rev.* **2015**, *44*, 159–181. [CrossRef]

37. Joudi, K.A.; Dhaidan, N.S. Application of solar assisted heating and desiccant cooling systems for a domestic building. *Energy Convers. Manag.* **2001**, *42*, 995–1022. [CrossRef]

38. Shariah, A.; Al-Akhras, M.A.; Al-Omari, I.A. Optimizing the tilt angle of solar collectors. *Renew. Energy* **2002**, *26*, 587–598. [CrossRef]

39. Elminir, H.K.; Ghitas, A.E.; El-Hussainy, F.; Hamid, R.; Beheary, M.M.; Abdel-Moneim, K.M.; Elminir, H.K. Optimum solar flat-plate collector slope: Case study for Helwan, Egypt. *Energy Convers. Manag.* **2005**, *47*, 624–637. [CrossRef]

40. Duffie, J.A.; Beckman, W.A. *Solar Engineering of Thermal Processes*; John Wiley & Sons: Hoboken, NJ, USA, 1991.

41. Duffie, J.A.; Beckman, W.A. *Solar Engineering of Thermal Processes*; John Wiley & Sons: Hoboken, NJ, USA, 2013.

42. Bahria, S.; Amirat, M.; Hamidat, A.; El Ganaoui, M.; Slimani, M.E.A. Parametric study of solar heating and cooling systems in different climates of Algeria–A comparison between conventional and high-energy-performance buildings. *Energy* **2016**, *113*, 521–535. [CrossRef]

43. Assilzadeh, F.; Kalogirou, S.A.; Ali, Y.; Sopian, K. Simulation and optimization of a LiBr solar absorption cooling system with evacuated tube collectors. *Renew. Energy* **2005**, *30*, 1143–1159. [CrossRef]

Article

Process Control Strategies in Chemical Looping Gasification—A Novel Process for the Production of Biofuels Allowing for Net Negative CO_2 Emissions

Paul Dieringer *, Falko Marx, Falah Alobaid, Jochen Ströhle and Bernd Epple

Institute for Energy Systems & Technology, Technical University Darmstadt, Otto-Berndt-Str. 2, 64287 Darmstadt, Germany; falko.marx@est.tu-darmstadt.de (F.M.); falah.alobaid@est.tu-darmstadt.de (F.A.); jochen.stroehle@est.tu-darmstadt.de (J.S.); bernd.epple@est.tu-darmstadt.de (B.E.)
* Correspondence: paul.dieringer@est.tu-darmstadt.de; Tel.: +49-6151-16-22692

Received: 20 May 2020; Accepted: 12 June 2020; Published: 22 June 2020

Abstract: Chemical looping gasification (CLG) is a novel gasification technique, allowing for the production of a nitrogen-free high calorific synthesis gas from solid hydrocarbon feedstocks, without requiring a costly air separation unit. Initial advances to better understand the CLG technology were made during first studies in lab and bench scale units and through basic process simulations. Yet, tailored process control strategies are required for larger CLG units, which are not equipped with auxiliary heating. Here, it becomes a demanding task to achieve autothermal CLG operation, for which stable reactor temperatures are obtained. This study presents two avenues to attain autothermal CLG behavior, established through equilibrium based process simulations. As a first approach, the dilution of active oxygen carrier materials with inert heat carriers to limit oxygen transport to the fuel reactor has been investigated. Secondly, the suitability of restricting the air flow to the air reactor in order to control the oxygen availability in the fuel reactor was examined. Process simulations show that both process control approaches facilitate controlled and de-coupled heat and oxygen transport between the two reactors of the chemical looping gasifier, thus allowing for efficient autothermal CLG operation. With the aim of inferring general guidelines on how CLG units have to be operated in order to achieve decent synthesis gas yields, different advantages and disadvantages associated to the two suggested process control strategies are discussed in detail and optimization avenues are presented.

Keywords: chemical looping; biomass gasification; process control; process simulation

---

1. Introduction

The reduction of greenhouse gas emissions (GHGE) in order to reach the unilateral goals agreed upon in the UNFCCC Paris Agreement is one of the major challenges of civilization in the 21st century. While notable advances in the energy sector have been achieved in recent years [1,2], the de-carbonization of the transport sector, which is responsible for almost one quarter of the European GHGE emissions [3] and consumes 36% of the global final energy [1], signifies a key issue on the path to a closed carbon cycle. Especially the replacement of conventional fuels in the heavy freight transport and aviation industry, where electrification is currently not viable, remains a major hurdle. When considering the European Union's Renewable Energy Directive (RED II) [4], which set a target of a share of 14% renewable energy in the transport sector by 2030, while at the same time alleviating negative impacts on food availability and prices, it is clear that significant advances in renewable fuel generation are required.

The production of so-called advanced or second-generation biofuels through thermochemical conversion of biomass-based residues is an auspicious pathway to achieve these goals. Gasification is a mature thermochemical biomass conversion process, although its primary use is the generation of heat and electricity, while the synthesis of advanced biofuels through the gasification route has not been implemented in an industrial scale, yet [5].

Commonly, biomass gasification is achieved through utilizing air or pure oxygen in the gasifier. Albeit, pure oxygen is typically used in gasification processes embedded in biomass-to-biofuel process chains, since a nitrogen-free, high calorific value syngas is required for fuel synthesis [6]. The provision of this oxygen requires an air separation unit (ASU), which is associated with high capital and operational costs, hence adversely affecting the energetic plant efficiency and process economics [6,7]. Alternatively, steam [8–10] or carbon dioxide [10–12] can be deployed as the gasification medium. Yet, either of the two suffers from slow gasification kinetics [6,13,14] and strong process endothermicity [6,15], limiting the process efficiency. To circumvent this, the dual fluidized bed gasification (DFBG) technology achieves feedstock gasification in two connected reactors; a gasifier in which steam gasification of the deployed feedstock is attained, and a combustor in which the residual char is combusted facilitating full char conversion and the provision of heat, which is transported to the gasifier using an inert circulating bed material [16–18].

A similar gasification concept allowing for decent fuel conversions, without requiring an ASU is the chemical looping gasification (CLG) process, where biomass gasification is also carried out in two separate reactors (see Figure 1) [15,19–22]. Just as the related chemical looping combustion (CLC) process, CLG is realized using two coupled fluidized bed reactors, in order to attain good heat and mass transport characteristics [21,23,24]. Here, steam or carbon dioxide provide bed fluidization and gasification (see Equations (1) and (2)) of the feedstock in the fuel reactor (FR) [15,24]. Additional oxygen for the partial (see Equation (3)) or full (see Equations (4)–(6)) oxidation of gaseous hydrocarbon species, enhancing gasification kinetics and reducing the process endothermicity, is supplied through a circulating oxygen carrier (OC, Me_xO_y) [19,21,24]. Furthermore, the homogeneous water gas shift (WGS) reaction (Equation (7)) takes place inside the gas phase.

$$C + CO_2 \rightarrow 2\,CO \tag{1}$$

$$C + H_2O \rightarrow CO + H_2 \tag{2}$$

$$Me_xO_y + CH_4 \rightarrow Me_xO_{y-1} + 2\,H_2 + CO \tag{3}$$

$$4Me_xO_y + CH_4 \rightarrow 4\,Me_xO_{y-1} + 2\,H_2O + CO_2 \tag{4}$$

$$Me_xO_y + CO \rightarrow Me_xO_{y-1} + CO_2 \tag{5}$$

$$Me_xO_y + H_2 \rightarrow Me_xO_{y-1} + H_2O \tag{6}$$

$$CO + H_2O \leftrightarrow H_2 + CO_2 \tag{7}$$

The required oxygen transport to the FR is facilitated through a repeated regeneration of the OC (see. Equation (8)) in the air reactor (AR) with oxygen contained in the inlet air [15,20,24]. Moreover, unconverted char is combusted in the air reactor (see. Equation (9)), leading to a full conversion of the deployed feedstock [23,25].

$$Me_xO_{y-1} + 0.5\,O_2 \rightarrow Me_xO_y \tag{8}$$

$$C + O_2 \rightarrow CO_2 \tag{9}$$

The latter reaction is generally undesired, as a high carbon conversion is targeted inside the FR, in order to maximize the carbon capture efficiency of the process [23,26,27]. In literature, carbon capture efficiencies in the range of 90–99% are reported for CLC [26,28,29]. As approximately one third of the carbon contained in the feedstock is transferred into the valorized end-product (e.g., liquid Fischer-Tropsch fuels) in process chains employing CLG for syngas generation, this means that up to 65% of the carbon contained in the feedstock can be captured and stored, constituting negative emissions in case biogenic feedstocks are being employed. Yet, in reality figures falling short of this value can be expected, as a fraction of the feedstock carbon will be lost in the AR in the form of CO_2.

Apart from the oxygen transport, the continuous solid circulation between the two reactors provides the required heat transport from the AR, in which the exothermic re-oxidation of the OC occurs, to the FR, where the endothermic gasification reactions take place [15,19,23], thus allowing for stable elevated reactor temperatures.

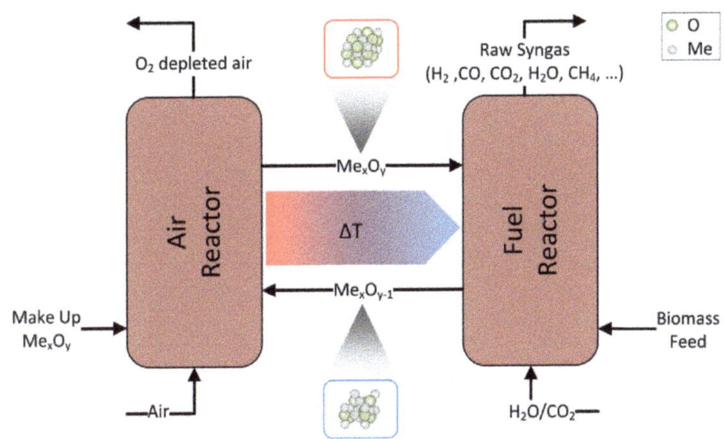

Figure 1. Schematic of chemical looping gasification (CLG) process.

CLG not only offers excellent characteristics in terms of feedstock flexibility [24], but is especially well suited for biomass-based feedstocks [30,31], commonly exhibiting a reactive char and containing a large fraction of volatiles. This means that high char conversions can be achieved through the gasification reaction with steam or CO_2, while volatiles are converted to the desired syngas species through their partial oxidation on the OC surface (see Equation (3)). Furthermore, it is reported that iron containing materials [32–35] can facilitate the cracking and oxidation of tars, which are known to be formed in significant amounts during biomass gasification [36].

While the role of the gasification agent is similar in CLC and CLG (i.e., char gasification), the oxygen carrier is meant to only partially oxidize the gaseous species in CLG, yielding a raw product gas with a high heating value [23,37], instead of a heat release from the AR, which is used for heat and power generation in CLC [24,38,39]. This shift from CLC to CLG is achieved through lowering the oxygen-to-fuel equivalence ratio in the FR to values below unity. An autothermal CLG process, maximizing the chemical energy contained in the raw syngas without relying on external heating, is obtained when the net heat release from the process equals zero (neglecting heat losses).

Although one might hence deduce that the transition from CLC to CLG is straightforward, there are major differences between the two processes. While large OC circulation rates are favorable in CLC, as they allow for a high oxygen availability in the FR, which favors fuel combustion [40–43] and provide for a large heat transport from the AR to the FR [41,44,45], the former is not desired in CLG. Here, the oxygen availability in the FR has to be limited in order to prevent the full oxidation of the employed feedstock. However, even more so than in CLC, CLG requires large heat transportation rates from the AR and FR due to the less pronounced occurrence of full oxidation reactions (Equations (4)–(6)), at the cost of highly endothermic partial oxidation reactions (Equation (3)) in the FR. This leads to a fundamental challenge in terms of process control, as both, heat and oxygen transfer between the two reactors, have to be controlled independently in order to attain an autothermal CLG process. Initial advances to reach this target were made by Ge et al. [37], diluting an active OC material with an inert, thus obtaining stable reactor temperatures for a lab-scale CLG unit. Yet, due to the significance of this inherent challenge, an in-depth analysis of this issue is required. Therefore, this work takes a holistic approach to this matter, employing process simulations in order to establish suitable process control measures to attain an autothermal CLG process. In the following, the developed process model will be introduced in Section 2, before general process control and optimization strategies are presented and discussed in detail in Section 3. To round off these elucidations, the most crucial findings and an outlook on future research topics are given in Section 4 of this article.

2. Modelling Methods

2.1. Description of the Process Model

The deployed Aspen Plus™ model, shown in Figure 2, is largely adopted from a previous study by Ohlemüller et al. [25]. Here, the chemical reactions occurring in the AR and FR are modelled in two separate reactors, whereas gas-solid and solid-solid separation is achieved through cyclones and separators, respectively. In order to reduce model complexity, the AR and FR were modelled as equilibrium RGIBBS reactors in this work, as this simplification allows for a basic description of the most crucial phenomena required for process control and obviates the necessity of accurate kinetic data. To account for the solid circulation in chemical looping processes, a constant mass stream of solids continuously cycles through the system (OCR-TOAR/OCO-TOFR), after being added to the system after initiation of the simulation (INIT).

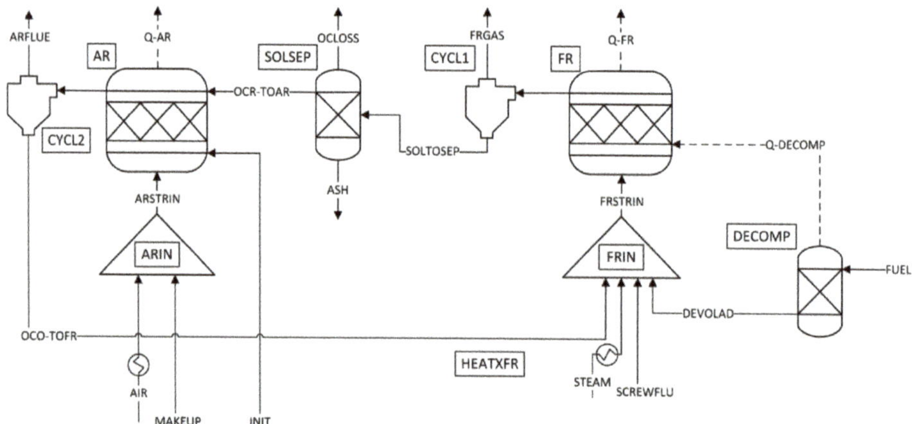

Figure 2. Flow sheet of the Aspen Plus™ CLG process model.

For completeness and comprehensibility reasons, all components and streams are briefly described in the following:

- Prior to any calculation, an initial solid mass flow is given into the system (INIT), to model the circulating solid OC mass. Instead of estimating the actual solid loss, the approach of Ohlemüller et al. [25], setting the total OC loss (OCLOSS) to 1% of the circulating mass to achieve fast flowsheet conversion, was adopted. The same amount of fresh solids was constantly fed to the AR (MAKEUP), to achieve constant solid circulation.
- For both reactors, cyclones are employed to achieve solid-gas separation. The FR products are separated into a gas (FRGAS) and solid (SOLTOSEP) stream, via CYCL1 (separation efficiency 100%). Similarly, the AR products are separated into a gas (ARFLUE) and solid stream (OCO-TOFR) in CYLC2 (separation efficiency 100%).
- All streams entering the process are fed at ambient temperature ($T_0 = 25\ °C$), except for the stream STEAM, which is fed as saturated steam (120 °C).
- The steam and the air entering the FR/AR are preheated to a designated inlet temperature ($T_{air,AR}$, $T_{H2O,FR}$). If not stated otherwise, the inlet temperature of both streams (STEAM, AIR) when entering the FR/AR is set to 400 °C.

- As Aspen Plus™ is not equipped to handle solid fuels, the biomass feedstock (FUEL) is fed to the decomposer (DECOMP), where it is decomposed into its pyrolysis products (DEVOLAD). The heat of pyrolysis (Q-DECOMP) is transferred to the fuel reactor. A detailed description of the decomposer block is given in Section 2.2.
- The pyrolysis products (DEVOLAD), the gasification agent (STEAMX), the OC recycled from the AR (OCO-TOFR), and the CO_2 required for solid feeding and loop seal fluidization (SCREWFLU) are mixed (FRIN) before entering the fuel reactor.
- Subsequently, the educts entering the fuel reactor (FR) are converted into reaction products according to the chemical equilibrium at the given boundary conditions (T_{FR}, P_{FR} = 1 atm).
- The solids leaving CYCL1 are separated into the OC fed to the AR (OCR-TOAR) and a stream containing carbon and ash (SOL) in the solids separation (SOLSEP). This separation signifies the removal of bed material (i.e., OC, ash and unconverted feedstock) from the FR via sluicing during operation. Additionally, a fraction of the oxygen carrier material is removed from the system (OCLOSS), to model OC losses via sluicing and attrition.
- The OC makeup stream (MAKEUP) and the inlet air (AIRX) are mixed (ARIN) before being fed to the AR.
- Inside the air reactor (AR) the reduced OC and the unreacted char react with the oxygen contained in the air according to the chemical equilibrium at the given boundary conditions (T_{AR}, P_{AR} = 1 atm).

2.2. Decomposer

Generally, the conversion of a fuel during gasification is described by three subsequent mechanisms: drying, pyrolysis and gasification [6]. While the gasification step is modelled in the FR, the former two mechanisms are modelled in the decomposer block in this study. As drying solely encompasses the release of moisture from the fuel [6,46], the main focus of this section is placed on fuel pyrolysis. Ohlemüller et al. [25] applied the pyrolysis model of Matthesius et al. [47] to predict the pyrolysis product composition from coal proximate and ultimate analysis parameters. Although it is reported that the basic mechanism of coal and biomass pyrolysis are similar [6,7], it was decided to employ a pyrolysis model specifically tailored for biomass feedstocks, as this study is focused on the conversion of biomass-based fuels. Neves et al. [48] devised a pyrolysis model for biomass feedstock built on the basis of an extensive experimental database. Similar to the pyrolysis model by Matthesius et al. [47], this model solely requires information on the feedstock composition (C, H, O and char content) to estimate the final chemical composition of the organics after pyrolysis, allowing for its straight forward implementation into the existing Aspen Plus™ model. Cuadrat et al. [49] found that the formation of tar and larger hydrocarbons (>C1) is negligible in the presence of ilmenite and steam/CO_2. Therefore, the assumption by Ohlemüller et al. [25] and Mendiara et al. [50] that tars and larger hydrocarbons are directly converted to methane and carbon monoxide was also adopted in this study. Moreover, oxygen and hydrogen contained in the char were converted to syngas, resulting in a char solely consisting of carbon. As the FR is modelled based on chemical equilibrium, these simplifications do not have an impact on the final simulation results.

By applying these assumptions, the product compositions after pyrolysis were calculated on the basis of the proximate and ultimate analysis of wood pellets, being the model feedstock for all subsequent considerations (see Table 1).

Table 1. Summary of the Ultimate and Proximate analysis for industrial wood pellets.

Ultimate Analysis	wt-%	Proximate Analysis	wt-%
C (d.a.f.)	50.8	Moisture	6.5
H (d.a.f.)	6	Ash (d.b.)	0.7
N (d.a.f.)	0.07	Volatile matter (d.b.)	85.1
O (d.a.f.)	43.2	Fixed carbon (d.b.)	14.2
S (d.a.f.)	0.008		
Cl (d.a.f.)	0.006		
Net calorific value [MJ/kg]	17.96		

Since the pyrolysis product composition is highly temperature dependent [6,7,48], a constant temperature representing the FR temperature during CLG was selected as the input for the pyrolysis model ($T_{devol.} = 900\ °C$). A summary of the final product composition after de-volatilization, which was implemented into the process model, is given in Table 2.

Table 2. Mass yields [wt-%] for DECOMP Aspen Plzus® block for industrial wood pellets according to pyrolysis model of Neves et al. [48] (T = 900 °C).

Component	wt-%	Component	wt-%
ASH	0.65	H_2O	14.06
CO	55.20	N_2	0.06
C	11.92	CO_2	3.11
CH_4	13.55	H_2S	0.01
H_2	1.43		

2.3. Boundary Conditions

For all subsequent simulations, the biomass input was selected in such a way, that the thermal load, P_{th}, of the chemical looping gasifier amounted to 1 MW. In terms of the circulating solid materials, the deployed oxygen carrier material is ilmenite, for which is has been established that the major redox stages are FeO + TiO_2, Fe_3O_4, TiO_2 and Fe_2TiO_5 [51]. These redox stages were modelled as $FeTiO_3$ (for FeO + TiO_2), Fe_3O_4, TiO_2, and Fe_2O_3 + TiO_2 (for Fe_2TiO_5). Deeper redox stages (e.g., FeO) were also considered in the process model, yet were not found to be formed in notable amounts. The inert solid sand was modelled through pure SiO_2. The FR and AR are operated under atmospheric pressure. Moreover, the air reactor temperature was set to 1050 °C, if not stated otherwise. The fuel reactor temperature results from the energy balance of the process, requiring that both reactors are in heat balance ($\dot{Q}_{FR} = 0$, $\dot{Q}_{AR} \geq 0$). As the kinetic syngas inhibition of char gasification reactions [8,12] is not considered in the RGIBBS equilibrium calculation, full char conversion is attained inside the FR for all temperatures considered in this study. Although this simplification signifies a deviation from reality, it does not impact the general inferences which will be elaborated on hereinafter. For the steam to biomass ratio in the FR a value of 0.9, reported for a 2–4 MW_{th} chemical looping gasifier in literature [52], was selected if not stated otherwise. During CLC/CLG operation CO_2 is required for fuel feeding and inerting. This stream of CO_2, entering the fuel reactor, was selected in such a way that the CO_2 to biomass ratio amounts to 0.2, to take into account that the CO_2 input through the feeding section increases with increased thermal load. The two remaining process variables, the air mass flow entering the AR and the circulating oxygen carrier mass, were adjusted in such a way that autothermal CLG operation was achieved. A summary of all boundary conditions is given in Table A1 in Appendix A.

3. Results and Discussion

3.1. Attaining CLG Behavior

Generally, shifting from a combustion to a gasification process is achieved through lowering the air/oxygen-to-fuel ratio of the process, thereby decreasing the ratio of fully to partially oxidized gas species leaving the process and hence increasing the heating value of the product gas [6,53,54]. Here, the critical parameter is the so called air-to-fuel equivalence ratio given by the ratio of oxygen fed to the AR, $\dot{m}_{O,AR}$, and the oxygen required for full feedstock combustion, $\dot{m}_{O,stoich}$:

$$\lambda = \frac{\dot{m}_{O,AR}}{\dot{m}_{O,stoich}}. \tag{10}$$

According to this definition, (close to) full combustion of the feedstock is attained for air-to-fuel equivalence ratios larger than unity ($\lambda > 1$), while gasification processes require sub-stoichiometric oxygen feeding (i.e., $\lambda < 1$).

Due to the dissection of the gasification/combustion reaction into two separate reactors in chemical looping processes, there is no direct contact between the air entering the AR and the fuel entering the FR. Hence, the application of an alternative parameter, the oxygen-carrier-to-fuel equivalence ratio, ϕ', relating the amount of oxygen carried by the OC to the FR to the oxygen required for stoichiometric combustion, has been suggested [43]:

$$\phi' = \frac{R_{OC}\cdot\dot{m}_{OC}}{\dot{m}_{O,stoich}}. \tag{11}$$

Here, R_{OC} denotes the oxygen transport capacity of the given oxygen carrier material. While this parameter accurately relates the two quantities for CLC, where the OC always leaves the AR in a (close to) fully oxidized state, this is not necessarily the case in CLG. Therefore, a slightly altered oxygen-carrier-to-fuel equivalence ratio, ϕ, considering the possibility of a partially reduced OC leaving the AR, has been proposed for gasification applications [35]:

$$\phi = \frac{R_{OC}\cdot\dot{m}_{OC}\cdot X_{s,AR}}{\dot{m}_{O,stoich}}, \tag{12}$$

where $X_{s,AR}$ signifies the oxidation degree of the oxygen carrier at the AR outlet, given by [24,35]:

$$X_{s,AR} = \frac{m_{OC,AR} - m_{OC,red}}{R_{OC}\cdot m_{OC,ox}}. \tag{13}$$

Here, $m_{OC,red}$ and $m_{OC,ox}$ are the mass of an OC sample in a fully reduced and oxidized state respectively, while $m_{OC,AR}$ is the mass of the OC sample leaving the AR. For ilmenite the fully reduced oxygen carrier is approximated by $FeTiO_3$, the fully oxidized state is approximated by $Fe_2O_3 + 2TiO_2$, and $Fe_3O_4 + 3TiO_2$ denotes an intermediate redox state ($X_s = 0.67$).

In order to assess how λ and ϕ have to be adjusted in order to obtain an efficient CLG process, one should first assess the general impact of these two parameters on the process. Due to the relative fast kinetics of the OC re-oxidation [55–57], the oxygen carrier is often assumed to leave the AR in a (close to) fully oxidized state for $\lambda > 1$ in chemical looping processes. In contrast, sub-stoichiometric air-to-fuel equivalence ratios ($\lambda < 1$) only lead to a partial re-oxidation of the OC in the AR. Following the same logic, the OC material can be assumed to leave the FR in a (close to) fully reduced state in case $\phi < 1$, whereas partial reduction is attained for $\phi > 1$. From these deductions, it becomes clear that "standard" CLC operation is attained for $\lambda > 1$ and $\phi > 1$, [42,43]. Here, a highly oxidized OC leaves the AR, before being partially reduced in the FR, which is illustrated in Figure 3a.

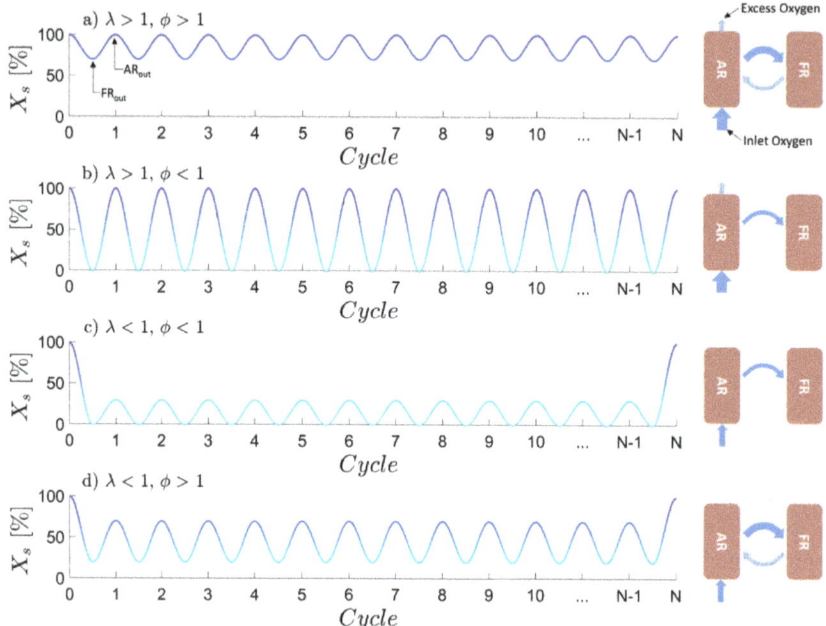

Figure 3. Different chemical looping modes (**a–d**) dependent on the air-to-fuel equivalence ratio λ and the oxygen-carrier-to-fuel equivalence ratio ϕ.

When targeting pronounced syngas formation, the oxygen release in the FR has to be limited, so that full feedstock oxidation is prevented [35,52]. The most obvious avenue that can be pursued to achieve this is lowering ϕ below unity. When doing so, the employed air-to-fuel equivalence ratio λ determines how much oxygen is transported between the two reactors per gram of OC. In case of $\lambda > 1$, which is illustrated in Figure 3b, the oxygen carrier undergoes a full redox cycle and hence the full oxygen transport capacity of the OC material (i.e., R_{OC}) is exploited. On the other hand, $\lambda < 1$ means that in equilibrium the OC leaves the AR in a partially reduced state, hence also reducing the mass specific oxygen transport of the OC (see Figure 3c). Lastly, one might also consider a process with $\lambda < 1$ and $\phi > 1$, as shown Figure 3d. In order to attain a steady-state process exhibiting these characteristics, full reduction of the oxygen carrier has to be prevented in the FR (e.g., kinetically), so that a fraction of oxygen is transported back to the AR. This means that in contrast to the former approaches, this case cannot be attained in equilibrium-like conditions. While this approach might also be feasible for CLG operation in theory, straight forward measures allowing for a controlled oxygen release in the FR are not at hand. Consequently, lowering the oxygen-to-fuel-ratio in the FR (i.e., $\phi < 1$) is the most promising avenue to attain CLG behavior. When aiming for large syngas yields, ϕ has to assume values below unity, while values exceeding unity are targeted in CLC [42,43]. In the following, different effective control strategies to achieve this reduction in ϕ, required for pronounced syngas formation in the FR, while at the same time achieving an autothermal process, will be investigated.

In order to simplify the subsequent considerations, a standard parameter to describe gasification processes, the cold gas efficiency (CGE), η_{CG}, will be deployed hereinafter. It describes which amount of chemical energy from the fuel is transferred to the gaseous FR product gas during gasification [6,7].

$$\eta_{CG} = \frac{\dot{n}_{gas,FR} \cdot (x_{CH4,FR} \cdot LHV_{CH4} + x_{CO,FR} \cdot LHV_{CO,FR} + x_{H2,FR} \cdot LHV_{H2})}{\dot{m}_{fuel} \cdot LHV_{fuel}} \tag{14}$$

Here, $\dot{n}_{gas,FR}$ and $\dot{m}_{fuel}$ denote the mole flow of the product gas stream and the fuel input into the FR, respectively. *LHV* is the lower heating value of the fuel (mass basis) and the gas species (molar basis) and x_i is the mole fraction of the gas species.

3.2. Reduction of OC Circulation

One approach to obtain CLG behavior, which has been suggested by Pissot et al. [52], is reducing the amount of OC cycled through the system ($\dot{m}_{OC}$), hence reducing ϕ. This approach can be deduced directly from Equation (12). Due to the resulting lower oxygen transport to the FR, syngas formation is favored, as less oxygen for full oxidation of the feedstock is provided by the OC. The simulation results for this approach are given in Figure 4. When considering the gas composition (Figure 4a) of the streams leaving the air and fuel reactor, various trends are visible. As expected, the syngas content in the gaseous FR products increases with decreasing OC circulation rate, which can directly be attributed to the lower oxygen/fuel ratio in the FR. Consequently, steam and CO_2 formation decrease. Yet, it has to be noted that substantial syngas concentrations are only attained for $\phi < 1$, which requires significant reductions in the OC circulation rate, when compared to CLC, where OC-to-fuel equivalence ratios as high as 8 [27] and 25 [40] are reported in literature for solid and gaseous fuels, respectively. For the gas concentrations leaving the AR, a strong impact of ϕ on the effluent oxygen is visible. As the inlet air mass flow was not varied ($\lambda = 1.2$), this observation is clear, as less O_2 is removed from the gas stream due to the lower OC circulation for $\phi < 1$. Furthermore, the CO_2 content in the AR product is predicted to be insignificant, indicating a complete char conversion, which is expected in chemical equilibrium. When considering Figure 4b, showing the solid composition after the fuel and air reactor, it can be seen that the OC leaves the AR and FR in a fully oxidized ($Fe_2O_3 + TiO_2$) and reduced ($FeTiO_3$) state, respectively for $\phi < 1$, whereas the OC is only partially reduced (indicated through the presence of Fe_3O_4) in the FR in case ϕ exceeds unity. Hence, the fraction of $FeTiO_3$ leaving the FR strongly increases with decreasing OC circulation, signifying a higher degree of reduction of the OC, due to the lower oxygen availability. As expected one consequently obtains chemical looping combustion behavior (see Figure 3a) for oxygen-carrier-to-fuel equivalence ratios greater than unity ($\phi > 1$), whereas chemical looping gasification behavior (see Figure 3b) is attained for $\phi < 1$.

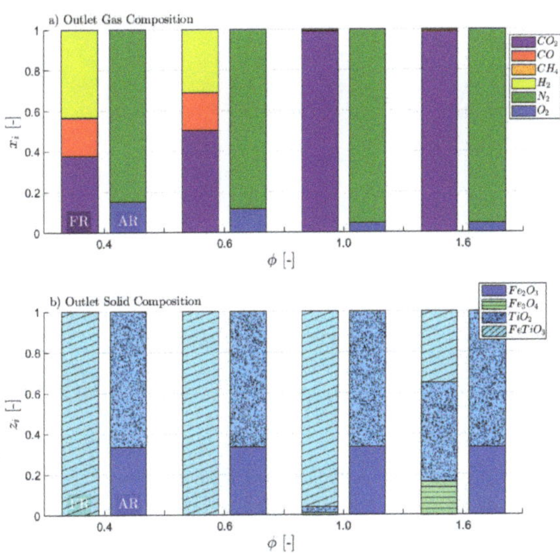

Figure 4. Simulation results for CLG operation through reduced oxygen carrier (OC) circulation. Dry molar gas composition (**a**) and molar solid composition (**b**) as a function of ϕ for varying OC circulation rates ($\lambda = 1.2$).

Based on these findings, one can conclude that a successful shifting from CLC to CLG for a given air-to-fuel ratio can be attained through a reduction in the OC circulation, which can also be seen in Figure 5a, showing a linear dependence between the two parameters. This means that for a change of ϕ from 1.0 to 0.5, the OC circulation rate has to be halved. However, lower solid circulation rates also result in a proportional decrease in the heat transport from the AR to the FR and hence a drop-off in FR temperatures [35,58]. While a moderate decrease in fuel reactor temperatures with decreasing OC circulation rate is visible for $\phi > 1$, for which complete feedstock conversion is attained in the FR, this decrease becomes more prominent for $\phi < 1$, where gasification reactions in the FR are dominant, hence increasing the endothermicity of reactions occurring in the FR. Consequently, FR temperatures fall below 800 °C for $\phi < 0.5$, where the availability of circulating OC material for sensible heat transport between the FR and AR is halved, when compared to $\phi = 1$ and more importantly the syngas content in the FR products is significant (see Figure 4a). This increase in syngas content also goes in hand with a decrease in the total net heat release from the CLG process ($\dot{Q}_{net}$), which can be calculated from the difference in the enthalpies of the streams entering ($_{in}$) and leaving ($_{out}$) the air and fuel reactor (see Equation (15)), as the enthalpy of the FR products increases.

$$\dot{Q}_{net} = \sum_{FR,in} \dot{m}_i \cdot h_i - \sum_{FR,out} \dot{m}_i \cdot h_i + \sum_{AR,in} \dot{m}_i \cdot h_i - \sum_{AR,out} \dot{m}_i \cdot h_i \tag{15}$$

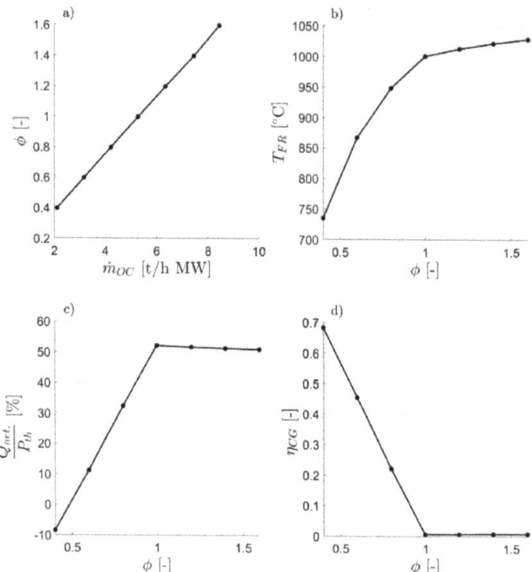

Figure 5. Simulation results for CLG operation through reduced OC circulation. OC-to-fuel ratio as a function of the OC circulation rate (**a**). Fuel reactor temperature (**b**), relative net process heat (**c**), and cold gas efficiency (**d**) for different values of ϕ ($\lambda = 1.2$).

The decrease in net process heat release with decreasing ϕ, indicating the retaining of chemical energy in the FR products, also becomes visible upon consideration of Figure 5c, depicting the relative net heat release of the process for the different OC-to-fuel ratios. For the given boundary conditions, an autothermal process, for which syngas yields are maximized without relying on external heat addition ($\dot{Q}_{net} = 0$) is attained for an OC to fuel ratio of approx. 0.5. The resulting cold gas efficiency for this operating point amounts to approx. 60% (see Figure 5d) at a FR temperature of 775 °C. Although the equilibrium model predicts full char and volatile conversions for these temperatures (see Figures 4a and 5d), char, volatile, and tar conversion are known to be kinetically governed processes in chemical looping systems [25,55,56,59], leading to product compositions deviating strongly from equilibrium composition [35,52]. Due to this reason, temperature differences in the range of 50 to 100 °C are generally targeted in dual fluidized bed gasification [16], in order to obtain sufficiently high gasifier temperatures, allowing for decent char, volatile, and tar conversions. Accordingly, FR temperatures in the range of 850–950 °C are desired in CLG, in order to attain high carbon capture efficiencies and cold gas efficiencies as well as low syngas tar loads [20,23,37,60,61].

These considerations underline that, although the desired reduction in ϕ is possible, attaining an efficient autothermal CLG process through a reduction in the OC circulation rate is not a recommendable strategy as it entails low fuel reactor temperatures, due to the dual-purpose of the OC circulation (i.e., oxygen and heat transport). Consequently, alternative approaches, allowing for a decoupling of oxygen and heat transport between the AR and FR and hence increased FR temperatures are required, in order to attain a CLG process exhibiting the desired characteristics.

3.3. Dilution of OC with Inert Bed Material

One strategy allowing for a decoupling of oxygen and heat transport between air and fuel reactor, which has been discussed in literature, is employing a mixture of an active OC material and a solid inert species (e.g., sand) [35,37,52]. Here, the inert fraction serves purely as a heat carrier, transferring sensible heat between the two reactors, without participating in the occurring reactions, while the active OC fraction fulfills its dual purpose of oxygen and heat transport. Consequently, this approach is a combination of CLG and dual fluidized bed gasification, which solely employs inert bed materials for heat transport. Following this logic, Ge et al. [37] found that through accurately tailoring the mixing ratio of inert silica sand and hematite, serving as an OC, FR temperatures can be stabilized at elevated levels (i.e., >900 °C), while at the same time ensuring a controlled oxygen transport to the FR, resulting in large syngas yields.

In terms of the impact of the variation in OC-to-fuel ratio on gas compositions achieved through this dilution of the OC material with an inert, similar observations are obtained (see Figure 6a). This means syngas formation increases steadily for $\phi < 1$. Moreover, the OC carrier composition, shown in Figure 3b, follows similar trends as observed for a plain reduction in the OC circulation rate (see Section 3.2), with a fully reduced OC leaving the FR for $\phi < 1$ (see Figure 3b), whereas only partial reduction is observed for $\phi > 1$ (see Figure 3a). Yet, the fraction of active OC material clearly decreases with decreasing ϕ, due to the dilution with silica sand.

As the total amount of circulating solids is kept constant, the mass of circulating OC material is inversely proportional to the dilution factor. This means that there exists a linear relationship between the solid fraction of the inert material (z_{SiO2}) and ϕ, which is visible in Figure 7a. Hence, for a given solid circulation rate, shifting from CLC to CLG can be attained through increased inert dilution. The positive effect of inert addition on FR temperatures becomes apparent upon consideration of Figure 7b. In contrast to a direct reduction in the OC circulation rate, the substitution of a fraction of the active metal oxide with an inert heat carrier allows for a sustaining of FR temperatures above 980 °C even for OC to fuel ratios as low as 0.5. Due to this increase in FR temperatures, the average temperature of the CLG process increases, leading to a slightly increased ϕ of approx. 0.55 for which autothermal operation is attained (see Figure 7c) (Higher process temperatures increase the heating demands of the educts entering the FR and AR and hence reduce the OC-to-fuel ratios for which autothermal operation can be obtained). Therefore, the cold gas efficiency obtained for autothermal operation for the given approach is also marginally reduced (see Figure 7d), when compared to the approach discussed in Section 3.2. Yet, it has to be noted that due to the intensified heat transport between the AR and FR, significantly smaller reactor temperature gradients are required for the given approach. Consequently, AR temperatures can be lowered without jeopardizing char conversions in the FR, thus reducing average process temperatures and allowing for strongly increased cold gas efficiencies (see also Section 3.5). Another advantage of this approach is that a catalytic material, not participating in oxygen transport (e.g., olivine), could be employed for OC dilution instead of sand, allowing for improved syngas characteristics with regard to tar content.

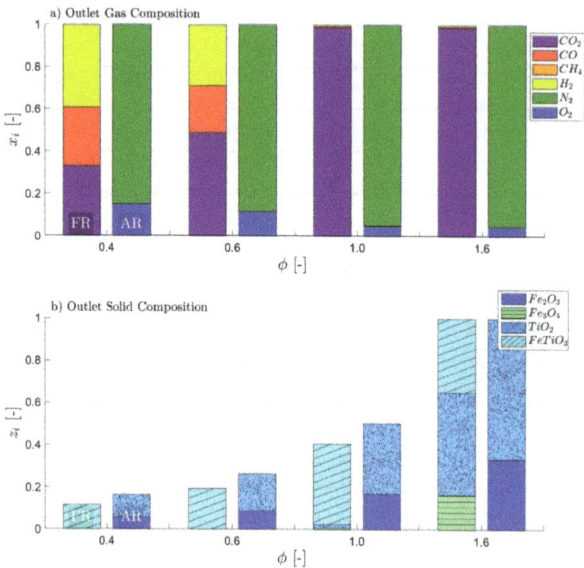

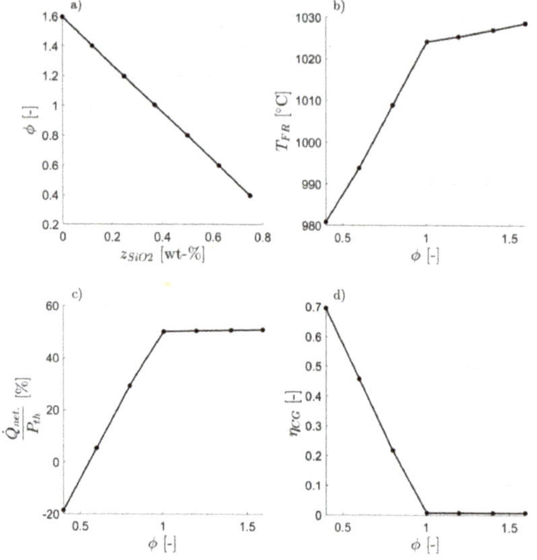

Figure 6. Simulation results for CLG operation through OC dilution with inert SiO_2 sand. Dry molar gas composition (**a**) and molar solid composition (**b**) as a function of ϕ for varying OC circulation rates ($\lambda = 1.2$, $\dot{m}_{OC} + \dot{m}_{SiO2} = const.$).

Figure 7. Simulation results for CLG operation through OC dilution with inert SiO_2 sand. OC-to-fuel ratio as a function of the inert concentration of the circulating solid mixture (**a**). Fuel reactor temperature (**b**), relative net process heat (**c**), and cold gas efficiency (**d**) for different values of ϕ ($\lambda = 1.2$, $\dot{m}_{OC} + \dot{m}_{SiO2} = const$).

Despite the presented advantages, Larsson et al. [35] found that, albeit slightly reducing tar loads, the addition of an active OC (ilmenite) to an inert circulating bed material in a dual-fluidized bed gasifier (for $\phi < 0.2$), entails a continuous drop in cold gas efficiency. This was explained by the fact that ilmenite addition does not enhance char conversion significantly, while its presence leads to a partial oxidation of the product gas. On the other hand, Pissot et al. [52] found that dilution of an active OC bed with up to 90% of an inert material does not entail visible enhancements in the cold gas efficiency of the CLG process, while it has a visible negative impact on carbon conversion. This shows that the mixing of an inert and an active OC material can have different effects on the process depending on the governing boundary conditions. Another drawback of this approach is that, albeit the addition of solids allows for an adjustment of ϕ during operation, it leads to a large system inertia, making it an arduous task to quickly react to disturbances. Moreover, a fraction of the solid material has to be removed from the system for ash removal in a continuously operated CLG unit. Economic considerations require a separation of these materials for further processing, recycling, and disposal. Clearly, the presence of a third component (i.e., sand, olivine) further complicates this task. Lastly, it is known that the operation of a fluidized bed with multiple bed materials of different characteristics brings about additional challenges in terms of material fluidization, entrainment, and attrition, as well as bed segregation [62]. Due to these reasons it was also suggested to employ materials of a low oxygen transport capability (R_O), such as LD-slag, containing a large inactive fraction not participating in the oxygen transport, which fulfills the purpose of the inert heat carrier [52]. Through this, oxygen carrier circulation rates providing sufficient heat transport between the reactors can be targeted, without obtaining OC-to-fuel equivalence ratios above unity. Yet, for this approach the main challenge is finding suitable OC materials exhibiting an oxygen transport capability in the desired range, high activity towards hydrocarbon conversion, and good chemical and mechanical stability.

3.4. Reduction of Air-to-Fuel Equivalence Ratio

To allow for a less restricted material selection and avoid solid inert addition, an alternative strategy to decouple oxygen and heat transport between the AR and FR is required. In order to achieve this, Larson et al. [35] suggested the deployment of a secondary system in which the OC is pre-reduced before entering the FR. This means that, as shown in Figure 3c, a partially reduced OC enters the FR ($X_s < 1$), thus entailing a lower OC-to-fuel ratio (see Equation (12)). Instead of employing a secondary reactor to accomplish this, one can also operate the AR in a sub-stoichiometric fashion ($\lambda < 1$), thereby preventing full re-oxidation of the OC in the AR. This means that in order to attain CLG conditions, the amount of air fed into the air reactor can be reduced, while retaining a constant OC circulation. As a consequence, the OC steadily reaches a lower degree of oxidation, hence lowering its oxygen release in the FR, until steady state is reached (more details see Appendix B). This approach has already been pursued in a 140 kW$_{th}$ chemical looping reforming unit, employing methane as a fuel [44]. The suggested concept becomes more lucid when considering the simulation results shown in Figure 8. Clearly, the amount of fully reduced ilmenite leaving the air and fuel reactor increases when decreasing the air input into the AR for $\phi < 1$ (see Figure 8b). While the same is true for the solids leaving the FR for all presented CLG approaches, a strong increase in the $FeTiO_3$ and Fe_3O_4 content in the AR products is obtained when reducing λ below unity. This can be explained by the fact that the oxygen available in the air reactor is insufficient to fully re-oxidize the OC, signified through an O_2–free product gas from the AR for $\phi < 1$ (see Figure 8a). Consequently, a pure stream of N_2 containing small concentrations of Argon and other minor compounds is produced in the AR [44]. Since substantial quantities of OC are cycled through the system in a fully reduced state, they effectively act as an inert, meaning that they transfer sensible heat, but do not participate in the occurring chemical reactions through oxygen release and uptake. However, in practice the reduced OC could potentially function as a catalytic site for tar cracking and methane reforming and favor the formation of syngas [32–35], thereby enhancing the process characteristics. Another advantage of the given approach is that an undiluted OC can be employed, which simplifies the required solid-gas and solid-solid (ash-OC-char) separation and the operation of the CLG unit with regard to the fluidization behavior. Moreover, the net heat duty of the process can be tailored promptly and easily through an adjustment of the air flow to the AR, allowing for quick responses to disturbances (e.g., variations in feedstock composition).

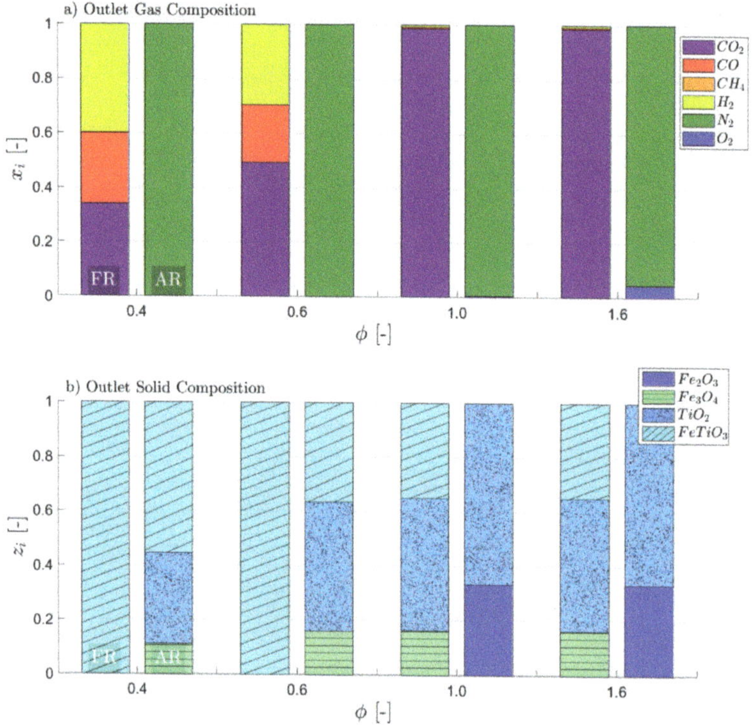

Figure 8. Simulation results for CLG operation through reducing λ. Dry molar gas composition (**a**) and molar solid composition (**b**) as a function of ϕ ($\dot{m}_{OC} = const.$).

The impact of the air-to-fuel equivalence ratio (λ) on ϕ is shown in Figure 9a. In CLC mode ($\lambda > 1$), where full OC oxidation is achieved in the AR (i.e., $X_{s,AR} = 1$), ϕ assumes a constant value, given by the amount of oxygen which is transported by a fully oxidized OC for a given circulation rate, regardless of the deployed air-to-fuel ratio (see Equation (12)). In contrast, lowering λ to values below unity to attain CLG operation means that ϕ and λ are equal, as the oxygen transport to the FR is limited by the oxygen availability in the AR:

$$\phi = \begin{cases} \lambda & \text{for } \lambda < 1 \\ \dfrac{R_{OC} \cdot \dot{m}_{OC}}{\dot{m}_{O,stoich}} = const. & \text{for } \lambda \geq 1 \end{cases} \tag{16}$$

The discontinuity of this relation for $\lambda = 1$ can be explained by the fact that when surpassing this value, a transient shift from CLC (see Figure 3a) to CLG (see Figure 3c) behavior (or vice versa) occurs, which goes in hand with a continuous decrease (resp. increase) in the oxidation degree of the oxygen carrier, before steady state sets in (more details see Appendix B).

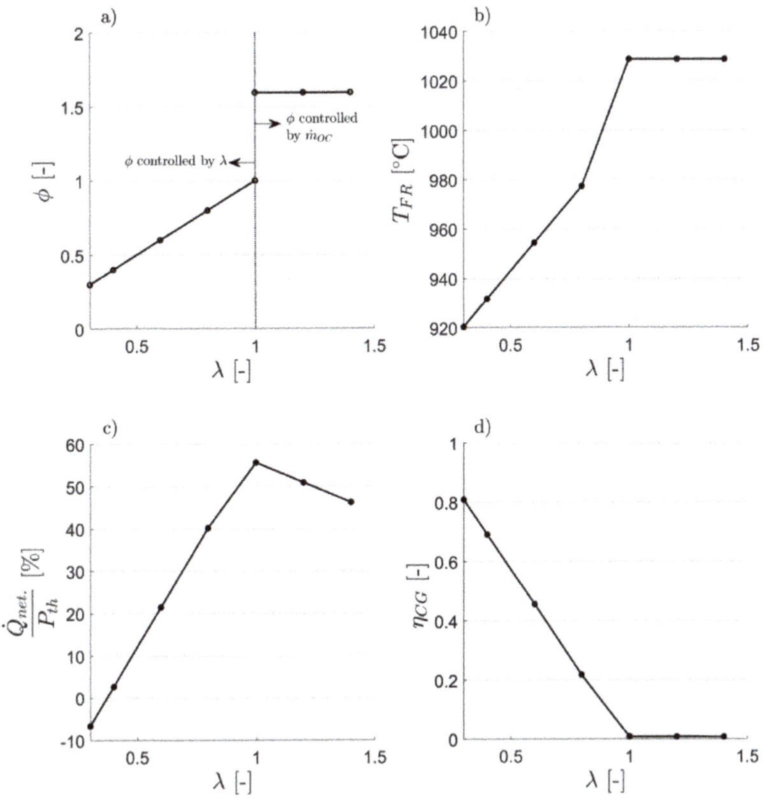

Figure 9. Simulation results for CLG operation through reducing λ. OC-to-fuel ratio as a function of the air-to-fuel equivalence ratio λ (**a**). Fuel reactor temperature (**b**), relative net process heat (**c**), and cold gas efficiency (**d**) for different values of λ ($\dot{m}_{OC} = const.$).

In terms of FR temperatures, Figure 9b shows that the given approach leads to a successful retaining of FR temperatures above 900 °C, even for ϕ-values as low as 0.4, due to the transportation of sensible heat by the OC. Moreover, the given approach yields more beneficial results in terms of the process heat balance, which can be seen in Figure 9c. Clearly, autothermal CLG operation is attained for $\phi = 0.37$, which means cold gas efficiencies exceeding 70% can be achieved (see Figure 9d). This is the case as in contrast to the previous approaches (see Sections 3.2 and 3.3), the AR is not operated in air excess during CLG operation, reducing the loss of sensible heat through the AR off-gases. This means that if one would reduce the air feed to the AR to the minimum extent required for full OC re-oxidation for the CLG approach employing inert dilution (see Section 3.3), enhanced cold gas efficiencies could be attained. Nonetheless, the given approach clearly shows advantages in terms of process control due to its flexibility, the possibility of freely selecting a suitable OC material (i.e., no specific limits on R_O), without having to consider material mixtures, and the availability of a catalytically active reduced OC material, instead of an inert solid, cycling through the system. Moreover, the chemical strain on the OC material is reduced as the change in oxidation degree for each redox cycle is lower, when compared to the former approaches, relying on full reduction and oxidation in the FR and AR, respectively (see Figure 3b,c), which should have beneficial effects on the OC lifetime.

However, one issue that might arise due to the operation of the AR in an sub-stoichiometric fashion is related to the fact that during operation a fraction of the feedstock char leaves the FR unconverted and hence travels to the AR with the circulating OC material [23,26,27]. This so called "carbon slip" leads to competing reactions between the OC material and the residual char, in case the AR is operated with $\lambda < 1$. Yet, simulations show that in an oxygen deficient atmosphere carbon conversion is favored to OC re-oxidation in chemical equilibrium. Moreover, CO formation shows to be negligible (more details see Appendix C). Due to the fast kinetics of both char conversion and OC re-oxidation, it can be expected that equilibrium-like conditions are attained in the AR and hence all residual char is fully oxidized to CO_2 in the AR. This hypothesis is also supported by chemical looping experiments in small scale fixed bed reactors, during which it was established that in the beginning of the re-oxidation stage oxygen preferentially reacts with deposited carbon before re-oxidizing the OC [21,63,64]. Nonetheless, experiments showing that this is also the case in a continuously operated CLG unit and that CO formation is negligible are required to establish that full char conversion without substantial CO formation in the AR can be attained for this approach. Another issue related to this approach is the potential deep reduction of the OC, which could potentially entail problems related to intensified OC attrition or bed agglomeration. Although the process model does not predict substantial formation of deeper reduction stages (e.g., FeO) in the FR, such phases, related to bed agglomeration, have been found to be formed in CLC under highly reducing conditions [51,65,66]. Therefore, the gravity of this issue should be further investigated in experimental studies.

3.5. Optimizing CLG Efficiency

In the previous section it was established that OC-to-fuel equivalence ratios smaller than unity are required in the FR. Moreover, it was demonstrated only when decoupling heat and oxygen transfer between the AR and FR, $\phi < 1$ and FR temperatures above 850 °C can be obtained for an autothermal CLG process. Thermodynamically speaking, it does not make a difference how this decoupling of heat and oxygen transport is attained, which is why the following considerations will focus on the CLG approach presented in Section 3.4, employing a reduction in the air-to-fuel equivalence ratio to achieve CLG behavior.

When optimizing gasification processes, the trade-off between maximizing the carbon conversion in the gasifier and at the same time attaining high cold gas efficiencies is at the core of many optimization strategies. This is also the case in CLG, where $\eta_{CGE} = 1$ and complete char conversion is desired, yet not attainable. While large carbon capture efficiencies are obtained in cases where the char is gasified in the fuel reactor to a large extent, which is promoted by high FR temperatures [20,23,37], large steam/biomass ratios [20,37], and high OC-to-fuel ratios (if sufficient char residence times are provided) [27,52], cold gas efficiencies are maximized by the minimization of the oxidation of H_2 and hydrocarbons in the FR [35]. Although full oxidation of syngas in the FR should be limited to achieve large CGEs, formation of steam and CO_2 in the FR is required to a certain extent to obtain autothermal CLG conditions. The degree to which this formation of fully oxidized gas species is required is determined by the criterion of the CLG process being in heat balance ($\dot{Q}_{net} = 0$). This means that the heat release attained through full feedstock oxidation has to balance the heat demand of pre-heating of all inlet streams to the given reactor temperatures, the heat of reaction for endothermic gasification reactions, and the heat losses of the CLG unit. This has also been shown in the previous sections where despite assuming chemical equilibrium (i.e., full feedstock conversion), cold gas efficiencies deviating strongly from unity were obtained for autothermal boundary conditions (see Figures 5, 7 and 9).

Therefore, one approach to enhance the cold gas efficiency in CLG is a reduction in the inlet gas flows entering the air and fuel reactor. Since the air mass flow entering the AR is required to control ϕ, this leaves the steam mass flow entering the FR as a free variable which can be altered to enhance cold gas efficiencies. The effect of a reduction in the steam to biomass ratio on the net heat release of the process is shown in Figure 10a. It is visible that, with a decreasing steam to biomass ratio, the air-to-fuel equivalence ratio for which an autothermal process is attained decreases. Due to the direct correlation between the oxygen availability and cold gas efficiency in CLG (see Figure 10b), this also means that the CGE obtained for autothermal operation increases with decreasing steam/biomass ratio, so that the CGE is raised from 72.5 to 77.1%, when decreasing the steam/biomass ratio from 0.9 to 0.3. However, it is obvious that the reduction of the steam to biomass ratio would also entail a drop in carbon capture efficiencies of the process, as less steam is available for char gasification and the kinetic inhibition effect of syngas increases with decreasing steam concentrations (entailing larger syngas partial pressures) in the FR [8,12,67,68]. This becomes most obvious for a steam to biomass ratio of 0, for which char conversions in the FR would be diminutive in a real gasifier, due to the slow kinetics of heterogeneous solid-solid OC-feedstock reactions [67–69]. As this drop in char conversion is not predicted by the equilibrium model, the negative effect on process efficiency with decreasing steam to biomass ratio cannot be evaluated in this study. However, sufficient steam availability clearly is a prerequisite in CLG, when targeting large char conversions and hence carbon capture efficiencies.

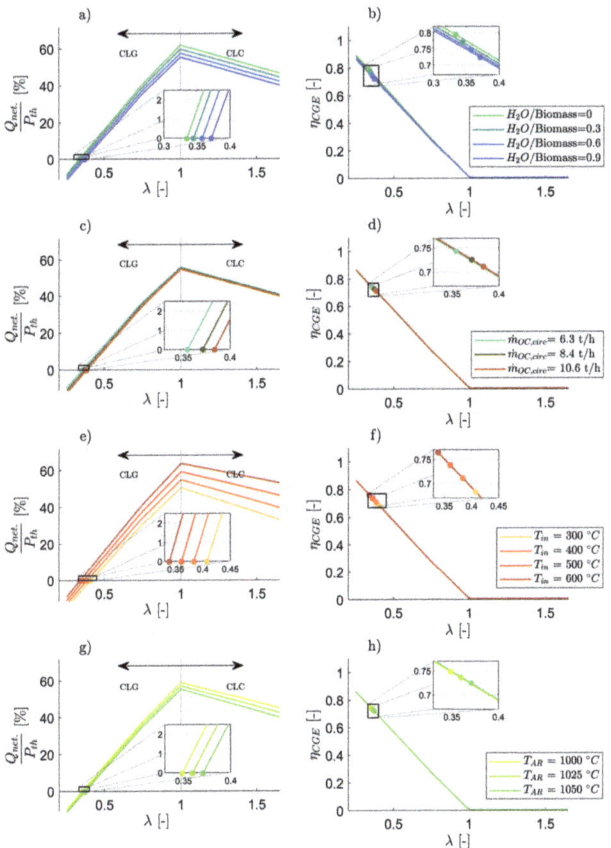

Figure 10. Net heat release and cold gas efficiency for CLC/CLG process as a function of the air to fuel equivalence ratio for different steam to biomass ratios (**a,b**), OC circulation rates (**c,d**), gas inlet temperatures (**e,f**), and air reactor temperatures (**g,h**). Circles mark the cold gas efficiency for autothermal CLG operation ($\dot{m}_{OC} = const.$, so $\phi = \lambda$ for $\lambda < 1$ and $\phi = const. > 1$ for $\lambda > 1$).

Another possible measure to enhance CGEs are variations in the circulation rate of the OC, which is shown in Figure 10c,d. Clearly, larger solid circulation rates enhance the heat transport between the reactors and hence entail higher FR temperatures [16]. However, due to material attrition, solid loss, which necessitates continuous make-up feeding, also scales with the circulation rate. As shown in Figure 10d, the effect of this material loss on the process heat balance is comparatively small, thus its effect on the cold gas efficiency is low. However, the model predicts an increase in FR temperatures from 892 to 951 °C, when increasing the circulation rate from 6.3 to 10.6 t/h. This means that generally, large solid circulation rates are desired in CLG units, as large FR temperatures are beneficial for volatile and char conversion [20,23,37]. Yet, it has to be kept in mind that the solid circulation in dual fluidized bed systems requires solid entrainment from the fluidized bed riser, which can be increased through an increase in gas velocities (i.e., increase in steam/biomass ratio), smaller particle diameters or smaller reactor diameters [16]. Moreover, intensified solid circulation also increases the occurrence of a "carbon slip" to the AR, due to the lower residence times of the char particles in the FR [27,28,70]. This means that the OC circulation rate can only be varied within a given range.

Increasing the inlet temperature of the steam and air entering the FR and AR respectively, thereby decreasing the heat demand for heating up of the gases inside the reactor, is a further strategy to boost cold gas efficiencies. As shown in Figure 10e, this approach allows for a reduction of the air-to-fuel equivalence ratio from 0.38 to 0.34 when increasing inlet temperatures from 400 °C to 600 °C. Hence, maximizing inlet gas temperatures through heat recuperation is a key task in CLG in order to optimize the process efficiency, which is illustrated by the increase in the CGE from 68.3 to 76.5%, when increasing gas inlet temperatures from 300 to 600 °C (see Figure 10f). Due to the absence of corrosive compounds and the high process temperatures, the hot off-gases leaving the AR are ideal for steam generation and heat recuperation. On the other hand, special syngas coolers are being used to recuperate sensible heat from syngas streams for steam production [71–73], highlighting that efficient gas pre-heating using heat from process off-gases is possible in CLG.

Furthermore, variations in the AR temperatures can be considered, in order to enhance CLG process efficiencies. Generally speaking, a reduction in average process temperatures is beneficial for the process heat release, as pre-heating demands for all educts (i.e., inlet gases & feedstock material) are being reduced as a consequence, thus allowing for intensified heat extraction for a given air-to-fuel ratio (see Figure 10g). As visible in Figure 10h, a slight increase in the CGE by 2.4 percentage points can be attained for autothermal CLG operation when lowering AR temperatures from 1050 to 1000 °C. Yet, it has to be kept in mind that in chemical looping processes, air and fuel reactor temperatures are coupled, which means that a drop in FR temperatures is an inevitable effect of reduced AR temperatures. For the given boundary conditions, FR temperatures are projected to directly correlate with AR temperatures, which means that for the given reduction in AR temperatures from 1050 to 1000 °C, a corresponding drop in FR temperatures from 928 to 880 °C entails. This means that when attempting to prevent the ensuing drop in FR temperatures, related to negative effects on volatile and carbon conversion, OC circulation rates have to be increased accordingly as a counter-measure.

Although these insights allow for a first glimpse on process optimization approaches, it becomes clear that a detailed consideration of reaction kinetics and reactor hydrodynamics is quintessential, when aiming for a holistic optimization of the CLG process, as both phenomena have a pronounced effect on the process parameters. As it is well known that the conversion of char and other hydrocarbons is kinetically governed [25,55,56,59], the impact of reactor temperature, residence time, and gas concentrations on reaction kinetics need to be established in detail, allowing for accurate predictions of the governing reactions in a realistic environment. Moreover, reactor hydrodynamics are a crucial factor in chemical looping systems [74,75], making it a pre-requisite to consider them in advanced CLG process models. Through considering these phenomena, it thus becomes feasible to assess to which extent the preceding approaches can be utilized to obtain a CLG process exhibiting not only a high cold gas efficiency, but also excellent carbon capture efficiencies. Nonetheless, the preceding explanations offer valuable insights on the fundamental challenges associated with the autothermal CLG process, which require catering to, when implementing the technology in large scale.

4. Conclusions

In the course of this study, an equilibrium process model for the chemical looping gasification of biomass, using ilmenite ore as the oxygen carrier, was deployed to establish adequate process control techniques to attain autothermal behavior for gasifiers of any scale. It was shown that pursuing continuous CLG operation leads to unique challenges in terms of the OC circulation, which is responsible for both, oxygen and heat transport between the air and fuel reactor. While high OC circulation is generally beneficial in CLC to achieve complete fuel conversion in the FR and prevent a drop in FR temperatures, CLG faces an essential dilemma. Here, large OC circulation rates are necessary to fulfill the process heat balance (i.e., retain constant temperatures in the FR), whereas significantly lower circulation rates are required in terms of the necessary oxygen transport. Hence, heat and oxygen transport have to be de-coupled. Based on model calculations, two strategies to achieve autothermal CLG behavior through a de-coupling of oxygen and heat transport were presented. One eligible option is the dilution of the OC with an inert solid (e.g., sand), allowing for an accurate tailoring of the mixture's heat capacity and oxygen transport capability through its composition. As an alternative, the oxygen transport to the FR can be controlled through the oxygen availability (i.e., air supply) in the AR, leading to a deeply reduced oxygen carrier cycling through the system, not being fully re-oxidized in the AR. While both approaches lead to stable autothermal CLG behavior with sufficiently high FR temperatures, the latter strategy possesses certain advantages in terms of process control and fuel reactor chemistry, based on which it was deemed more suitable for large-scale operation. Regardless of the deployed approach, it was shown that restricting oxygen release in the FR is key in controlling CLG operation, where large cold gas efficiencies are desired. As partial oxidation of the feedstock is necessary in order to fulfill the heat balance of an autothermal process, this means that heat losses and heat sinks in the chemical looping gasifier have to be minimized, so that the oxygen input into the FR can be reduced, thus boosting syngas yields. Possible strategies to achieve this are gas pre-heating, variations in the OC circulation, alterations in the average CLG process temperature, and a reduction in the H_2O/biomass ratio in the FR.

Certainly, the presented findings encourage a deeper investigation of the chemical looping gasification of biomass on a numerical level, as only through the deployment of elaborate models considering hydrodynamics and reaction kinetics in-depth inferences regarding the process efficiency are facilitated. Moreover, they also call for experimental investigations of the suggested process control strategies. Especially the suggested continuous CLG operation with a deeply reduced OC, not being fully re-oxidized in the AR, means setting foot on a new terrain. Here, the suitability of the presented approach is decided by the fact whether positive (e.g., pronounced methane reforming ability, increased syngas selectivity & tar cracking) or negative effects (e.g., intensified attrition, reactivity loss, particle agglomerations) prevail.

Author Contributions: Conceptualization, J.S. and P.D.; methodology, P.D.; writing—original draft preparation, P.D.; writing—review and editing, F.M., J.S. and F.A.; visualization, P.D.; supervision, B.E. All authors have read and agreed to the published version of the manuscript.

Funding: This work has received funding of the European Union's Horizon 2020-Research and Innovation Framework Programme under grant agreement No. 817841 (Chemical Looping gasification foR sustainAble production of biofuels-CLARA).

Acknowledgments: The authors would like to thank the Technical University of Darmstadt, enabling the open-access publication of this paper.

Nomenclature

Symbol	Explanation	Unit
h_i	Enthalpy of stream i	kJ/kg
$\dot{m}_i$	Mass flow of component/element i	kg/h
M_i	Molar mass of component/element i	g/mole
$\dot{n}_i$	Mole flow of component/element i	kmole/h
P	Power	kW
p	Pressure	bar
R_{OC}	Oxygen transport capacity of oxygen carrier	-
$\dot{m}_{air,AR}$	Mass flow of air entering the AR	kg/h
T	Temperature	°C
x_i	Mass/mole fraction in gas phase	-
X_i	Conversion of component i	-
$Y_{i,j}$	Mass yield of component/element i from substance j	-
z_i	Mass/mole fraction in solid phase	-
η_{CC}	Carbon capture efficiency	-
η_{CGE}	Cold gas efficiency	-
λ	Air-to-fuel equivalence ratio	-
ϕ	Oxygen carrier-to-fuel equivalence ratio	-

Subscript	Explanation
AR	Air reactor
devol.	Devolatilization.
FR	Fuel reactor
init	Initial
net	net
O	Oxygen
OC	Oxygen Carrier
ox	Oxidation
red	Reduction
s	Solid
stoich	Stoichiometric
th	Thermal

Abbreviation	Explanation
AR	Air Reactor
ASU	Air Separation Unit
CGE	Cold Gas Efficiency
CLC	Chemical Looping Combustion
CLG	Chemical Looping Gasification
FR	Fuel Reactor
GHGE	Greenhouse Gas Emissions
LHV	Lower Heating Value
OC	Oxygen Carrier
RED II	European Union Renewable Energy Directive
WGS	Water-Gas-Shift

Appendix A. Boundary Conditions for CLG Process Model

A summary of all model boundary conditions employed for the simulations presented in Section 3.2, Section 3.3 and Section 3.4 is given in Table A1.

Table A1. Boundary conditions for 1 MW$_{th}$ CLC/CLG process model for different CLG approaches.

Parameter	Approach 1 *	Approach 2 *	Approach 3 *	Unit
T_{FR}	730–1030	980–1030	930–1030	°C
T_{AR}	1050	1050	1050	°C
p_{FR}/p_{AR}	1.013	1.013	1.013	bar
$\dot{m}_{fuel}$	200.4	200.4	200.4	kg/h
$\dot{m}_{H2O,FR}$	180.4	180.4	180.4	kg/h
$\dot{m}_{CO2,FR}$	40.1	40.1	40.1	kg/h
$\dot{m}_{air,AR}$	1362.6	1362.6	454–1590	kg/h
$T_{CO2,FR}$	25	25	25	°C
$T_{H2O,FR}/T_{air,AR}$	400	400	400	°C
$\dot{m}_{OC,init}$	2.11–8.45	8.45	8.45	t/h
z_{SiO2}	0	0–75	0	wt-%

* CLG approach I: Reduction in OC circulation rate (see Section 3.2), CLG approach 2: Dilution with solid inert (see Section 3.3), CLG approach 3: Reduction of air inlet into AR (see Section 3.4).

Appendix B. Shifting from CLC to CLG Operation through Variations in the Air-to-Fuel Equivalence Ratio

As described in Section 3.4, the oxygen availability in the FR is solely dependent on the circulation rate of the OC and the oxygen transport capability of the OC material (R_O), when operating the AR in air excess ($\lambda > 1$) in CLC, as the OC material is fully oxidized inside the AR. When subsequently reducing λ to values below unity from a steady state CLC operating point (see Figure A1a), the limited air availability in the AR leads to a transient phase during which the OC undergoes a continuous drop in the oxidation degree with each redox cycle, as more oxygen is consumed in the FR (combustion conditions) than is being supplied in the AR. As soon as the oxidation degree in the FR approaches 0, the oxygen availability in the subsequent redox cycle is determined by the oxygen supply in the AR. Hence, ϕ is equal to λ from this point onwards. As indicated in Figure A1a, this means that steady state CLG conditions are attained as a consequence. When on the other hand starting off with steady state CLG operation ($\lambda < 1$) before increasing λ beyond unity, the OC undergoes a transient phase during which its oxidation degree increases with each redox cycle, since more oxygen is supplied in the AR than is being consumed in the FR. As soon as the amount of oxygen transported by the OC is sufficient to fully oxidize the deployed feedstock, CLC conditions are attained. It has to be noted that this can be the case before steady state is reached (see Figure A1b). This means that despite the described discontinuity in the relation between λ and ϕ for $\lambda = 1$, a rapid switch in the OC-to-fuel ratio will not occur during operation, as the transition from CLC to CLG or vice versa will occur smoothly via a transient phase during which the oxidation degree of the OC adapts to the newly set boundary conditions.

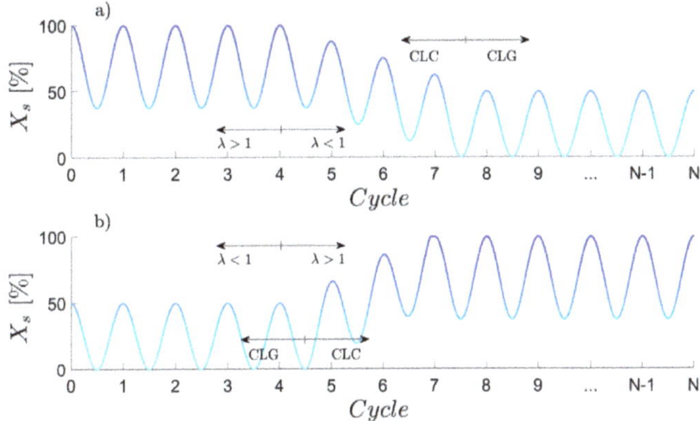

Figure A1. Progression of the OC oxidation degree when shifting from CLC ($\lambda > 1$) to CLG ($\lambda < 1$) mode through variations of the air-to-fuel equivalence ratio. (**a**) Shift from CLC to CLG, (**b**) shift from CLG to CLG.

Appendix C. Char Conversion in an Sub-Stoichiometrically Operated AR

In order to establish how a mixture of unconverted char and a fully reduced OC behaves in an sub-stoichiometric oxygen containing atmosphere in the AR, a mixture of char (5 mole-%) and a reduced

OC (78 mole-% $FeTiO_3$, 6 mole-% Fe_2O_3 and 11 mole-% TiO_2) were reacted with different amounts of air in an RGIBBS reactor of varying temperature (900–1100 °C). The results for an AR temperature of 1000 °C are shown in Figure A2. It is visible that char conversion occurs prior to OC re-oxidation, as the char fraction is zero regardless of the deployed air-to-fuel ratio. Moreover, the chemical equilibrium predicts a further reduction of the OC in case the amount of oxygen contained in the inlet air is insufficient for char conversion. Certainly, this behavior can only be observed in case of sufficiently long reaction times (rarely given in a fluidized bed), since solid-solid reactions between OC and char particles are known to exhibit slow kinetics [67–69]. This means that when attempting full char conversion, the inlet air entering the AR has to be sufficient to provide full carbon combustion. When this is the case, it can be assumed that full char conversion is attained inside the AR. In terms of the CO content at the reactor outlet it can be seen that full CO conversion to CO_2 is achieved regardless of the utilized air-to-fuel ratio, indicated by negligible concentrations of CO in the AR outlet (see Figure A2).

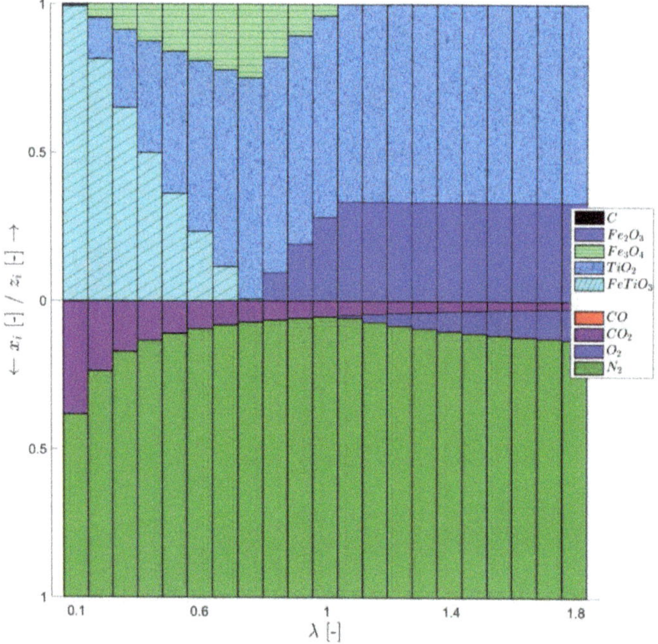

Figure A2. Solid and gas composition at chemical equilibrium for TAR = 1000 °C at varying air-to-fuel equivalence ratio λ (Inlet solid composition: 78 mole-% $FeTiO_3$, 6 mole-% Fe_2O_3 and 11 mole-% TiO_2).

References

1. International Energy Agency (IEA). *Key World Energy Statistics 2018*; International Energy Agency: Paris, France, 2018.
2. Energy Information Administration. Statistics Data Browser—Electricity Generation from Renewables by Source. Available online: https://www.iea.org/statistics/ (accessed on 23 August 2019).
3. European Commission. Transport Emissions—A European Strategy for Low-Emission Mobility. Available online: https://ec.europa.eu/clima/policies/transport_en (accessed on 23 August 2019).
4. EUR-Lex. Directive (EU) 2018/2001 of the European Parliament and of the Council of 11 December 2018 on the Promotion of the Use of Energy from Renewable Sources (Text with EEA Relevance). Available online: https://eur-lex.europa.eu/legal-content/en/TXT/?uri=CELEX:32018L2001 (accessed on 12 June 2020).
5. Carrasco, J.E.; Monti, A.; Tayeb, J.; Kiel, J.; Girio, F.; Matas, B.; Santos Jorge, R. *Strategic Research and Innovation Agenda 2020*; EERA: Brussels, Belgium, 2020; p. 82.
6. De, S.; Agarwal, A.K.; Moholkar, V.S.; Thallada, B. *Coal and Biomass Gasification: Recent Advances and Future Challenges*; Springer: Berlin/Heidelberg, Germany, 2018.
7. Higman, C.; Van der Burgt, M. *Gasification*, 2nd ed.; Gulf Professional Publishing: Houston, TX, USA; Elsevier Science: Amsterdam, The Netherlands, 2008.

8. Barrio, M.; Gbel, B.; Rimes, H.; Henriksen, U.; Hustad, J.E.; Srensen, L.H. Steam gasification of wood char and the effect of hydrogen inhibition on the chemical kinetics. In *Progress in Thermochemical Biomass Conversion*; Bridgwater, A.V., Ed.; Blackwell Science Ltd.: Oxford, UK, 2001; pp. 32–46.

9. Hansen, L.K.; Rathmann, O.; Olsen, A.; Poulsen, K. *Steam Gasification of Wheat Straw, Barley Straw, Willow and Giganteus*; Risø National Laboratory: Roskilde, Denmark, 1997.

10. Klose, W.; Wolki, M. On the intrinsic reaction rate of biomass char gasification with carbon dioxide and steam. *Fuel* **2005**, *84*, 885–892. [CrossRef]

11. Ollero, P.; Serrera, A.; Arjona, R.; Alcantarilla, S. The CO_2 gasification kinetics of olive residue. *Biomass Bioenergy* **2003**, *24*, 151–161. [CrossRef]

12. Barrio, M.; Hustad, J.E. CO_2 Gasification of birch char and the effect of CO inhibition on the calculation of chemical kinetics. In *Progress in Thermochemical Biomass Conversion*; Bridgwater, A.V., Ed.; Blackwell Science Ltd.: Oxford, UK, 2001; pp. 47–60.

13. Basu, P. *Biomass Gasification and Pyrolysis*; Elsevier: Amsterdam, The Netherlands, 2010.

14. Ge, H.; Zhang, H.; Guo, W.; Song, T.; Shen, L. System simulation and experimental verification: Biomass-based integrated gasification combined cycle (BIGCC) coupling with chemical looping gasification (CLG) for power generation. *Fuel* **2019**, *241*, 118–128. [CrossRef]

15. Huang, Z.; Zhang, Y.; Fu, J.; Yu, L.; Chen, M.; Liu, S.; He, F.; Chen, D.; Wei, G.; Zhao, K.; et al. Chemical looping gasification of biomass char using iron ore as an oxygen carrier. *Int. J. Hydrogen Energy* **2016**, *41*, 17871–17883. [CrossRef]

16. Karl, J.; Pröll, T. Steam gasification of biomass in dual fluidized bed gasifiers: A review. *Renew. Sust. Energ. Rev.* **2018**, *98*, 64–78. [CrossRef]

17. Xu, G.; Murakami, T.; Suda, T.; Matsuzawa, Y.; Tani, H. The superior technical choice for dual fluidized bed gasification. *Ind. Eng. Chem. Res.* **2006**, *45*, 2281–2286. [CrossRef]

18. Aigner, I.; Pfeifer, C.; Hofbauer, H. Co-gasification of coal and wood in a dual fluidized bed gasifier. *Fuel* **2011**, *90*, 2404–2412. [CrossRef]

19. Huang, Z.; He, F.; Feng, Y.; Zhao, K.; Zheng, A.; Chang, S.; Wei, G.; Zhao, Z.; Li, H. biomass char direct chemical looping gasification using NiO-modified iron ore as an oxygen carrier. *Energy Fuels* **2014**, *28*, 183–191. [CrossRef]

20. Ge, H.; Guo, W.; Shen, L.; Song, T.; Xiao, J. Biomass gasification using chemical looping in a 25 kW$_{th}$ reactor with natural hematite as oxygen carrier. *Chem. Eng. J.* **2016**, *286*, 174–183. [CrossRef]

21. Guo, Q.; Cheng, Y.; Liu, Y.; Jia, W.; Ryu, H.-J. Coal chemical looping gasification for syngas generation using an iron-based oxygen carrier. *Ind. Eng. Chem. Res.* **2014**, *53*, 78–86. [CrossRef]

22. Huang, Z.; He, F.; Feng, Y.; Zhao, K.; Zheng, A.; Chang, S.; Li, H. Synthesis gas production through biomass direct chemical looping conversion with natural hematite as an oxygen carrier. *Bioresour. Technol.* **2013**, *140*, 138–145. [CrossRef] [PubMed]

23. Huseyin, S.; Wei, G.; Li, H.; He, F.; Huang, Z. Chemical-looping gasification of biomass in a 10 kW$_{th}$ interconnected fluidized bed reactor using Fe_2O_3/Al_2O_3 oxygen carrier. *J. Fuel Chem. Technol.* **2014**, *42*, 922–931. [CrossRef]

24. Adanez, J.; Abad, A.; Garcia-Labiano, F.; Gayan, P.; De Diego, L.F. Progress in chemical-looping combustion and reforming technologies. *Prog. Energy Combust. Sci.* **2012**, *38*, 215–282. [CrossRef]

25. Ohlemüller, P.; Alobaid, F.; Abad, A.; Adanez, J.; Ströhle, J.; Epple, B. Development and validation of a 1D process model with autothermal operation of a 1 MW$_{th}$ chemical looping pilot plant. *Int. J. Greenh. Gas Control* **2018**, *73*, 29–41. [CrossRef]

26. Markström, P.; Linderholm, C.; Lyngfelt, A. Chemical-looping combustion of solid fuels—Design and operation of a 100 kW unit with bituminous coal. *Int. J. Greenh. Gas Control* **2013**, *15*, 150–162. [CrossRef]

27. Cuadrat, A.; Abad, A.; García-Labiano, F.; Gayán, P.; De Diego, L.F.; Adánez, J. Effect of operating conditions in Chemical-Looping Combustion of coal in a 500 Wth unit. *Int. J. Greenh. Gas Control* **2012**, *6*, 153–163. [CrossRef]

28. Pérez-Vega, R.; Abad, A.; García-Labiano, F.; Gayán, P.; De Diego, L.F.; Adánez, J. Coal combustion in a 50 kWth chemical looping combustion unit: Seeking operating conditions to maximize CO_2 capture and combustion efficiency. *Int. J. Greenh. Gas Control* **2016**, *50*, 80–92. [CrossRef]

29. Ströhle, J.; Orth, M.; Epple, B. Chemical looping combustion of hard coal in a 1 MW$_{th}$ pilot plant using ilmenite as oxygen carrier. *Appl. Energy* **2015**, *157*, 288–294. [CrossRef]

30. Cao, Y.; Casenas, B.; Pan, W.-P. Investigation of chemical looping combustion by solid fuels. 2. Redox reaction kinetics and product characterization with coal, biomass, and solid waste as solid fuels and CuO as an oxygen carrier. *Energy Fuels* **2006**, *20*, 1845–1854. [CrossRef]

31. Leion, H.; Mattisson, T.; Lyngfelt, A. Solid fuels in chemical-looping combustion. *Int. J. Greenh. Gas Control* **2008**, *2*, 180–193. [CrossRef]

32. Virginie, M.; Adánez, J.; Courson, C.; De Diego, L.F.; García-Labiano, F.; Niznansky, D.; Kiennemann, A.; Gayán, P.; Abad, A. Effect of Fe–olivine on the tar content during biomass gasification in a dual fluidized bed. *Appl. Catal. B Environ.* **2012**, *121–122*, 214–222. [CrossRef]

33. Kuhn, J.N.; Zhao, Z.; Felix, L.G.; Slimane, R.B.; Choi, C.W.; Ozkan, U.S. Olivine catalysts for methane and tar-steam reforming. *Appl. Catal. B Environ.* **2008**, *81*, 14–26. [CrossRef]

34. Mendiara, T.; Johansen, J.M.; Utrilla, R.; Geraldo, P.; Jensen, A.D.; Glarborg, P. Evaluation of different oxygen carriers for biomass tar reforming (I): Carbon deposition in experiments with toluene. *Fuel* **2011**, *90*, 1049–1060. [CrossRef]

35. Larsson, A.; Israelsson, M.; Lind, F.; Seemann, M.; Thunman, H. Using ilmenite to reduce the tar yield in a dual fluidized bed gasification system. *Energy Fuels* **2014**, *28*, 2632–2644. [CrossRef]

36. Milne, T.A.; Evans, R.J.; Abatzaglou, N. *Biomass Gasifier "Tars": Their Nature, Formation, and Conversion*; National Renewable Energy Laboratory: Golden, CO, USA, 1998.

37. Ge, H.; Guo, W.; Shen, L.; Song, T.; Xiao, J. Experimental investigation on biomass gasification using chemical looping in a batch reactor and a continuous dual reactor. *Chem. Eng. J.* **2016**, *286*, 689–700. [CrossRef]

38. Leion, H.; Jerndal, E.; Steenari, B.-M.; Hermansson, S.; Israelsson, M.; Jansson, E.; Johnsson, M.; Thunberg, R.; Vadenbo, A.; Mattisson, T.; et al. Solid fuels in chemical-looping combustion using oxide scale and unprocessed iron ore as oxygen carriers. *Fuel* **2009**, *88*, 1945–1954. [CrossRef]

39. Fan, L.-S. *Chemical Looping Systems for Fossil Energy Conversions*; Wiley-AIChE: Hoboken, NJ, USA, 2010.

40. Mayer, K.; Penthor, S.; Pröll, T.; Hofbauer, H. The different demands of oxygen carriers on the reactor system of a CLC plant—Results of oxygen carrier testing in a 120 kW$_{th}$ pilot plant. *Appl. Energy* **2015**, *157*, 323–329. [CrossRef]

41. Ohlemüller, P.G. *Untersuchung von Chemical-Looping-Combustion im Megawatt-Maßstab*; Cuvillier: Göttingen, Germany, 2019.

42. De Diego, L.F.; Garcı́a-Labiano, F.; Gayán, P.; Celaya, J.; Palacios, J.M.; Adánez, J. Operation of a 10 kW$_{th}$ chemical-looping combustor during 200h with a CuO–Al$_2$O$_3$ oxygen carrier. *Fuel* **2007**, *86*, 1036–1045. [CrossRef]

43. Adánez, J.; Gayán, P.; Celaya, J.; De Diego, L.F.; García-Labiano, F.; Abad, A. Chemical Looping Combustion in a 10 kW$_{th}$ Prototype Using a CuO/Al$_2$O$_3$ Oxygen Carrier: Effect of Operating Conditions on Methane Combustion. *Ind. Eng. Chem. Res.* **2006**, *45*, 6075–6080. [CrossRef]

44. Pröll, T.; Bolhàr-Nordenkampf, J.; Kolbitsch, P.; Hofbauer, H. Syngas and a separate nitrogen/argon stream via chemical looping reforming—A 140 kW pilot plant study. *Fuel* **2010**, *89*, 1249–1256. [CrossRef]

45. Ohlemüller, P.; Busch, J.-P.; Reitz, M.; Ströhle, J.; Epple, B. Chemical-Looping Combustion of Hard Coal: Autothermal Operation of a 1 MW$_{th}$ Pilot Plant. *J. Energy Resour. Technol.* **2016**, *138*, 042203. [CrossRef]

46. Mallick, D.; Mahanta, P.; Moholkar, V.S. Co-gasification of coal and biomass blends: Chemistry and engineering. *Fuel* **2017**, *204*, 106–128. [CrossRef]

47. Matthesius, G.A.; Morris, R.M.; Desai, M.J. Prediction of the volatile matter in coal from ultimate and proximate analyses. *J. S. Afr. Inst. Min. Metall.* **1987**, *5*, 157–161.

48. Neves, D.; Thunman, H.; Matos, A.; Tarelho, L.; Gómez-Barea, A. Characterization and prediction of biomass pyrolysis products. *Prog. Energy Combust. Sci.* **2011**, *37*, 611–630. [CrossRef]

49. Cuadrat, A.; Abad, A.; Gayán, P.; De Diego, L.F.; García-Labiano, F.; Adánez, J. Theoretical approach on the CLC performance with solid fuels: Optimizing the solids inventory. *Fuel* **2012**, *97*, 536–551. [CrossRef]

50. Mendiara, T.; Pérez-Astray, A.; Izquierdo, M.T.; Abad, A.; De Diego, L.F.; García-Labiano, F.; Gayán, P.; Adánez, J. Chemical Looping Combustion of different types of biomass in a 0.5 kW$_{th}$ unit. *Fuel* **2018**, *211*, 868–875. [CrossRef]

51. Leion, H.; Lyngfelt, A.; Johansson, M.; Jerndal, E.; Mattisson, T. The use of ilmenite as an oxygen carrier in chemical-looping combustion. *Chem. Eng. Res. Des.* **2008**, *86*, 1017–1026. [CrossRef]

52. Pissot, S.; Vilches, T.B.; Maric, J.; Seemann, M. Chemical looping gasification in a 2–4 MWth dual fluidized bed gasifier. In Proceedings of the 23rd International Conference on Fluidized Bed Conversion, Seoul, South Korea, 13 May 2018.

53. Li, K.; Zhang, R.; Bi, J. Experimental study on syngas production by co-gasification of coal and biomass in a fluidized bed. *Int. J. Hydrogen Energy* **2010**, *35*, 2722–2726. [CrossRef]

54. Narváez, I.; Orío, A.; Aznar, M.P.; Corella, J. Biomass gasification with air in an atmospheric bubbling fluidized bed. effect of six operational variables on the quality of the produced raw gas. *Ind. Eng. Chem. Res.* **1996**, *35*, 2110–2120. [CrossRef]

55. Abad, A.; Adánez, J.; Cuadrat, A.; García-Labiano, F.; Gayán, P.; De Diego, L.F. Kinetics of redox reactions of ilmenite for chemical-looping combustion. *Chem. Eng. Sci.* **2011**, *66*, 689–702. [CrossRef]

56. Zafar, Q.; Abad, A.; Mattisson, T.; Gevert, B. Reaction kinetics of freeze-granulated $NiO/MgAl_2O_4$ oxygen carrier particles for chemical-looping combustion. *Energy Fuels* **2007**, *21*, 610–618. [CrossRef]

57. Mattisson, T.; Lyngfelt, A.; Cho, P. The use of iron oxide as an oxygen carrier in chemical-looping combustion of methane with inherent separation of CO_2. *Fuel* **2001**, *80*, 1953–1962. [CrossRef]

58. Ohlemüller, P.; Ströhle, J.; Epple, B. Chemical looping combustion of hard coal and torrefied biomass in a 1 MW_{th} pilot plant. *Int. J. Greenh. Gas Control* **2017**, *65*, 149–159. [CrossRef]

59. Dennis, J.S.; Scott, S.A. In situ gasification of a lignite coal and CO_2 separation using chemical looping with a Cu-based oxygen carrier. *Fuel* **2010**, *89*, 1623–1640. [CrossRef]

60. He, F.; Huang, Z.; Li, H.; Zhao, Z. Biomass Direct Chemical Looping Conversion in a Fluidized Bed Reactor with Natural Hematite as an Oxygen Carrier. In Proceedings of the Asia-Pacific Power and Energy Engineering Conference (IEEE), Wuhan, China, 28–31 March 2011; pp. 1–7.

61. Zhao, H.; Guo, L.; Zou, X. Chemical-looping auto-thermal reforming of biomass using Cu-based oxygen carrier. *Appl. Energy* **2015**, *157*, 408–415. [CrossRef]

62. Kunii, D.; Levenspiel, O. *Fluidization Engineering*, 2nd ed.; Butterworth-Heinemann: Boston, MA, USA, 1991.

63. Song, Q.; Xiao, R.; Deng, Z.; Zhang, H.; Shen, L.; Xiao, J.; Zhang, M. Chemical-looping combustion of methane with $CaSO_4$ oxygen carrier in a fixed bed reactor. *Energy Convers. Manag.* **2008**, *49*, 3178–3187. [CrossRef]

64. Ryu, H.-J.; Bae, D.-H.; Jin, G.-T. Effect of temperature on reduction reactivity of oxygen carrier particles in a fixed bed chemical-looping combustor. *Korean J. Chem. Eng.* **2003**, *20*, 960–966. [CrossRef]

65. Cuadrat, A.; Abad, A.; Adánez, J.; De Diego, L.F.; García-Labiano, F.; Gayán, P. Behavior of ilmenite as oxygen carrier in chemical-looping combustion. *Fuel Process. Technol.* **2012**, *94*, 101–112. [CrossRef]

66. Cho, P.; Mattisson, T.; Lyngfelt, A. Carbon Formation on Nickel and Iron Oxide-Containing Oxygen Carriers for Chemical-Looping Combustion. *Ind. Eng. Chem. Res.* **2005**, *44*, 668–676. [CrossRef]

67. Leion, H.; Lyngfelt, A.; Mattisson, T. Effects of Steam and CO_2 in the Fluidizing Gas when Using Bituminous Coal in Chemical-Looping Combustion. In Proceedings of the 20th International Conference on Fluidized Bed Combustion, Xi'an, China, 18–21 May 2009; pp. 608–611.

68. Leion, H.; Mattisson, T.; Lyngfelt, A. The use of petroleum coke as fuel in chemical-looping combustion. *Fuel* **2007**, *86*, 1947–1958. [CrossRef]

69. Brown, T.A.; Dennis, J.S.; Scott, S.A.; Davidson, J.F.; Hayhurst, A.N. Gasification and chemical-looping combustion of a lignite char in a fluidized bed of iron oxide. *Energy Fuels* **2010**, *24*, 3034–3048. [CrossRef]

70. Mendiara, T.; De Diego, L.F.; García-Labiano, F.; Gayán, P.; Abad, A.; Adánez, J. Behaviour of a bauxite waste material as oxygen carrier in a 500 Wth CLC unit with coal. *Int. J. Greenh. Gas Control* **2013**, *17*, 170–182. [CrossRef]

71. Herdel, P.; Krause, D.; Peters, J.; Kolmorgen, B.; Ströhle, J.; Epple, B. Experimental investigations in a demonstration plant for fluidized bed gasification of multiple feedstock's in 0.5 MW_{th} scale. *Fuel* **2017**, *205*, 286–296. [CrossRef]

72. Weidenfeller, D.J.; Kulik, R.; Rothenpieler, K.; Stückrath, K.; Hetzer, J. Design, Simulation and Practical Experience of the Largest Syngas Cooler in Operation for Coal Gasification. In Proceedings of the 8th International Freiberg Conference, Cologne, Germany, 12–16 June 2016.

73. Schmidtsche Schack, ARVOS GmbH. *Schmidtsche Schack®Solutions for Gasification Plants*. Available online: https://www.schmidtsche-schack.com/products/syngas-cooler#c255 (accessed on 17 June 2020).

74. Bischi, A.; Langørgen, Ø.; Morin, J.-X.; Bakken, J.; Ghorbaniyan, M.; Bysveen, M.; Bolland, O. Hydrodynamic viability of chemical looping processes by means of cold flow model investigation. *Appl. Energy* **2012**, *97*, 201–216. [CrossRef]

Appl. Sci. **2020**, *10*, 4271

75. Markström, P.; Lyngfelt, A. Designing and operating a cold-flow model of a 100 kW chemical-looping combustor. *Powder Technol.* **2012**, *222*, 182–192. [CrossRef]

Article

Influence of Pressure on Gas/Liquid Interfacial Area in a Tray Column

Adel Almoslh *, Falah Alobaid, Christian Heinze and Bernd Epple

Institut Energiesysteme und Energietechnik, Technische Universität Darmstadt, Otto-Berndt-Straße 2, 64287 Darmstadt, Germany; falah.alobaid@tu-darmstadt.de (F.A.); christian.heinze@tu-darmstadt.de (C.H.); bernd.epple@tu-darmstadt.de (B.E.)

* Correspondence: adel.almoslh@tu-darmstadt.de; Tel.: +49-6151/16-23004; Fax: +49-(0)-6151/16-22690

Received: 28 May 2020; Accepted: 1 July 2020; Published: 3 July 2020

Abstract: The influence of pressure on the gas/liquid interfacial area is investigated in the pressure range of 0.2–0.3 MPa by using a tray column test rig. A simulated waste gas, which consisted of 30% CO_2 and 70% air, was used in this study. Distilled water was employed as an absorbent. The temperature of the inlet water was 19 °C. The inlet volumetric flow rate of water was 0.17 m^3/h. Two series of experiments were performed; the first series was performed at inlet gas flow rate 15 Nm^3/h, whereas the second series was at 20 Nm^3/h of inlet gas flow rate. The results showed that the gas/liquid interfacial area decreases when the total pressure is increased. The effect of pressure on the gas/liquid interfacial area at high inlet volumetric gas flow rates is more significant than at low inlet volumetric gas flow rates. The authors studied the effect of decreasing the interfacial area on the performance of a tray column for CO_2 capture.

Keywords: CO_2 capture; CO_2 absorption; process simulation; validation study; experimental study

---

1. Introduction

The global atmospheric concentration of carbon dioxide and its contribution to global warming requires technical measures to limit the emission of this greenhouse gas. One possibility is by separating the carbon dioxide from the waste gas stream of fossil power plants by applying absorption technologies. The absorption process of CO_2 can be performed using solvents with different additives. Amine solvents such as monoethanolamine (MEA), diethanolamine (DEA), and methyldiethanolamine (MDEA) are common for CO_2 capture (Wilk et al. 2017) [1]. The reversible reactions and the moderate reactivity between CO_2 and amine solutions enable efficient CO_2 capture (Yamada 2016) [2]. Barzagli et al. (2014) [3] studied nonaqueous amine solvents such as 2-(isopropylamino)ethanol (IPMEA), 2-(tert-butylamino)ethanol (TBMEA) and N-methyl-2,2'-iminodiethanol(MDEA) for CO_2 capture and found that CO_2 removal efficiency was in the range of 87–95% at equilibrium, depending on the operational conditions. Although amine absorption processes are widely used for CO_2 capture, they have some disadvantages related to equipment corrosion, amine degradation (Sanna 2014) [4], and high energy consumption (Wilk et al. 2017) [1]. The absorption technology for CO_2 capture consists mainly of the absorber column and the regeneration unit. The absorber column can be a tray or packed column. The absorbent enters the absorber from the top, and the waste gas, which contains CO_2, enters the absorber from the bottom. The gas and the liquid phases contact each other on the trays or the packing materials. The trays or the packed material enhance the gas/liquid interfacial area, which increases the mass and heat transfer between the contact phases. The CO_2 component transforms from the gas phase to a liquid phase and then is absorbed.

Studying the Effect of Pressure on Gas/Liquid Interfacial Area

Literature reviews reveal that there are various studies concerning the influence of pressure on the hydrodynamics and mass transfer in gas/liquid systems. Some of the studies are devoted to the influence of pressure on the creation of bubbles. These studies used a capillary tube or single orifices connected to a gas chamber. Other studies have investigated the influence of pressure on interfacial areas in a bubble column, packed column, or tray column. Kling et al. (1962) [5] were the first to observe that an increase in pressure at a single gas inlet orifice and constant superficial gas velocity creates a decrease in the initial bubble volume (Oyevaar 1989) [6]. Kling et al. (1962) [5] suggested that the increase in energy content causes the gas to enter further into the liquid, causing elongated bubbles, which separate more easily from the orifice, leading to smaller bubbles at higher pressures. LaNauze et al. (1974) [7] studied the influence of pressure and gas flow rates on the creation of CO_2 bubbles in the water at different diameters of orifices photographically. They published the results of the behavior of bubble volume over pressure up to 2.1 MPa at different gas flow rates. They found that the bubble volume is increased when the gas flow rate is increased. Furthermore, it was shown that the bubble volume decreased significantly when the pressure was increased between 0.1–1 MPa, whereas it slightly decreased when the pressure was increased between 1–2.1 MPa. Bier et al. (1978) [8] studied the influence of operating pressure on an initial bubble volume, by sparging N_2 or He through a capillary tube into the water or ethanol. The authors concluded that the influence of the operating pressure is much smaller compared with sparging the gas through an orifice connected to a gas chamber since the gas chamber limits pressure vibrations that happen in close gas supply lines (Oyevaar 1989) [6]. Idogawa et al. (1987) [9] noted that the diameter of the initial bubble decreases to 25% when pressure is increased from 0.1 to 15 MPa. Oyevaar et al. (1989) [10] determined interfacial areas at pressures up to 1.85 MPa in a bubble column and a packed column. The authors found that the interfacial areas are unaffected by pressure in the packed bubble column, but that the influence of the pressure on the interfacial areas in the bubble column arises from the generation of smaller bubbles at the gas distributor. Badssi et al. (1988) [11] investigated the effect of the pressure and superficial velocity of gas and liquid on the interfacial area in a laboratory column equipped with cross-flow sieve trays; they checked that each tested variable has an independent influence on the interfacial area. They investigated the effect of the pressure on the total interfacial area in two different gas–liquid systems, CO_2-DEA and CO_2-NaOH. They reviewed that the total interfacial area decreases when the pressure is increased. Benadda et al. (1996) [12] studied the effect of pressure on the interfacial area in a counter-current packed column. Their experiment conditions were conducted at specific gas mass flow rates of 0.1 kg/m^{2}s, and specific liquid mass flow rates of 5.52 kg/m^{2}s. They concluded that the interfacial area decreases when the pressure is increased between 0.1 and 1.2 MPa. Molga et al. (1996) [13] determined the gas–liquid interfacial during CO_2 absorption by using DEA as well as DEA-ETG aqueous solutions. Their experiment device was a bubble column reactor with an inner diameter of 156 mm. The results obtained by Molga et al. (1996) [13] are different; they reviewed that there is no observed influence of pressure on the measured interfacial areas.

Most available studies in the literature are interested in the proportional correlation between the pressure and the absorbed amount of gas in a liquid. In contrast, studies on the effect of pressure on the gas/liquid interfacial area are still limited. One can summarize the objectives of this study as follows:

(1) To experimentally investigate the effect of the pressure on gas/liquid interfacial area and its effect on the performance of the absorber; an absorber test rig was constructed and operated.

(2) To assemble a property package for a rate-based model by applying Aspen Plus software for simulation of CO_2 absorption using water as an absorbent.

(3) To validate the mathematical model against experimental data at different operation points.

(4) To investigate the effect of pressure on the gas/liquid interfacial area for mass transfer and its impact on the performance of a sieve tray absorber for capturing CO_2 by using water as an absorbent.

2. Experimental

2.1. Test Rig Setup

Figures 1 and 2 illustrate an absorber test rig at Technische Universität Darmstadt. The absorber test rig consists of four main parts: a gas mixing unit, an absorber column, a regeneration unit, and a gas analysis unit. The mixing unit consists of two lines, one is connected with a compressed-air source, and the other is combined with a CO_2 gas cylinder. The mixing unit is connected to a manifold upstream of the absorber. The absorber is made of a glass column that has a height of 1500 mm, and its internal diameter is 152 mm. The column has 12 glass nozzles to which metal flanges can be attached, ten nozzles used to measure pressures and temperatures in the absorber, and two nozzles for introducing the inlet gas and liquid to the absorber. At the bottom and the top, it is closed by suited metal flanges as well. The top flange contains the exit of the gas, and the bottom flange contains the exit of liquid. Five sieve trays are fixed by threaded rods and inserted inside the absorber. The diameter of the tray is 150 mm, the diameter of the hole is 2 mm, the fraction of the sieve hole area to active area is 0.071, and the weir height is 15 mm. The tray spacing is 240 mm, and the space between the tray and glass wall is closed with rubber seals. A gas analysis unit is connected at the gas outlet line to measure the volumetric fraction of CO_2 at the outlet of the absorber.

Figure 1. Side view of the absorber test rig: 1, control panel; 2, absorber column; 3, recycling pump; 4, liquid level control valve; 5, make-up pump; 6, pressure difference transmitter; 7, Coriolis device; 8, packed column; 9, reboiler; 10, gas analysis unit; 11, pressure control valve; 12, gas outlet.

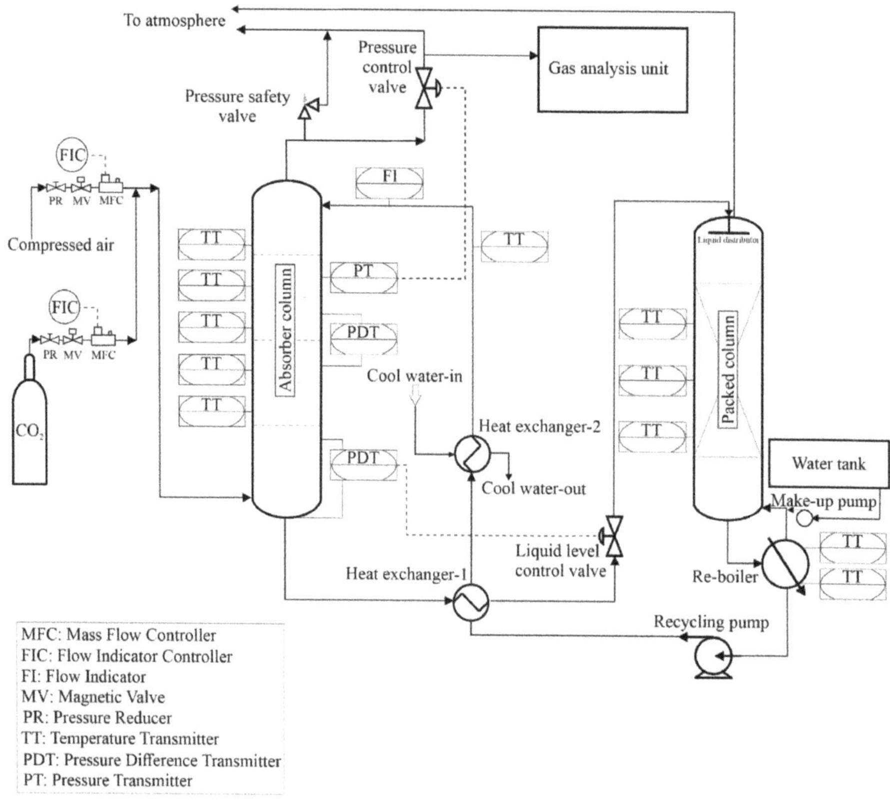

Figure 2. Schematic diagram of the absorber test rig.

The objective of the absorbent regeneration unit is to regenerate the absorbent and recycle it to the absorber as a lean absorbent. It consists of two heat exchangers, a reboiler, a packed column, a make-up pump, and a recycle pump. The packed column is a glass column with a diameter of 152 mm and a high of 1300 mm, which is filled to a height of 1 m with a metal packing material from type Pall-Ring 15 mm, with a specific surface of 360 m^2/m^3, and free volume 95 %. The rich absorbent enters the packed column through a liquid distributor on the top of the packed column. The purpose of the liquid distributor is to spray the absorbent uniformly on the top of the packings; the manufactured liquid distributor is from a spray type that contains 10 holes distributed uniformly on the liquid distributor. The packed column is placed on the reboiler, which has a thermal power of 4.5 kW. The recycle pump is connected directly to the reboiler, which pumps the water from the reboiler to the absorber. The lean hot absorbent is precooled in the first heat exchanger by exchanging heating with the absorbent, which comes out from the absorber. After that, the precooled absorbent cools down through the second heat exchanger, which is installed before the absorber, by exchanging heating with cool water.

2.2. Instrumentation and Control Equipment of the Test Rig

The test rig is provided with various devices and control circuits, which aim at measuring the essential parameters of the absorption process, as well as safe operation. A pressure reducer is fixed on every line of the gas mixer, adjusting the maximum pressure of gas entering the absorber. A magnetic valve is fixed after a pressure reducer, enabling the possibility of opening and closing the gas supply. A mass flow controller (MFC) is used to control the volumetric flow rate for every gas entering the absorber. A temperature sensor is fixed at each absorber tray to measure the temperature of the fluid on each tray. A Coriolis device and temperature sensor are attached at the inlet of the liquid to measure the inlet flow rate of water as well as the temperature of the inlet water. To estimate the pressure drop due to a tray and liquid holdup, the pressure difference before and after the tray is determined. For this purpose, a pressure difference device, illustrated in Figure 2, is fixed at the absorber.

Five control circles control the test rig. The first control circuit is aimed at controlling the pressure to the set point so that it does not exceed 0.6 MPa (permissible internal pressure of the glass absorber). The pressure control circuit consists of a pressure sensor and a control valve. The pressure sensor is fixed at the top of the tray column, whereas the control valve is installed at the absorber's gas outlet. The pressure control circuit starts controlling the pressure after the gas enters the absorber, resulting in a pressure increase. The pressure sensor sends a signal with the actual value of the pressure to a PID controller. The PID controller compares the set point of pressure and the actual value of pressure and sends a signal to the control valve to open or close, maintaining the pressure at its set point. For safety reasons, in particular, a safety pressure valve at the outlet of the absorber is installed, releasing the pressure inside the glass absorber when it exceeds the value of 0.45 MPa. By this procedure, the glass absorber is protected from any sudden increase of pressure above 0.45 MPa.

The second control circuit controls the liquid level at the bottom part of the absorber to its set point. Controlling the liquid level is essential since it prevents the gas from exiting from the liquid outlet and prevents the accumulation of the liquid inside the absorber to a high level. Without a level controller, the accumulated liquid causes the closing of the inlet of the gas or the possibility of immersing the trays of the absorber with the liquid, which leads to a high-pressure drop and low efficiency of the process. The liquid level control circuit consists of a pressure difference measurement and a control valve. The pressure difference is installed at the bottom of the tray column, whereas the control valve is arranged at the liquid outlet of the absorber. The third control circuit adjusts the level of absorbent inside the reboiler to its set point since there is an absorbent loss due to the absorbent being exposed to stripping and evaporation during the operation. The control circuit consists of a level sensor and a make-up pump. If the level of absorbent decreases below the set point, the level sensor sends a signal to the make-up pump to supply a new absorbent inside the reboiler for the compensation of absorbent loss.

For regeneration of the absorbent, heat is required to increase the absorbent temperature, breaking the bond between the CO_2 and the absorbent. For this purpose, a fourth control circuit is used to control the absorbent's temperature inside the reboiler. This control circuit consists of a heater and two temperature sensors that are installed at the top and the bottom of the reboiler. The temperature sensors send signals to the controller that commands the heater. The controller compares the actual value of the temperature inside the reboiler and the set point and sends a signal to the heater to heat the absorbent if the temperature of the absorbent is below the set point. In some cases, for example, the level control circuit does not work correctly. As a result, the level of absorbent inside the reboiler decreases, and it may damage the heater inserted in the reboiler. As a safety procedure, a control circuit (the fifth control circuit) is installed at the reboiler, which aims to shut-down the heater when the liquid level inside the reboiler exceeds the set point of the absorbent level. Accordingly, the heater is protected from any sudden decrease in the absorbent level during the operation.

2.3. Test Procedure

The CO_2 gas was mixed with air by the gas mixing unit; the air plays a role as a carrier gas. The volume fraction of CO_2 was 0.3 in all the experiments. Table 1 illustrates the operating conditions of experiments that have been performed in our study. Two separate experimental series were conducted to investigate the influence of pressure on the gas/liquid interfacial area. The first series of experiments was performed at the inlet gas flow rate of 15 Nm^3/h, the second series of experiments were run at the inlet gas flow rate of 20 Nm^3/h. In a series of experiments, the pressure was varied in the ranges of 0.2–0.3 MPa. The volumetric flow rate of the inlet water was almost constant at 0.17 m^3/h, and its temperature was controlled at 19 °C. The duration of every experiment was 15 min. The regeneration unit was operated with thermal power of 4.5 kW over time.

Table 1. The operation conditions of performed experiments in this study with a CO_2 volume fraction of 0.3, inlet water volumetric flow rate of 0.17 m^3/h, and inlet temperature of the inlet water of 19 °C.

Number of the Experiment	The Total Inlet Gas Flow Rate (Nm^3/h)	Pressure (MPa)
1	15	0.2
2	15	0.21
3	15	0.22
4	15	0.23
5	15	0.24
6	15	0.25
7	15	0.26
8	15	0.27
9	15	0.28
10	15	0.29
11	15	0.3
12	20	0.2
13	20	0.21
14	20	0.22
15	20	0.23
16	20	0.24
17	20	0.25
18	20	0.26
19	20	0.27
20	20	0.28
21	20	0.29
22	20	0.3

3. Modeling Approach for Tray Column

3.1. Rate-Based Modelling for CO_2 Absorption

The two-film theory developed by Lewis and Whitman (1924) [14] explains the mass transfer between gas and liquid. The two-film theory is based on the hypothesis that the gas and liquid phases form a thin layer of fluid on each side of the interface (Whitman 1962) [15]. Figure 3 illustrates the situation of the concentration of component i (in our study, i represent CO_2) at gas/liquid interfacial area. It should be noted that the concentration of component i decreases from $C_{i,G}$ in gas bulk to $C_{i,G*}$ in the interface. The difference of concentration creates a driving force for component i to shift from the gas bulk to gas film and then from the gas film to liquid film. The growth of component i in the liquid film forms a concentration difference between the liquid film and the liquid bulk; likewise, this concentration difference generates a driving force for component i to shift it from liquid film to liquid bulk (Whitman 1962) [15].

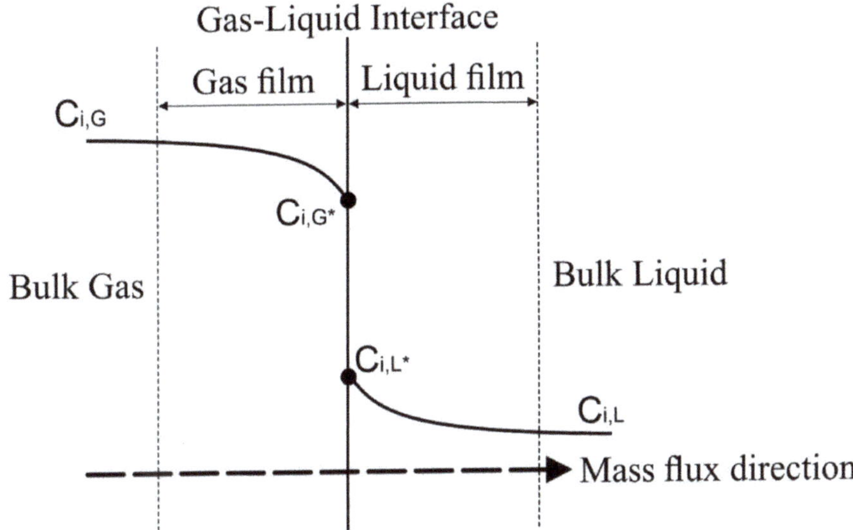

Figure 3. Illustration for the situation of the concentration of component *i* at the gas/liquid interfacial area (Doran 2013) (with permission from [16], Copyright Elsevier, 2013).

The rate of mass transfer of component *i* through the gas boundary layer is calculated as follows (Doran 2013) [16]:

$$N_{i,G} = k_G a^*(C_{iG} - C_{iG*})$$ (1)

The rate of mass transfer of component *i* through the liquid boundary layer is calculated as follows (Doran 2013) [16]:

$$N_{i,L} = k_L a^*(C_{iL*} - C_{iL})$$ (2)

where k_G, k_L are the mass transfer coefficient of the gas-phase and liquid-phase, respectively, a^* is the interfacial area.

At steady state, the flux of component *i* from bulk gas to the interface must be equal to the flux of component *i* from the interface to the bulk liquid (Ngo 2013) [17]:

$$N_{i,G} = N_{i,L}$$ (3)

For modeling the tray column, the rate-based model developed by Pandya (1983) [18] was applied. The rate-based model consists of a set of correlations that calculate the mass and energy transfer across the interfacial area using mass transfer coefficients (Afkhamipour and Mofarahi 2013) [19]. By applying the rate-based model, the absorber column is divided into stages. Figure 4 shows a stage of the column, which represents a tray in the column.

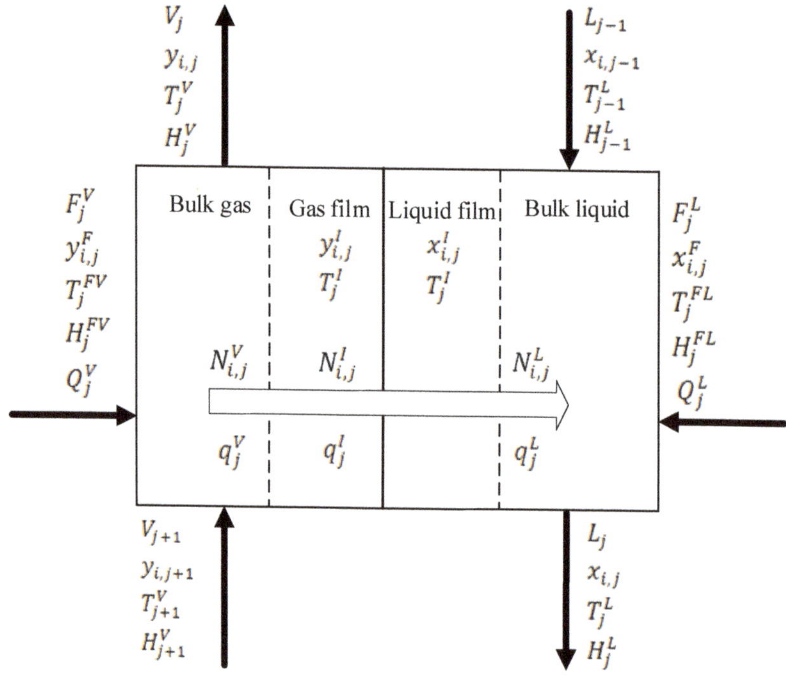

Figure 4. Rate-based stage model (reproduced from the ASPEN PLUS software manual).

The material and energy balances around a stage are conducted by applying the MERSHQ equations presented by Taylor and Krishna (1993) [20] as follows:

Material balance for bulk vapor :
$$F_j^V y_{i,j}^F + V_{j+1} y_{i,j+1} + N_{i,j}^V + r_{i,j}^V - V_j y_{i,j} = 0 \tag{4}$$

Material balance for bulk liquid :
$$F_j^L x_{i,j}^F + L_{j-1} x_{i,j-1} + N_{i,j}^L + r_{i,j}^L - L_j x_{ij} = 0 \tag{5}$$

Material balance for vapor film :
$$N_{i,j}^V + r_{i,j}^{fV} - N_{i,j}^I = 0 \tag{6}$$

Material balance for liquid film :
$$N_{i,j}^I + r_{i,j}^{fL} - N_{i,j}^L = 0 \tag{7}$$

Energy balance for bulk vapor :
$$F_j^V H_j^{FV} + V_{j+1} H_{j+1}^V + Q_j^V - q_j^V - V_j H_j^V = 0 \tag{8}$$

Energy balance for bulk liquid :
$$F_j^L H_j^{FL} + L_{j-1} H_{j-1}^L + Q_j^L + q_j^L - L_j H_j^L = 0 \tag{9}$$

Energy balance for vapor film :
$$q_j^V - q_j^I = 0 \tag{10}$$

Energy balance for liquid film :
$$q_j^I - q_j^L = 0 \tag{11}$$

Phase equilibrium at the interface :
$$y_{i,j}^I - K_{i,j} x_{i,j}^I = 0 \tag{12}$$

Summation for bulk vapor :
$$\sum_{i=1}^{n} y_{i,j} - 1 = 0 \tag{13}$$

Summation for bulk liquid :
$$\sum_{i=1}^{n} x_{i,j} - 1 = 0 \tag{14}$$

Summation for vapor film :
$$\sum_{i=1}^{n} y_{i,j}^I - 1 = 0 \tag{15}$$

$$\text{Summation for liquid film}: \sum_{i=1}^{n} x_{i,j}^{I} - 1 = 0 \tag{16}$$

where F is the molar flow rate of feed, L and V are the molar flow rate of liquid and vapor, respectively, N molar transfer rate, K is equilibrium ratio, r is reaction rate, H is enthalpy, Q is heat input to a stage, q is heat transfer rate, x_i , y_i are mole fraction of component i in liquid and vapor phases, respectively.

The empirical correlation, according to AICHE (1958) [21], is used for the estimation of the binary mass transfer coefficient for the liquid phase, where Chan and Fair's (1984) [22] correlation is applied for the estimation of the binary mass transfer coefficient for the gas phase. For the estimation of the interfacial area, Zuiderweg's (1982) [23] correlation is used.

In the following, empirical correlations are used in the rate-based model for the estimation of binary mass transfer coefficients and interfacial area a^I (AICHE 1958) [21], (Chan and Fair 1984) [22], (Zuiderweg 1982) [23], (Švandová 2011) [24].

Binary mass transfer coefficient for the liquid

$$k_{i,k}^{L} = \frac{\left(4.127 * 10^{8} D_{i,k}^{L}\right)^{0.5} (0.21313 F_s + 0.15) L t_L}{\overline{\rho}^{L} a^I} \tag{17}$$

Binary mass transfer coefficient for the gas

$$k_{i,k}^{V} = \left(10300 - 8670 F_f\right) F_f \left(D_{i,k}^{V}\right)^{0.5} \left(\frac{1-a}{a}\right) \frac{h_{cl}^{0.5} A_b}{a^I} \tag{18}$$

Interfacial area

$$a^I = \frac{40 A_b}{\phi^{0.3}} \left[\frac{\left(u_s^V\right)^2 \rho_t^V h_{cl} FP}{\sigma} \right]^{0.37} \tag{19}$$

For Spray regime where $FP \leq 3.0 l_w h_{cl} / A_b$,
Interfacial area

$$a^I = \frac{43 A_b}{\phi^{0.3}} \left[\frac{\left(u_s^V\right)^2 \rho_t^V h_{cl} FP}{\sigma} \right]^{0.53} \tag{20}$$

For a mixed froth-emulsion regime where $FP > 3.0 l_w h_{cl} / A_b$ Superficial F-factor

$$F_s = u_s^V \left(\rho_t^V\right)^{0.5} \tag{21}$$

The superficial velocity of vapor

$$u_s^V = \frac{Qv}{A_b} \tag{22}$$

Fractional approach to flooding

$$F_f = \frac{u_s^V}{u_{sf}^V} \tag{23}$$

Clear liquid height

$$h_{cl} = 0.6 h_w^{0.5} \left(FP \frac{P A_b}{l_w N_p} \right)^{0.25} \tag{24}$$

Flow parameter

$$FP = \frac{Q_L}{Qv} \left[\frac{\rho_t^L}{\rho_t^V} \right]^{0.5} \tag{25}$$

where $k_{i,K}^{L}$ is the binary mass transfer coefficient for the liquid which is predicted by using the AICHE 1958 correlation [21]; $k_{i,K}^{V}$ is the binary mass transfer coefficient for the vapor that is predicted by using

Chan and Fair's (1984) correlation [22]; $D_{i,k}^L$, $D_{i,k}^V$ are diffusivity of the liquid and vapor, respectively; Fs is the superficial F-factor; L is the total molar flow rate of the liquid; t_L is the average residence time for the liquid (per pass); $\overline{\rho}^L$ is the molar density of the liquid; a^I is the total interfacial area for mass transfer that is calculated according to the Zuiderweg's (1982) correlation [23] ; F_f is a fractional approach to flooding; a is the relative froth density ; h_{cl} is the clear liquid height; A_b is the total active bubbling area on the tray; u_s^V is the superficial velocity of vapor; ρ_t^L, ρ_t^V are the density of the liquid and vapor, respectively; FP is the flow parameter; $\varnothing$ is the fractional hole area per unit bubbling area; σ liquid surface tension; l_w is the average weir length (per liquid pass); Q_L, Q_V are volumetric flow rate for the liquid and vapor, respectively; u_{sf}^V is the superficial velocity of vapor at flooding; l_w is average weir height; P is sieve tray hole pitch; N_p is the number of liquid flows.

The heat transfer through the interfacial area is estimated using the Chilton–Colburn analogy. The heat transfer coefficient h is calculated as follows (Simon, Elias et al. 2011) [25]:

$$h = k_G \left(\frac{\rho_G \left(c_p / M_{w,L} \right) \lambda^2}{D^2} \right)^{1/3} \tag{26}$$

where k_G is the mass transfer coefficient for the gas phase, ρ_G is the density of the gas, c_p is specific molar heat capacity, $M_{w,L}$ is the molecular weight of the liquid phase, λ is thermal conductivity, and D is the diffusion coefficient.

3.2. Tray Column Simulation

For simulating the tray absorber column, ASPEN PLUS software was used due to its extensive property databanks and rigorous equation solvers. The rate-based model was applied due to its empirical correlations for the estimation of mass transfers. For building the rate-based model by using Aspen PLUS, the column was divided into five stages or trays. The inlet gas consists of air and CO_2; the mole fraction of CO_2 was 0.3. The water was used as an absorbent; the inlet volumetric flow of water was 0.17 m^3/h. Two series of the simulation were run: one at 15 Nm^3/h of inlet gas flow rate and the other at 20 Nm^3/h. The pressure was increased in the range of 0.2–0.3 MPa. The AICHE (1958) correlation [21] was applied for the prediction of the binary mass transfer coefficient for the liquid. Chan and Fair's (1984) correlation [22] was used to predict the binary mass transfer coefficient for the gas. The correlation of Chilton–Colburn was applied to estimate the heat transfer coefficient. The nonrandom two-liquid model (NRTL) was applied for the prediction of phase behavior properties.

Assumptions for the rate-based model approach are summarised below:

- The absorption column is assumed to be adiabatic.
- There is diffusion resistance in film.
- There are no reactions in the film.
- The liquid is well mixed, and the vapor is a plug flow; changes in concentration and temperature in the radial direction are negligible.
- The interfacial surface area is identical for both heat and mass transfer.
- The interface temperature is identical to the bulk liquid temperature since the liquid-side heat transfer resistance is small and balanced to the gas phase (Afkhamipour and Mofarahi 2013) [19].
- Both liquid and gas phases are formally handled as non-ideal mixtures.

4. Results and Discussion

4.1. Model Validation

The absorber test rig is operated at specified conditions for 15 min, which gives results in time-dependent values for each measured parameter (i.e., temperature, pressure, and gas concentrations). The standard deviation that displays the amount of variation of each measured parameter is then calculated to estimate the random error. The systematic error of the measurement devices is constant throughout the experiments and is, therefore, not additionally shown in this chapter. Generally, the measurement uncertainty of directly measured values (e.g., pressure, temperature, and flue gas concentrations) depends only on the relative uncertainty of the measuring devices and is given by the relative error. For indirectly measured parameters or calculated values (e.g., volumetric flow rate, where the pressure difference and temperature are applied for its calculation), the Gaussian error propagation method is applied, with the assumption of normally distributed uncertainties. In this work, the volumetric concentrations are detected using the gas analysis unit, and the maximum relative error for CO_2 in the different process streams is approximately 3%.

Figure 5 shows a comparison between the outlet volume fraction of CO_2 obtained by the experimental model (including the random error bars) and by the Aspen PLUS simulation model as a function of pressure at the inlet gas flow rates of 15 and 20 Nm^3/h. It can be noted from Figure 5 that there is an agreement between the simulation results and experimental data at 15 Nm^3/h. There is good agreement between the results up to 0.24 MPa at an inlet flow rate 20 Nm^3/h; the deviation between the results increase when the pressure is increased between 0.24–0.3 MPa at the inlet gas flow of 20 Nm^3/h.

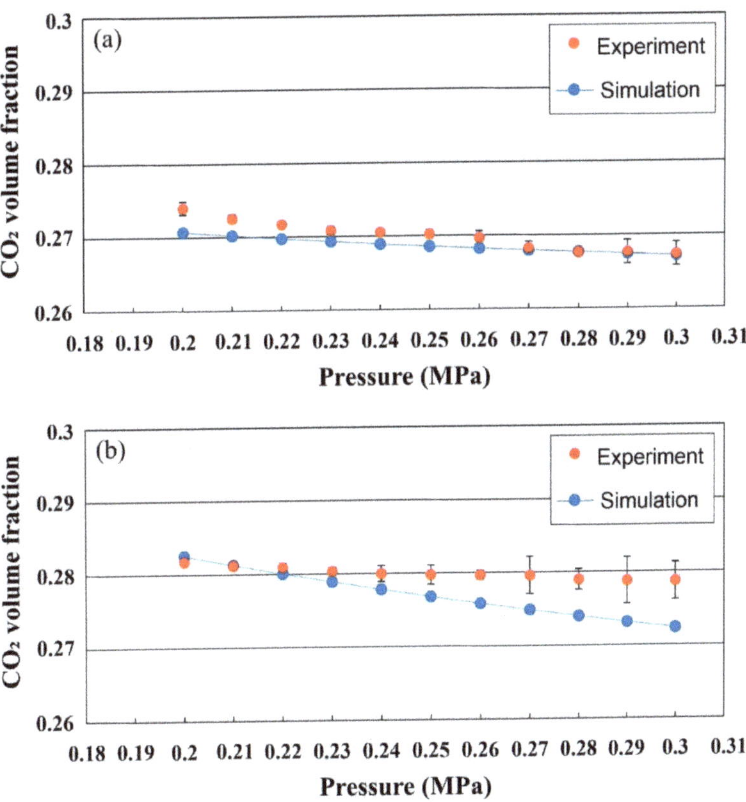

Figure 5. Effect of pressure on the outlet volume fraction of CO_2 at the inlet gas flow rate of 15 (**a**) and 20 Nm^3/h (**b**). The error bars are obtained by the standard errors of measurement.

For assessing the accuracy of the rate-based simulation model, the relative error is calculated between the concentrations estimated using the rate-based simulation model and those obtained from the experimental model. The relative error values are shown in Table 2. As can be seen in Table 2, there is good agreement between the experimental model and the Aspen PLUS simulation model obtained, with a maximum relative error of 2.27%.

Table 2. The relative error between the values calculated using the rate-based model and those obtained from empirical measurement at inlet gas flow rates of 15 and 20 Nm^3/h.

Pressure (MPa)	Inlet Gas Flow Rate (15 Nm^3/h)	Inlet Gas Flow Rate (20 Nm^3/h)
0.2	1.195%	0.262%
0.21	0.856%	0.018%
0.22	0.711%	0.308%
0.23	0.551%	0.511%
0.24	0.591%	0.764%
0.25	0.6182%	1.063%
0.26	0.528%	1.379%
0.27	0.104%	1.678%
0.28	0.008%	1.785%
0.29	0.085%	2.017%
0.3	0.062%	2.271%

4.2. Estimation of the Gas/Liquid Interfacial Area

It was observed during the performance of the experiments that the foam created above trays decreased when the pressure was increased. The effect of pressure increase on the reduction of the foam increased when the inlet gas flow rate was increased. For the estimation of the gas/liquid interfacial area, the validated rate-based model was used. Figure 6 shows the estimated gas/liquid interfacial area calculated by Zuiderweg's (1982) correlation [23] when the pressure is increased. As can be seen in Figure 6, at 15 Nm^3/h of the inlet gas flow rate, the gas/liquid interfacial area distinctly decreases when pressure is increased up 0.23 MPa, whereas the gas/liquid interfacial area slightly decreases when the pressure is increased between 0.23–0.3 MPa. In contrast, the gas/liquid interfacial area significantly decreases when pressure is increased between 0.2–0.3 MPa at 20 Nm^3/h of the inlet gas flow rate. The trend may be explained when consideration is taken of the fact that the gas/liquid interfacial area depends on superficial velocity, according to Zuiderweg's (1982) correlation [23]. The gas/liquid interfacial area increases when the superficial velocity of the gas is increased, and vice versa. With a constant inlet mass flow rate of the gas, the superficial velocity decreases when pressure is increased (Benadda 1996) [12], which leads to a decrease in the interfacial area according to Zuiderweg (1982) [23].

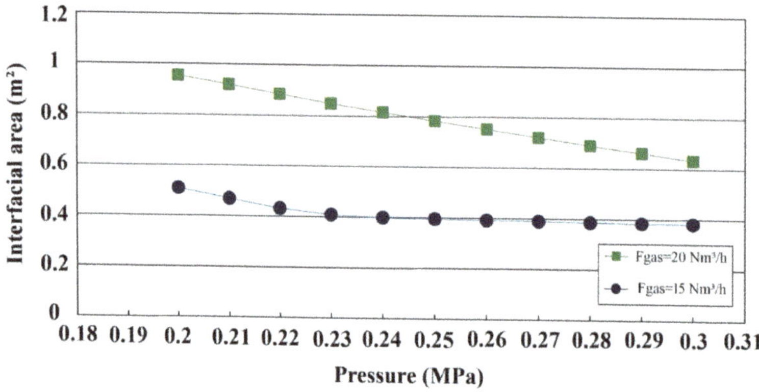

Figure 6. Effect of the pressure on the gas/liquid interfacial area at inlet gas flow rates of 15 and 20 Nm^3/h.

4.3. Studying the Effect of Pressure on the Performance of the Absorber

The performance of the absorber for CO_2 capture was measured by estimation of the absorbed rate of CO_2 per unit Nm^3/h of the inlet flow rate of CO_2. The absorbed rate N_{CO2} of CO_2 per unit Nm^3/h of inlet CO_2 was calculated using the following equation:

$$N_{CO_2}\left[Nm^3/Nm^3\right] = \left[\left(y_{CO_2,in} - y_{CO_2,out}\right)F_{gas,in}\left[Nm^3/h\right]\right]/F_{CO_2,\,in}\left[Nm^3/h\right] \tag{27}$$

where $F_{gas,in}$ is the inlet gas flow rate, $F_{CO2,in}$ is the inlet CO_2 flow rate, $y_{CO2,out}$ is the outlet volumetric fraction of CO_2 that was measured by the gas analysis unit, and $y_{Co2,in}$ is the inlet volumetric fraction of CO_2, which was calculated as follows:

$$y_{CO_2,in} = \frac{F_{CO_2,in}\left[Nm^3/h\right]}{F_{gas,in}\left[Nm^3/h\right]} = \frac{F_{CO_2,in}\left[Nm^3/h\right]}{F_{CO_2,in}\left[Nm^3/h\right] + F_{air,in}\left[Nm^3/h\right]} \tag{28}$$

$F_{air,in}$ is the inlet air flow rate.

Figure 7 illustrates the effect of the pressure on the absorbed rate of CO_2 at 15 and 20 Nm^3/h of inlet gas flow rates. It is clear that at the inlet gas flow rate of 15 Nm^3/h, increasing the pressure has a significant effect on the absorbed rate of CO_2. The absorbed rate of CO_2 increased significantly by increasing the pressure at 15 Nm^3/h of the inlet gas flow rate, whereas at the inlet gas flow rate of 20 Nm^3/h, the increase of pressure has a slight effect on the absorbed rate of CO_2 compared with the inlet gas flow rate of 15 Nm^3/h. Such trends of the curves can be explained when consideration is taken of the influence of pressure on the interfacial area (see Figure 6). It can be seen from Figure 6 that the effect of pressure on decreasing the interfacial area at 20 Nm^3/h is more significant than at 15 Nm^3/h of the inlet gas flow rate. Decreasing the interfacial area at 20 Nm^3/h leads to the reduction of the mass transfer of CO_2 from the gas phase to a liquid phase, according to Equations (1)–(3). This explains that there is no distinct increase of the absorbed rate of CO_2 per unit Nm^3/h of inlet CO_2 although the pressure is increased (as seen in Figure 7). In contrast, at 15 Nm^3/h, the absorbed amount of CO_2 increased when the pressure is increased since the interfacial area slightly decreases when the pressure is increased. As can be seen in Figure 7, at conditions of 0.2 MPa of pressure and 15 Nm^3/h of inlet gas flow rate, the absorbed rate of CO_2 per unit Nm^3/h is 0.087 $\left[Nm^3/Nm^3\right]$. At conditions of 0.3 MPa and 15 Nm^3/h, the absorbed rate increases to 0.109 $\left[Nm^3/Nm^3\right]$, with an increased rate of 0.022 $\left[Nm^3/Nm^3MPa\right]$, whereas at conditions 0.2 MPa of pressure and 20 Nm^3/h of inlet gas flow rate, the absorbed rate of CO_2 per unit Nm^3/h of inlet CO_2 is 0.061 $\left[Nm^3/Nm^3\right]$. At conditions 0.3 MPa and 20 Nm^3/h, this absorbed rate increases to 0.071 $\left[Nm^3/Nm^3\right]$, with an increased rate of 0.01 $\left[Nm^3/Nm^3MPa\right]$.

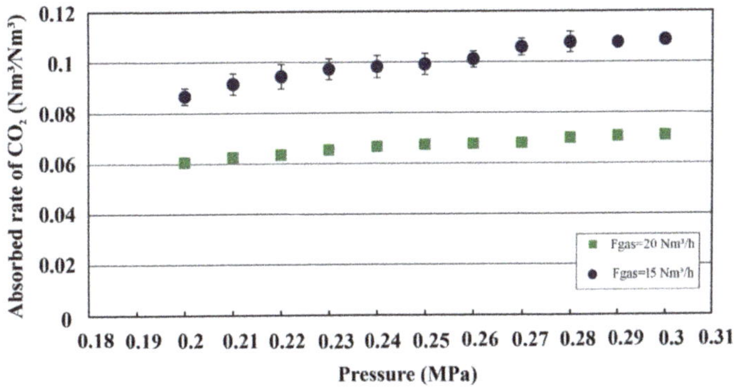

Figure 7. Effect of pressure on the absorbed rate of CO_2 per unit Nm3/h of inlet CO_2 at inlet gas flow rates of 15 and 20 Nm^3/h. The error bars are obtained by the standard errors of measurement.

As explained above, one can conclude that the pressure may have an obvious effect on the absorbed rate of CO_2 at a high inlet gas flow rate since it decreases the gas/liquid interfacial area, which reflects at the end of the absorbed rate of CO_2.

5. Conclusions

Our study contributes to the overall knowledge of CO_2 absorption under pressure. An absorber test rig was constructed and operated. Furthermore, a rate-based model was built by applying Aspen Plus software to simulate CO_2 absorption using water as an absorbent. The model was validated against experimental data at different operating points. The comparison between the results predicted by the rate-based model with the experimental data shows high agreement. The relative error was calculated between the gas concentration calculated using the model and those obtained from the results of the experiment to show the deviation of the model. An analytical study of the CO_2 absorption process has also been presented, which highlights the following points:

(1) It shows that the pressure influences the gas/liquid interfacial area. By increasing the pressure, the gas/liquid interfacial area slightly decreases at a low inlet gas flow rate. At a higher inlet gas flow rate, the gas/liquid interfacial area was significantly decreased.
(2) Decreased gas/liquid interfacial area when the pressure is increased has a significant influence on decreasing the absorption rate of CO_2.
(3) This study highlights the point of the effect of increasing pressure at a high inlet gas flow rate since it decreases the gas/liquid interfacial area.

This work is a contribution to the knowledge available for modeling studies for CO_2 absorption using water as an absorbent in the sieve tray column. Our results confirm other quotes in the literature which are still limited to this issue.

Author Contributions: A.A. is responsible for preparing the original draft, developing the applied methodology, and supporting the writing process with his reviews and edits. F.A. conducted the conceptualization and supported the writing process with his reviews and edits. C.H. conducted the conceptualization and supported the writing process with his reviews and edits. B.E. supervised the research progress and the presented work. All authors have read and agreed to the published version of the manuscript.

Funding: The authors received no specific funding for this work. The corresponding author would like to thank the Technical University of Darmstadt, enabling the open-access publication of this paper.

Nomenclatures

L	molar flow rate of liquid, [kmol/s]
V	molar flow rate of vapor [kmol/s]
F	molar flow rate of feed, [kmol/s]
N	molar transfer rate, [kmol/s]
K	equilibrium ratio, [−]
r	reaction rate, [kmol/s]
H	Enthalpy, [kmol/s]
Q	heat input to a stage, [J/s]
q	heat transfer rate, [J/s]
T	temperature, [K]
x	liquid mole fraction [−]
y	vapor mole fraction [−]
c_p	specific molar heat capacity, [J/kmol K]
$M_{w,L}$	molecular weight of the liquid phase
λ	Thermal conductivity, [w/m K]
h	Heat transfer coefficient $[w/m^2K]$
D	diffusion coefficient, m^2/s

$N_{i,G}$	rate of mass transfer of component i through the gas boundary, [kmol/s]
$N_{i,L}$	rate of mass transfer of component i through the liquid boundary, [kmol/s]
a	The interfacial area between gas and liquid phases, m^2
C_{iG}	concentration of component i in gas bulk, $\left[\frac{kmol}{m^3}\right]$
C_{iL}	concentration of component i in liquid bulk, $\left[\frac{kmol}{m^3}\right]$
C_{iG*}	concentration of component i in the interface from the gas side, $\left[\frac{kmol}{m^3}\right]$
C_{iL*}	concentration of component i in the interface from the liquid side, $\left[\frac{kmol}{m^3}\right]$
$k_{i,k}^V$	binary mass transfer coefficient for vapor, m/s
$k_{i,k}^L$	binary mass transfer coefficient for liquid, m/s
$D_{i,k}^V$	diffusivity of the vapor, m^2/s
$D_{i,k}^L$	diffusivity of the liquid, m^2/s
F_s	superficial F-factor, $kg^{0.5}/m^{0.5}s$
t_L	average residence time for the liquid (per pass), s
V	total molar flow rate of, liquid, vapor, kmol/s
L	total molar flow rate of, liquid, vapor, kmol/s
a^I	total interfacial area for mass transfer, m^2
$\bar{\rho}^V$	molar density of vapor, kmol/m^3
$\bar{\rho}^L$	molar density of liquid kmol/m^3
F_f	fractional approach to flooding, -
a	relative froth density, -, Equation (18)
A_b	total active bubbling area on the tray, m^2
u_s^V	superficial velocity for the vapor, m/s
u_s^L	superficial velocity for the liquid, m/s
u_{sf}^V	superficial velocity of vapor at flooding, m/s
ρ_t^V	density of the vapor, kg/m^3
ρ_t^L	density of the liquid, kg/m^3
l_w	average weir length (per liquid pass), m
Q_V	volumetric flow rate for the liquid, vapor, m^3/s
Q_L	volumetric flow rate for the liquid, vapor, m^3/s
σ	liquid surface tension N/m
FP	flow parameter, -
$\varnothing$	fractional hole area per unit bubbling area on the tray, -
h_{cl}	clear liquid height, m
h_w	average weir height, m
A_b	total active bubbling area on the tray, m^2
N_p	number of liquid flow passes, -
P	sieve tray hole pitch, m, Equation (24)

Subscripts

L	liquid
V	vapor
F	feed
G	gas
m^3/h	cubic meter per hour
i	component
n	number of components
I	interface
f	film
j	stage number

Abbreviations

NRTL	non-random two-liquid model
Nm3/h	cubic meter of gas per hour at the normal temperature and pressure

Aspen PLUS	simulation software program
MERSHQ	equations of material, energy balances, rate of mass and heat transfer, summation of composition, hydraulic, and equilibrium
MEA	2-aminoethano
DEA	2,2′-iminodiethanol
MDEA	N-methyl-2,2′-iminodiethanol
IPMEA	2-(isopropylamino) ethanol
TBMEA	2-(tert-butylamino) ethanol
PID controller	proportional–integral–derivative controller
MFC	mass flow controller

References

1. Wilk, A. Solvent selection for CO_2 capture from gases with high carbon dioxide concentration. *Korean J. Chem. Eng.* **2017**, *34*, 2275–2283. [CrossRef]
2. Yamada, H. Comparison of solvation effects on CO_2 capture with aqueous amine solutions and amine-functionalized ionic liquids. *J. Phys. Chem. B* **2016**, *120*, 10563–10568. [CrossRef] [PubMed]
3. Barzagli, F.; Lai, S.; Mani, F. Novel non-aqueous amine solvents for reversible CO_2 capture. *Energy Proced.* **2014**, *63*, 1795–1804. [CrossRef]
4. Sanna, A.; Vega, F.; Navarrete, B.; Maroto-Valer, M.M. Accelerated MEA degradation study in hybrid CO_2 capture systems. *Energy Proced.* **2014**, *63*, 745–749. [CrossRef]
5. Kling, G. Über die dynamik der Blasenbildung beim begasen von Flüssigkeiten unter druck. *Int. J. Heat Mass Transf.* **1962**, *5*, 211–223. [CrossRef]
6. Oyevaar, M.H.; Westerterp, K.R. Mass transfer phenomena and hydrodynamics in agitated gas—Liquid reactors and bubble columns at elevated pressures: State of the art. *Chem. Eng. Process. Process Intensif.* **1989**, *25*, 85–98. [CrossRef]
7. LaNauze, R.D. Gas bubble formation at elevated system pressures. *Trans. Inst. Chem. Eng.* **1974**, *52*, 337–348.
8. Bier, K.; Gorenflo, D.; Kemmade, J. Bubble formation and interfacial area during injection of gases into liquids through single orifices. *Wärme Und Stoffübertragung* **1978**, *11*, 217–228. [CrossRef]
9. Idogawa, K. Formation and flow of gas bubbles in a pressurized bubble column with a single orifice or nozzle gas distributor. *Chem. Eng. Commun.* **1987**, *59*, 201–212. [CrossRef]
10. Oyevaar, M.H.; De La Rie, T.; Van der Sluijs, C.L.; Westerterp, K.R. Interfacial areas and gas hold-ups in bubble columns and packed bubble columns at elevated pressures. *Chem. Eng. Process. Process Intensif.* **1989**, *26*, 1–14. [CrossRef]
11. Badssi, A.; Bugarel, R.; Blanc, C.; Peytavy, J.L.; Laurent, A. Influence of pressure on the gas-liquid interfacial area and the gas-side mass transfer coefficient of a laboratory column equipped with cross-flow sieve trays. *Chem. Eng. Process. Process Intensif.* **1988**, *23*, 89–97. [CrossRef]
12. Benadda, B.; Otterbein, M.; Kafoufi, K.; Prost, M. Influence of pressure on the gas/liquid interfacial area a and the coefficient k_La in a counter-current packed column. *Chem. Eng. Process. Process Intensif.* **1996**, *35*, 247–253. [CrossRef]
13. Molga, E.J.; Westerterp, K.R. Gas-liquid interfacial area and holdup in a cocurrent upflow packed bed bubble column reactor at elevated pressures. *Ind. Eng. Chem. Res.* **1997**, *36*, 622–631. [CrossRef]
14. Lewis, W.K.; Whitman, W.G. Principles of gas absorption. *Ind. Eng. Chem.* **1924**, *16*, 1215–1220. [CrossRef]
15. Whitman, W.G. The two film theory of gas absorption. *Int. J. Heat Mass Transf.* **1962**, *5*, 429–433. [CrossRef]
16. Doran, P.M. *Bioprocess Engineering Principles*, 2nd ed.; Elsevier: Amsterdam, The Netherlands, 2013; pp. 379–444.
17. Ngo, T.H. Gas Absorption into Emulsions. Ph.D. Thesis, Technical University of Braunschweig, Braunschweig, Germany, 5 March 2013.
18. Pandya, J. Adiabatic gas absorption and stripping with chemical reaction in packed towers. *Chem. Eng. Commun.* **1983**, *19*, 343–361. [CrossRef]
19. Afkhamipour, M.; Mofarahi, M. Comparison of rate-based and equilibrium-stage models of a packed column for post-combustion CO_2 capture using 2-amino-2-methyl-1-propanol (AMP) solution. *Int. J. Greenh. Gas Control* **2013**, *15*, 186–199. [CrossRef]

Appl. Sci. **2020**, *10*, 4617

20. Taylor, R.; Krishna, R. *Multicomponent Mass Transfer*; John Wiley & Sons: Hoboken, NJ, USA, 1993.
21. Contributors American Institute of Chemical Engineers. *Bubble-Tray Design Manual*; Contributors American Institute of Chemical Engineers: New York, NY, USA, 1958.
22. Chan, H.; Fair, J.R. Prediction of point efficiencies on sieve trays. 2. Multicomponent systems. *Ind. Eng. Chem. Process Des. Dev.* **1984**, *23*, 820–827. [CrossRef]
23. Zuiderweg, F.J. Sieve trays: A view on the state of the art. *Chem. Eng. Sci.* **1982**, *37*, 1441–1464. [CrossRef]
24. Švandová, Z.; Markoš, J.; Jelemenský, L. Impact of mass transfer on modelling and simulation of reactive distillation columns. In *Mass Transfer in Multiphase Systems and Its Applications*; Slovak University of Technology: Bratislava, Slovakia, 2011.
25. Simon, L.L.; Elias, Y.; Puxty, G.; Artanto, Y.; Hungerbuhler, K. Rate based modeling and validation of a carbon-dioxide pilot plant absorbtion column operating on monoethanolamine. *Chem. Eng. Res. Des.* **2011**, *89*, 1684–1692. [CrossRef]

Article

Energy and Exergy Analyses of an Existing Solar-Assisted Combined Cycle Power Plant

Ayman Temraz [1,2,*], Ahmed Rashad [2], Ahmed Elweteedy [2], Falah Alobaid [1] and Bernd Epple [1]

[1]   Institute for Energy Systems and Technology, Mechanical Engineering Department, Technical University of Darmstadt, 64287 Darmstadt, Germany; falah.alobaid@est.tu-darmstadt.de (F.A.); bernd.epple@est.tu-darmstadt.de (B.E.)

[2]   Mechanical Power and Energy Department, Military Technical College, Cairo 11766, Egypt; rashad@mtc.edu.eg (A.R.); aelweteedy@mtc.edu.eg (A.E.)

*   Correspondence: ayman.temraz@est.tu-darmstadt.de

Received: 12 June 2020; Accepted: 15 July 2020; Published: 20 July 2020

Abstract: Solar-assisted combined cycle power plants (CCPPs) feature the advantages of renewable clean energy with efficient CCPPs. These power plants integrate a solar field with a CCPP. This integration increases the efficiency of solar power plants while decreasing the CO_2 emissions of the CCPPs. In this paper, energy and exergy analyses were performed for an existing solar-assisted CCPP. The overall thermal efficiency and the exergetic efficiency of each component in the power plant were calculated for different solar field capacities. Also, a parametric study of the power plant was performed. The analysis indicated that the exergetic efficiency of the power plant components has its lowest value in the solar field while the condenser has the lowest exergetic efficiency in the combined cycle regime of operation. Further, a parametric study revealed that the thermal efficiency and the exergetic efficiency of the power plant as a whole decrease with increasing ambient temperature and have their highest values in the combined cycle regime of operation. Owing to these results, an investigation into the sources of exergy destruction in the solar field was conducted.

Keywords: solar energy; energy analysis; exergy analysis; CSP; PTC; ISCC; power plant

---

1. Introduction

The world society actively supports measures aimed at facilitating flexible and low-carbon energy economy. These actions mainly include the promotion of renewable energy sources for power generation with possible electrification of the heating and transport sectors.

Concentrated solar power (CSP) plants use steam to produce energy, similar to the conventional steam power plants. They consist of a solar field and a power block, and they may have an energy storage system (optional).

Many types of collectors have been developed to be used in CSP technology. Parabolic trough collectors (PTCs) have been used in many CSP plants. Parabolic trough power plants are ready for use today because they were tested on a commercial basis [1]. This has been proven in California since 1985 with parabolic trough power plants, with 354 MW of installed capacity [2], which have succeeded in commercial operation and generate electricity using a steam turbine connected to a generator as conventional power plants.

However, the increasing share of renewables is raising awareness of a critical challenge. Renewables normally provide fluctuating feed-in into the electricity grid so that energy reserves, e.g., energy storage systems or conventional thermal power plants such as nuclear power plants and combined cycle power plants (CCPPs), are required to achieve a balance between current electricity supply and demand. Also, to increase the profitability of the CSP plants, CSP technologies have been integrated with conventional plants.

The high thermodynamic efficiency of the CCPPs has attracted the construction of these power plants worldwide. On the one hand, the nominal process efficiency of a large-scale CCPP with a net electrical power of about 605 MW_{el} per unit can reach levels greater than 60% [3–6]. On the other hand, state-of-the-art coal-fired power plants reach a net thermal process efficiency of about 46% with single reheat and several low-pressure and high-pressure feedwater preheaters [7]. According to the International Energy Agency (IEA) in 2018, gas-fired power generation accounted for approximately 24% of the total share of worldwide electricity generation, dominated by CCPPs.

The CCPPs have the advantage of absorption of the waste heat in the flue gas of a gas turbine using a heat recovery steam generator (HRSG) installed downstream of the gas turbine. The integration of solar energy into this technology is an effective method for cleaner and cheaper power generation.

CSPs integrated within CCPPs are known as integrated solar combined cycle (ISCC) power plants. ISCC power plants consist of a solar field and a solar steam generator integrated into a conventional CCPP. This ISCC system improves the solar-to-electricity conversion efficiency [8–10] and the economic feasibility of the CSP plants. In addition, it increases the solar share, which leads to saving the fossil fuels used in these plants [10] while decreasing the CO_2 emissions [11].

B. Kelly et al. [12] demonstrated that the ISCC power plant concept presents an effective path for the continued development of PTC technology regarding the solar thermal-to-electric conversion efficiencies and the solar energy levelized energy cost (LEC). J. Dersch et al. [11], in collaboration with the International Energy Agency SolarPACES (Solar Power and Chemical Energy Systems) organization, studied the advantages and disadvantages of ISCC systems compared with solar electric generation systems (SEGS) and conventional CCPPs. The study showed the environmental and economic benefits of each ISCC configuration.

G. Bonforte et al. [13], P. Iora et al. [14], M. Mehrpooya et al. [15], and A. Baghernejad and M. Yaghoubi [16] implemented exergetic analyses of ISCC power plants. G. Bonforte et al. [13] developed an exergo-environmental and exergo-economic model to analyze an ISCC power plant in Southern Poland under the design conditions. The results showed that the CO_2 emissions were reduced by 9%. P. Iora et al. [14] presented a novel allocation method for the electricity produced in an ISCC based on the exergy loss approach by implementing internal exergy balances. They showed that this method is reliable and as good as the conventional Separate Production Reference method. M. Mehrpooya et al. [15] constructed a model using ASPEN HYSYS simulation software and MATLAB code to exergetically analyze an ISCC with a high-temperature energy storage system. It was found that the largest exergy losses were at the solar collector, the energy storage system, and the combustor.

A. Baghernejad and M. Yaghoubi [16] carried out energy and exergy analyses for an ISCC in Yazd, Iran using the design data of the power plant. The results showed that the energy and exergy efficiencies of this power plant are higher than those for a simple CCPP without a solar contribution and those for steam power plants with PTC technology. O. Behar et al. [17] simulated the performance of the first ISCC in Algeria, under HassiR'mel climate conditions. The results showed that the output power and the thermal efficiency increased at daytime than at night by 17% and 16.5%, respectively.

S. Wang et al. [18] analyzed the performance variation of the solar field and overall ISCC using advanced exergy analysis methods and hourly analysis within a typical day. The results showed that increasing the solar energy input to the ISCC system decreases the exergy destruction of the Brayton cycle and increases the exergy destruction of the Rankine cycle.

In the literature, one can find several papers regarding the investigation of ISCC power plants applied to different atmospheric conditions. The originality of this work is the parametric study of the energy and exergy analyses regarding an existing ISCC power plant in Egypt, under Kureimat climate conditions, as a whole and for the main components in the ISCC power plants. This work aims to identify the sites of major exergy destruction, clarify the reasons for exergy destruction in these sites, and attempt to clarify how to decrease the exergy destruction in this type of power plant. Besides this, given the challenges for the electricity market with the continuing expansion of intermittent renewables, in this work, we investigate the operational flexibility of ISCC power plants.

In this paper, we start with a description of the ISCC power plant under investigation, and we provide the method and the equation for the calculation of energy and exergy parameters. Then, we show the influence of ambient temperature and solar heat input on the plant performance. Finally, on the basis of these results, we investigate the sources of exergy destruction in the solar field and the combustion chamber to identify the possibility to enhance the performance of these components.

2. Plant Description

The ISCC power plant in Kureimat, Egypt is located at a northern latitude of $29°16'$ and eastern longitude of $31°15'$. It has 135 MW total power capacity, comprising a solar field with an electrical output of 20 MW and a combined cycle field with a power of 115 MW [19]. The integration of a combined cycle with the solar field ensures the delivery of the required electricity contribution to the grid regardless of solar radiation conditions.

2.1. Solar Field

The solar field comprises parallel rows of solar collector arrays and typical glass mirrors of 61 MW installed thermal capacity. The solar field comprises 40 loops, each loop having four parabolic trough collectors (type Skal-ET 150 designed by TSK Flagsol Engineering GmbH), and each collector has an aperture area of 817.15 m^2.

The solar heat transfer from the solar field collectors (PTCs) to the steam cycle uses a heat transfer fluid (HTF) system. The HTF is Therminol VP-1 from Solutia (ultra-high-temperature, liquid/vapor phase fluid) and operates between 12 °C and 400 °C (54–750°F) [20,21]. The HTF system is designed for an HTF mass flow of 250 kg/s at full load (61 MW of solar field thermal power output). Hot HTF returning from the solar field at 393 °C is pumped through the solar heat exchanger. The HTF leaves the solar heat exchanger at 293 °C and is pumped back into the solar field, as shown in Table 1.

Table 1. The solar field design parameters [15,16,22].

Solar Field Operation Parameter	Unit	Value
Solar Field Total Aperture Area	m^2	130,800
Number of Collectors	N°	160
Number of Collector Loops	N°	40
Design Irradiation	W/m^2	700
Maximum Solar Field Thermal Power Output	MW	61
Output Temperature of the HTF	°C	393
Input Temperature of the HTF	°C	293

2.2. Combined Cycle

The combined cycle field consists of one MS6001FA heavy-duty gas turbine with a generator of a rated electric power output of 70 MW at 20 °C ambient dry bulb temperature. It has one HRSG that receives about 206 kg/s flue gas at about 600 °C from the gas turbine at full load operation. The flue gas leaves the HRSG at about 100 °C. Under the rated conditions of the gas turbine and HRSG, full load operation and solar heat input of 50 MW and 20 °C ambient dry bulb temperature, the steam turbine generator output is about 65 MW.

The HRSG of the ISCC power plant in Kureimat comprises three high-pressure economizers (HP ECO), a high-pressure evaporator (HP EV), a high-pressure steam drum (HP Drum at a pressure about 80 bar), and five high-pressure superheaters (HP SH) for a feed of the high-pressure section of the steam turbine, as shown in Figure 1. Besides this, it includes a low-pressure evaporator (LP EV), low-pressure steam drum (LP Drum at a pressure about 11 bar), and low-pressure superheater (LP SH) for a feed of the steam turbine low-pressure section. The solar-generated steam is injected into the high-pressure drum.

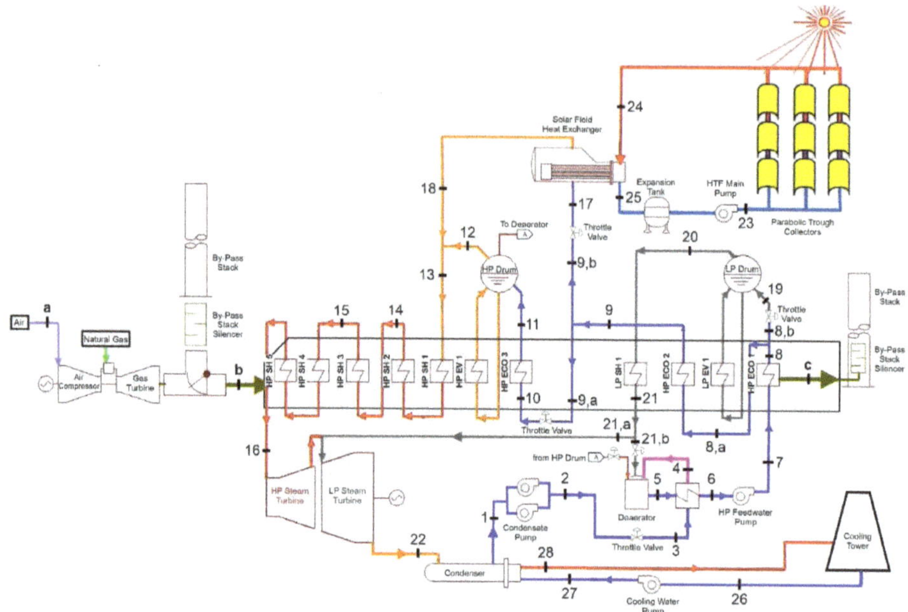

Figure 1. A flow diagram of the integrated solar combined cycle (ISCC) power plant in Kureimat with state points illustration.

3. Thermodynamic Analysis

The ISCC was evaluated assuming steady-state operation, with no accounting for thermal capacitance. This assumption works fairly well through the majority of the operating day but creates some problems in the morning when the solar field is warming up.

3.1. Energy Balance

The net rate of heat input to the ISCC ($\dot{Q}_{ISCC,in}$) is given by

$$\dot{Q}_{ISCC,in} = \dot{Q}_{fuel} + \dot{Q}_{inc} \tag{1}$$

where $\dot{Q}_{fuel}$ is the rate of heat addition to the ISCC from the fuel combustion and $\dot{Q}_{inc}$ is the absorbed incident solar radiation. The heat from the fuel combustion ($\dot{Q}_{fuel}$) as a function of the fuel flow rate ($\dot{m}_{fuel}$) and the fuel heating value (H_v) is given by

$$\dot{Q}_{fuel} = \dot{m}_{fuel} * H_v \tag{2}$$

and the heat from the absorbed incident solar radiation ($\dot{Q}_{inc}$) as a function of the direct normal insolation (*DNI*), incidence angle (θ), incidence angle modifier (*IAM*), and solar collectors' aperture area ($A_{mirrors}$) is given by

$$\dot{Q}_{inc} = DNI * \cos\theta * IAM * A_{mirrors}. \tag{3}$$

Here, the incidence angle modifier (*IAM*) is the correlation of the losses from the collectors due to additional reflection and absorption by the glass envelope, and it can be calculated as follows:

$$IAM = 1 + \frac{0.000884 * \theta}{\cos\theta} - \frac{0.00005369 * \theta^2}{\cos\theta} \tag{4}$$

and the solar collectors' aperture area ($A_{mirrors}$) is calculated from the number of solar field collectors ($N_{collectors}$) and the width ($W_{collector}$) and length ($L_{collector}$) of the collectors as follows:

$$A_{mirrors} = N_{collectors} * W_{collector} * L_{collector} \tag{5}$$

The electric power output of the ISCC ($\dot{W}_{elec,ISCC}$) is equal to the sum of the electric power outputs of the gas turbine ($\dot{W}_{elec,GT}$) and the steam turbine ($\dot{W}_{elec,ST}$) as follows:

$$\dot{W}_{elec,ISCC} = \dot{W}_{elec,GT} + \dot{W}_{elec,ST} \tag{6}$$

As a result, the overall first law efficiency of the ISCC power plant is

$$\eta_{I,cycle} = \frac{\dot{W}_{elec,ISCC}}{\dot{Q}_{ISCC,in}} \tag{7}$$

3.2. Exergetic Efficiencies

Exergy-based performance analysis is the performance study of a system based on the second law of thermodynamics, which overcomes the limitations of studying the system based on the first law of thermodynamics. Exergy is a measure of the maximum useful work of a system as it proceeds to a specified final state in equilibrium with its surroundings (dead state). Exergy is destroyed in the system, not conserved as energy is.

Two different approaches are generally used to calculate the exergy efficiency of a system, one is called "brute force", while the other is called "functional" [16].

The brute force form of exergy efficiency is used in this paper. The brute force form requires accuracy and an explicit definition of each input and output exergy term before calculating the exergy efficiency as shown in Table 2. The input exergy terms of the ISCC represent the chemical exergy of the fuel and the exergy associated with the solar thermal energy input.

Table 2. Definitions of the exergy destruction and second law efficiency.

Component	Exergy Destruction	Second Law Efficiency (η_{II})
Pumps	$\dot{I}_{pump} = \dot{X}_{in} - \dot{X}_{out} + \dot{W}_{pump}$	$\eta_{II,pump} = 1 - \frac{\dot{I}_{pump}}{\dot{W}_{pump}}$
Heaters	$\dot{I}_{heater} = \dot{X}_{in} - \dot{X}_{out}$	$\eta_{II,heater} = 1 - \frac{\dot{I}_{heater}}{\dot{X}_{in}}$
Turbine	$\dot{I}_{turbine} = \dot{X}_{in} - \dot{X}_{out} - \dot{W}_{turbine}$	$\eta_{II,turbine} = 1 - \frac{\dot{I}_{turbine}}{\dot{X}_{in} - \dot{X}_{out}}$
Condenser	$\dot{I}_{condenser} = \dot{X}_{in} - \dot{X}_{out}$	$\eta_{II,condenser} = \frac{\dot{X}_{out}}{\dot{X}_{in}}$
Cycle	$\dot{I}_{cycle} = \sum_{all\ components} \dot{I}$	$\eta_{II,cycle} = \frac{\dot{W}_{net,out}}{\dot{X}_{cycle,in}}$

So, the second law efficiency (η_{II}) (exergetic efficiency) is given by

$$\eta_{II} = \frac{Exergy\ output}{Exergy\ input} \tag{8}$$

The net exergy transfer by heat ($\dot{X}_{heat}$) at the source temperature (T_s) and dead state temperature (T_0) is given by

$$\dot{X}_{heat} = \sum (1 - \frac{T_0}{T_s})\dot{Q}_s \tag{9}$$

and the specific exergy (Ψ) is given by

$$\Psi = (h - h_0) - T_0(s - s_0) \tag{10}$$

where h, h_0, s, and s_0 are the specific enthalpy, the specific enthalpy under the dead state condition, the specific entropy, and the specific entropy under the dead state condition, respectively.

Then, the total exergy rate associated with a fluid stream ($\dot{X}$) at the mass flow rate ($\dot{m}$) becomes

$$\dot{X} = \dot{m} * \Psi = \dot{m}[(h - h_0) - T_0(s - s_0)]. \tag{11}$$

The exergy destruction rate in the ISCC as a whole ($\dot{I}_{ISCC}$) was obtained from

$$\dot{I}_{ISCC} = \dot{I}_{compressor} + \dot{I}_{CC} + \dot{I}_{GT} + \dot{I}_{SF} + \dot{I}_{SFHex} + \dot{I}_{condenser} + \dot{I}_{CP} + \dot{I}_{FWP} + \dot{I}_{ST} + \dot{I}_{HRSG}. \tag{12}$$

3.2.1. The Exergetic Efficiency of the Solar Field

The solar heat input from the HTF ($\dot{Q}_{HTF}$) to the water in the solar field heat exchanger is given by

$$\dot{Q}_{HTF} = \dot{m}_{23}(h_{23} - h_{24}), \tag{13}$$

$$\dot{Q}_{water} = \dot{m}_{17}(h_{18} - h_{17}). \tag{14}$$

The exergy destruction rate in the solar field ($\dot{I}_{SF}$) is calculated from

$$\dot{I}_{SF} = \dot{X}_{SF,in} - \dot{X}_{SF,gain} \tag{15}$$

$$\dot{X}_{SF,gain} = \dot{X}_{23} - \dot{X}_{24} \tag{16}$$

$$\dot{X}_{SF,in} = \dot{Q}_{inc}\left[1 - \left(\frac{T_0}{T_{sun}}\right)\right] \tag{17}$$

where T_{sun} is the sun temperature, which equals 5777 K. The exergetic efficiency of the solar field ($\eta_{II,SF}$) is given by

$$\eta_{II,SF} = \frac{\dot{X}_{SF,gain}}{\dot{X}_{SF,in}}. \tag{18}$$

3.2.2. The Exergetic Efficiency of the ISCC

The fuel chemical exergy per unit time ($\dot{X}_{fuel}$) equals

$$\dot{X}_{fuel} = \zeta * \dot{Q}_{fuel} \tag{19}$$

where ζ is the ratio of the chemical exergy to the net calorific value, which equals 1.04 for natural gas [23].

The exergetic efficiency of the ISCC ($\eta_{II,Cycle}$) is given as

$$\eta_{II,Cycle} = \frac{\dot{W}_{elec,ISCC}}{\dot{X}_{ISCC,in}}, \tag{20}$$

$$\dot{X}_{ISCC,in} = \dot{X}_{SF,in} + \dot{X}_{fuel}. \tag{21}$$

4. Results and Discussion

The performance of the ISCC power plant was analyzed under different design conditions. The analyses were performed for different solar field thermal outputs (0 MW, 50 MW, and 75 MW) and different ambient temperatures (5, 20, and 35 °C). All calculations were made based on design condition data.

The energy efficiency (first law efficiency) and the exergetic efficiency (second law efficiency) were calculated based on the heat input to the plant by the fuel and the sun.

4.1. The Overall Thermal Efficiency of the ISCC Power Plant

The overall thermal efficiency of the ISCC power plant in Kureimat at different ambient temperatures for solar heat inputs of 0 MW, 50 MW, and 75 MW is shown in Figure 2.

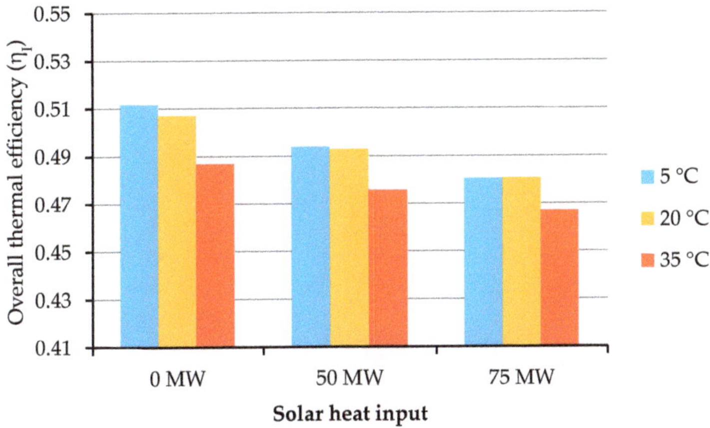

Figure 2. The overall thermal efficiency of the power plant at different ambient temperatures for different solar heat inputs.

The overall thermal efficiency of the power plant at different ambient temperatures for solar heat input equal to 0 MW, which represents the combined cycle regime, is shown in Figure 2. At no solar heat input (combined cycle regime), the thermal efficiency of the plant was reduced from 51.14% at ambient temperature 5 °C to 48.67% at 35 °C.

Figure 2 shows that the overall thermal efficiency of the ISCC decreases with increasing ambient temperature at different solar heat inputs (0, 50, 75 MW), and that appears most distinctly at ambient temperature 35 °C. This may be due to the direct effect of the ambient temperature increase on the efficiency of the condenser and the gas turbine: the condenser and gas turbine efficiency decreases with increasing ambient temperature.

The overall thermal efficiency of the ISCC is lower than the overall thermal efficiency of the plant in the combined cycle regime in all cases (different solar heat inputs and ambient temperatures). Figure 2 shows that the integration of the solar field with the combined cycle (i.e., ISCC) reduced the thermal efficiency of the power plant at all ambient temperatures. This may be because the target of the ISCC is not to increase the overall thermal efficiency of the Brayton cycle, like the combined cycle, but to increase the economic feasibility of the solar power plants. Elimination of the thermal storage system reduces the cost of the power plant [24–26].

4.2. Exergy Destruction in Each Component of the ISCC as a Percentage of the Total Exergy Destruction in the Whole ISCC

The exergy destruction in each component of the ISCC and the exergy destruction in the whole ISCC were calculated for different solar heat inputs and ambient temperatures 5 °C, 20 °C, and 35 °C.

The percentages of exergy destruction in each component of the ISCC out of the total exergy destruction of the power plant at different ambient temperatures for solar heat inputs 0 MW, 50 MW, and 75 MW are shown in Figures 3–5, respectively.

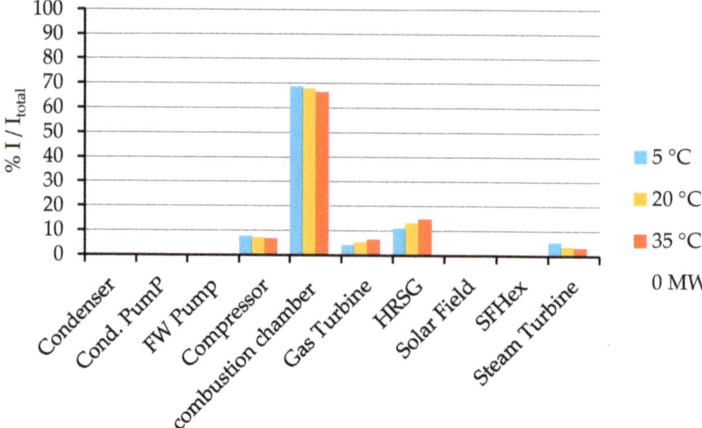

Figure 3. Percentage of exergy destruction in each component of the ISCC out of the total exergy destruction of the plant at different ambient temperatures for solar heat input equal to 0 MW.

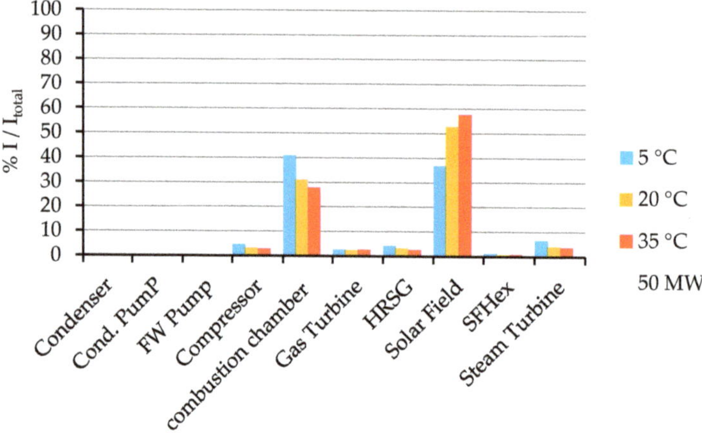

Figure 4. Percentage of exergy destruction in each component of the ISCC out of the total exergy destruction of the plant at different ambient temperatures for solar heat input equal to 50 MW.

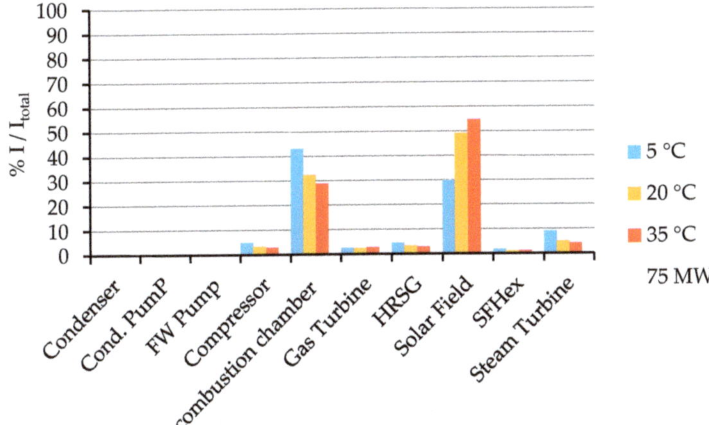

Figure 5. Percentage of exergy destruction in each component of the ISCC out of the total exergy destruction of the plant at different ambient temperatures for solar heat input equal to 75 MW.

It is revealed in Figure 3 that the combustion chamber (CC) has the highest percentage of exergy destruction, and this value is higher in the combined cycle regime than in the ISCC regime. This may ensure that the solar field has high irreversibility weight, which affects the percentage of exergy destruction in the combustion chamber compared to its value in the combined cycle regime.

However, Figure 3 shows that the exergy destruction in the combustion chamber decreases slightly with the increase of the ambient temperature under the combined cycle regime (0 MW solar heat input).

It can be observed from Figures 4 and 5 that the exergy destruction in the combustion chamber decreases significantly with the increase of the ambient temperature in the case of ISCC. This may account for the weight of exergy destruction in the solar field. Also, the exergy destruction in the solar field increases with increasing ambient temperature, in contrast to the exergy destruction in the combustion chamber.

Figures 3–5 show that the combustion chamber and the solar field have the highest exergy destruction among all the subsystems. This is valid for all cases of solar heat input. It was also revealed from the values at different ambient temperatures that the exergy destruction of the solar field decreases with increasing solar thermal input.

4.3. The Exergetic Efficiency of the Main Components of the ISCC

The exergetic efficiency of different components of the ISCC at different ambient temperatures for solar heat inputs 0 MW, 75 MW, and 50 MW is shown in Figures 6–8, respectively.

Figure 6 depicts the exergetic efficiency of different components of the ISCC at different ambient temperatures in the absence of the solar field (solar heat input equal to 0 MW), i.e., under the combined cycle regime. Under the combined cycle regime, the condenser has the lowest exergetic efficiency except at ambient temperature 5 °C. That may be due to the decrease in the low-temperature reservoir which increases the heat dissipated to the condenser cooling water.

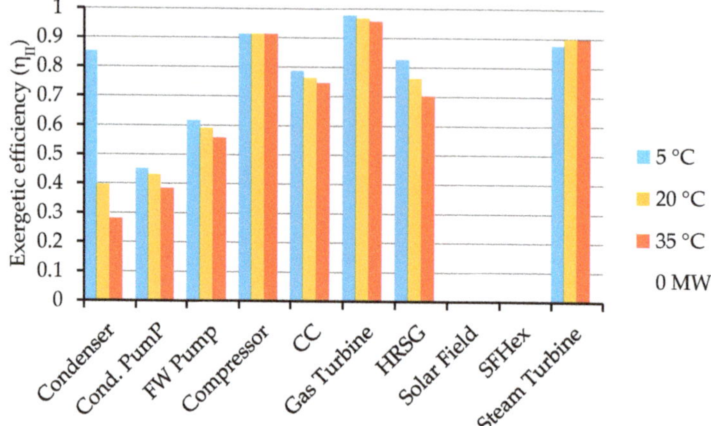

Figure 6. Exergetic efficiency of different components of the ISCC at different ambient temperatures for solar heat input equal to 0 MW.

The exergetic efficiency of the solar field decreased from 31.3% to 14.5% when the ambient temperature increased from 5 °C to 35 °C, as shown in Figure 7. The condenser exergetic efficiency also decreased from 75.5% to 19.3% when the ambient temperature increased from 5 °C to 35 °C for solar heat input equal to 50 MW. This may be due to the decrease in the temperature difference between the exhausted steam from the low-pressure turbine and the cooling water from the cooling tower.

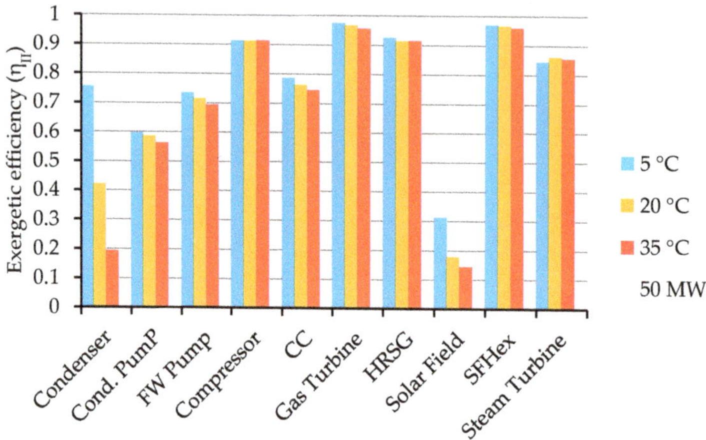

Figure 7. Exergetic efficiency of different components of the ISCC at different ambient temperatures for solar heat input equal to 50 MW.

Figure 8 shows that the exergetic efficiency of the solar field decreased from 47% to 21.7% when the ambient temperature increased from 5 °C to 35 °C. The condenser exergetic efficiency also decreased from 65.8% to 19.3% when the ambient temperature increased from 5 °C to 35 °C for solar heat input equal to 75 MW.

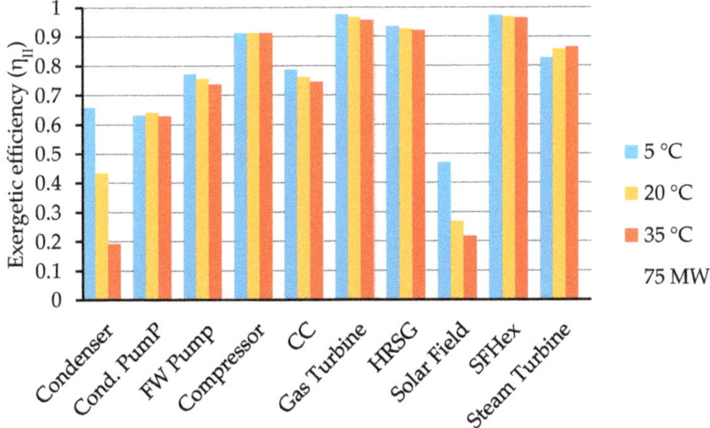

Figure 8. Exergetic efficiency of different components of the ISCC at different ambient temperatures for solar heat input equal to 75 MW.

As shown in Figures 6–8, the exergetic efficiency of the HRSG decreased with increasing ambient temperature, and this may be due to the existence of the attemperators in the HRSG which limit the steam temperature to the setpoint value. In the HRSG installed in the Kureimat power plant, attemperators were installed at the surface of the superheaters to control the temperature at the inlet of the high-pressure steam turbine. These attemperators use water directly from the main feedwater pump of the power plant. An increase in the ambient temperature may lead to an increase in the flue gas temperature of exhaust from the gas turbine into the HRSG, and the attemperators limit the effect of this temperature increase on the temperature of the superheated steam going into the steam turbine using water directly from the main feedwater pump. This may be a reason for the decreasing exergetic efficiency of the HRSG with increasing ambient temperature as shown in Figures 6–8.

Unlike the thermal efficiency [27], the exergetic efficiency of the solar field explicitly decreased with increasing ambient temperature, as shown in Figures 7 and 8. This may be due to the increase of the exergy destruction in the solar field with increasing ambient temperature, as shown in Figures 4 and 5.

4.4. The Exergetic Efficiency of the ISCC Power Plant

The exergetic efficiency of the ISCC power plant was calculated for different solar heat power inputs. The comparison was implemented at three different ambient temperatures: 5, 20, and 35 °C.

The ISCC power plant exergetic efficiency for solar heat inputs 0 MW, 50 MW, and 75 MW at different ambient temperatures is depicted in Figure 9. The exergetic efficiency of the ISCC power plant was calculated based on the design condition data for the different solar heat power inputs.

Figure 9 reveals that the exergetic efficiency of the ISCC power plant is inversely proportional to the ambient temperature, where it decreased from 47.2% to 46% with increasing ambient temperature from 5 °C to 35 °C for solar heat input equal to 75 MW. In addition, it decreased from 48.2% to 46.58% when the ambient temperature increased from 5 °C to 35 °C for solar heat input equal to 50 MW.

Figure 9 also illustrates the exergetic efficiency of the combined cycle regime (solar heat input equal to 0 MW) at different ambient temperatures. In the absence of the solar field, the exergetic efficiency of the plant reached 49.18% and 47.21% at ambient temperatures 5 °C and 35 °C, respectively. This demonstrates that the exergetic efficiency of the ISCC power plant in Kureimat has higher efficiency under the combined cycle regime than under the ISCC regime, as shown in Figure 9. This may be due to the existence of the solar field, which needs precise design optimization of solar energy integration in a CCPP.

Like the overall thermal efficiency, the exergetic efficiency of the ISCC power plant decreased with increasing ambient temperature, mainly at ambient temperature 35 °C. This may be due to the sharp decrease in the exergetic efficiency of the condenser and the solar field with increasing ambient temperature, as shown in Figures 6–8. These figures also show that the exergetic efficiency of the gas turbine and the HRSG decreased with increasing ambient temperature, and that also affected the exergetic efficiency of the ISCC power plant, as shown in Figure 9.

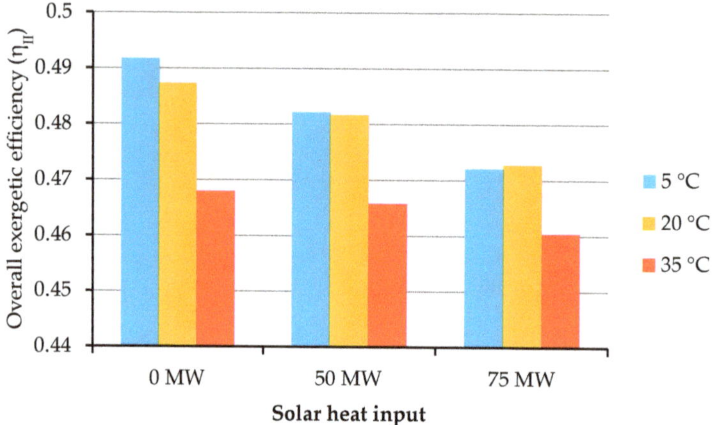

Figure 9. ISCC power plant exergetic efficiency at different ambient temperatures for different solar heat inputs.

4.5. Investigating the Sources of Exergy Destruction

From the attained results, it is clear that the amount of exergy destruction in the various components of the ISCC is altered. This variation is assumed to be due to different reasons such as the type of device, the process, etc.

Moreover, the results showed that the combustion chamber and the solar field represent the sites of highest exergy destruction in the ISCC. In this section, an attempt is made to explore and clarify the sources of exergy destruction in the solar field and the combustion chamber to identify the possibility of enhancing the performance of these components.

4.5.1. Irreversibility in the Solar Field

The exergy destruction in the solar field is due to heat transfer between the sun and the absorber, heat transfer between the absorber and the HTF, and the friction of the viscous HTF. The exergy loss is due to the optical efficiency (the ratio of sunlight capture to incident sunlight) and the heat transfer to the surroundings.

The solar collector is considered to be the main source of exergy destruction in the solar field due to the high temperature difference in the collector. The major contribution to the exergy destruction in the solar collector is due to the heat transfer between the sun and the absorber, while the major exergy loss occurs due to optical errors [28].

It was reported that exergy destruction due to heat transfer between the sun and the absorber accounts for 35% to 40% of the total exergy destroyed. Exergy losses to the surroundings account for 5% to 10% of the total exergy destroyed [28].

It is thought that to decrease the exergy losses from the solar collector (i.e., increase the collector energetic efficiency), attention should be pointed toward improving the optical parameters of the collector (such as mirror reflectivity, transmissivity of the glass envelope, absorptivity of the heat collection element selective coating, focal length of the collectors etc.).

Regarding the exergy destruction due to heat transfer, improving that part may involve great challenges because of the existence of the finite temperature differences which are essential for the heat transfer process and cannot be avoided.

4.5.2. Irreversibility in the Combustion Chamber

The combustion process is complex. Thus, the entropy generation during the combustion process is rather high due to the complexity of that process. It was reported that oxidation of fuel during the combustion process utilizes around 1/3 of the usable fuel energy [29]. This feature of the combustion process causes it to have the highest exergy destruction. The combustion process includes diffusion, chemical reaction, heat transfer, friction, and mixing. To implement all of these subprocesses, a considerable amount of the available energy is consumed. Most of this energy is unreachable (combustion activation energy, mixing, and diffusion).

There are three major physicochemical subprocesses responsible for entropy production during the combustion process [29]:

- Diffusion of reactants (mixing of fuel and air molecules) and chemical reaction (fuel oxidation) where energy is consumed to overcome the activation energy.
- Heat transfer between combustion products and other neighbors of particles; this is called "internal thermal energy exchange".
- Mixing of combustion products with other constituents.

These processes cause exergy consumption (destruction) and thus result in a reduction in the system exergy. On one hand, all these processes destroy up to 40% of the useful exergy of the fuel. On the other hand, it was found that the dominant process of exergy destruction is the internal thermal energy exchange process. It was found that more than 2/3 of the exergy destruction in the combustion process occurs at the internal thermal exergy exchange process, while fuel oxidation is responsible for up to 30% of the exergy destruction, and the exergy destruction due to the mixing process is about 3% of the total exergy destruction of the combustion process [29].

The thermodynamically irreversible combustion process is path-dependent. To get a quantitative solution for the total entropy production during the combustion process, correct information of the sequence of the combustion process and reactions must be offered.

Many factors affect the exergy destruction in the combustion chamber. For example, the exergy destruction decreases with decreasing excess air and increasing preheating temperature. Mixing at a large temperature difference leads to high exergy destruction [30]. Also, the exergy destruction of the combustion chamber is affected by the molecular structure of the fuel, where the exergy destruction of the combustion chamber increases with the increase of the hydrocarbon chain length [31].

An attempt was made to avoid this heat transfer by introducing the concept of reversible combustion, where it was proposed theoretically to preheat the reactants to the equilibrium temperature and partial pressures without a reaction, but it could not be achieved in practice [29].

The major exergy destruction in the combustion chamber occurs during the phase of the internal thermal energy exchange between the system particles [29]. The unavoidability of the internal thermal energy exchange makes reducing the exergy destruction during the combustion process very difficult.

5. Conclusions

The objective of this study was to investigate the performance of an existing 135 MW ISCC power plant in Kureimat. The ISCC power plant was thermodynamically studied under Kureimat climatic conditions. Energy and exergy analyses were performed for the ISCC power plant as a whole at different ambient temperatures (5, 20, and 35 °C) and different solar heat inputs (0, 50, 75 MW). Moreover, the exergy destruction and the exergetic efficiency for the main components of the ISCC power plant were calculated and investigated regarding the influence of the ambient temperature and the solar heat input to identify the causes and locations of the highest thermodynamic irreversibility. The integration of solar energy into a natural gas CCPP was analyzed as a power-boosting mode.

The main conclusions of this study are as follows:

- The solar field has the lowest exergetic efficiency (17.8%), followed by the condenser (42.2%), at ambient temperature 20 °C and solar heat input 50 MW.
- The exergy destruction in the solar field is the largest part of the exergy destruction in the ISCC power plant (52.9% at ambient temperature 20 °C and solar heat input 50 MW).
- The thermal efficiency and the exergetic efficiency of the ISCC decrease with increasing solar field thermal input, where it has its highest values (51.14% and 49.18%, respectively) at no solar field thermal input (combined cycle regime) and ambient temperature 5 °C.
- The thermal efficiency and the exergetic efficiency of both the ISCC and the combined cycle (i.e., at no solar field heat input) decrease with increasing ambient temperature at different solar heat inputs (0, 50, 75 MW). This is due to the decrease of the exergetic efficiency of the gas turbine, the solar field, the condenser, and the HRSG with increasing ambient temperature.
- The integration of a solar field with a combined cycle (i.e., ISCC) reduced the thermal and exergetic efficiencies of the power plant under the combined cycle regime due to the low thermal and exergetic efficiencies of the solar field because the solar fuel cost was considered in this study.
- The target of the ISCC power plants is not to increase the overall thermal efficiency of the Brayton cycle, like the combined cycle, but to increase the economic feasibility of solar power plants. Elimination of the thermal storage system reduces the cost of the power plant [24–26]. So, this integration of a solar field is recommended regarding the given challenges for the electricity market with the continuing expansion of intermittent renewables.

Author Contributions: A.T. is responsible for the technical, analysis, writing and funding acquisition parts. A.E. has contributed strongly to the design and preparation of the investigation. A.T., A.R., A.E., F.A., and B.E. discussed the formulation and the results and contributed to the review and editing of the manuscript. All authors have read and agreed to the published version of the manuscript.

Funding: The authors appreciate the Technical University of Darmstadt for funding the publication costs for the article in an open-access journal. The corresponding author appreciates the Egyptian Government for offering the PhD scholarship.

Conflicts of Interest: The authors declare no conflict of interest.

References

1. Zhang, H.L.; Baeyens, J.; Degrève, J.; Cacères, G.; Cac, G.; Zhang, H.L.; Baeyens, J.; Degrève, J.; Cacères, G. Concentrated solar power plants: Review and design methodology. *Renew Sustain. Energy Rev.* **2013**, *22*, 466–481. [CrossRef]
2. Mills, D. Advances in solar thermal electricity technology. *Sol. Energy* **2004**, *76*, 19–31. [CrossRef]
3. Ratliff, P.; Garbett, P.; Fischer, W. The new siemens gas turbine sgt5-8000h for more customer benefit. *VGB Powertech* **2007**, *87*, 128–132.
4. Scholz, C.; Zimmermann, H.; an der Ruhr, M. First Long-Term Experience with the Operational Flexibility of the SGT5-8000H. 2012. Available online: https://www.osti.gov/etdeweb/biblio/22090573 (accessed on 1 July 2020).
5. Vandervort, C.; Wetzel, T.; Leach, D. Engineering and validating a world record gas turbine. *Mech. Eng.* **2017**, *139*, 48–50. [CrossRef]
6. Vandervort, C. Advancements in h class gas turbines and combined cycle power plants. In Proceedings of the ASME Turbo Expo 2018: Turbomachinery Technical Conference and Exposition, Oslo, Norway, 11–15 June 2018.
7. Spliethoff, H. *Power Generation from Solid Fuels*; Springer Science & Business Media: Berlin, Germany, 2010; ISBN 364202856X.
8. Behar, O.; Khellaf, A.; Mohammedi, K.; Ait-kaci, S. A review of integrated solar combined cycle system (ISCCS) with a parabolic trough technology. *Renew. Sustain. Energy Rev.* **2014**, *39*, 223–250. [CrossRef]

9.   Montes, M.J.; Rovira, A.; Muñoz, M.; Martínez-val, J.M. Performance analysis of an integrated solar combined cycle using direct steam generation in parabolic trough collectors. *Appl. Energy* **2011**, *88*, 3228–3238. [CrossRef]

10. Rovira, A.; Montes, M.J.; Varela, F.; Gil, M. Comparison of heat transfer fluid and direct steam generation technologies for integrated solar combined cycles. *Appl. Therm. Eng.* **2013**, *52*, 264–274. [CrossRef]

11. Dersch, J.; Geyer, M.; Herrmann, U.; Jones, S.; Kelly, B.; Kistner, R.; Ortmanns, W.; Pitz-Paal, R.; Price, H. Trough integration into power plants—A study on the performance and economy of integrated solar combined cycle systems. *Energy* **2004**, *29*, 947–959. [CrossRef]

12. Kelly, B.; Herrmann, U.; Hale, M.J. Optimization studies for integrated solar combined cycle systems. In Proceedings of the Solar Forum 2001 Solar Energy: The Power to Choose, Washington, DC, USA, 21–25 April 2001; ASME: Washington, DC, USA, 2001; pp. 393–398.

13. Bonforte, G.; Buchgeister, J.; Manfrida, G.; Petela, K. Exergoeconomic and exergoenvironmental analysis of an integrated solar gas turbine/combined cycle power plant. *Energy* **2018**, *156*, 352–359. [CrossRef]

14. Iora, P.; Beretta, G.P.; Ghoniem, A.F. Exergy loss based allocation method for hybrid renewable-fossil power plants applied to an integrated solar combined cycle. *Energy* **2019**, *173*, 893–901. [CrossRef]

15. Mehrpooya, M.; Tosang, E.; Dadak, A. Investigation of a combined cycle power plant coupled with a parabolic trough solar field and high temperature energy storage system. *Energy Convers. Manag.* **2018**, *171*, 1662–1674. [CrossRef]

16. Baghernejad, A.; Yaghoubi, M. Exergy analysis of an integrated solar combined cycle system. *Renew. Energy* **2010**, *35*, 2157–2164. [CrossRef]

17. Behar, O.; Kellaf, A.; Mohamedi, K.; Belhamel, M. Instantaneous performance of the first integrated solar combined cycle system in algeria. *Energy Procedia* **2011**, *6*, 185–193. [CrossRef]

18. Wang, S.; Fu, Z.; Zhang, G.; Zhang, T. Advanced thermodynamic analysis applied to an integrated solar combined cycle system. *Energies* **2018**, *11*, 1574. [CrossRef]

19. Brakmann, G.; Mohammad, F.A.; Dolejsi, M.; Wiemann, M. Construction of the ISCC Kuraymat. In Proceedings of the International SolarPACES Conference, Berlin, Germany, 15–18 September 2009; pp. 1–8.

20. SOLUTIA. Data Sheet. p. 7. Available online: http://iapws.ru/MCS/Worksheets/HEDH/.%5C..%5C..%5C..%5CTTHB%5CHEDH%5CHTF-62.PDF (accessed on 17 January 2008).

21. Eastman Selection Guide High-Performance Fluids for Precise Temperature Control 10. Available online: http://www.synthec.com.pe/media_synthec/uploads/pdf_brochures/therminol_selection_guide.pdf (accessed on 14 May 2020).

22. Anderson, I.; Craig, A.D.; Kamal, J.A.; Veevers-Carter, P.; Govindarajalu, C.; Hassan, F. Report No: ICR2173 Implementation Completion and Results Report on a Grant In The Amount of US\$ 49.80 Million to the Arab Republic of Egypt for the Kureimat Solar Thermal Hybrid Project. 2012. Available online: http://documents1.worldbank.org/curated/en/739501468021865667/pdf/ICR21730P050560C0disclosed050100120.pdf (accessed on 10 April 2012).

23. Kotas, T.J. *The Exergy Method of Thermal Plant Analysis*, 1st ed.; Anchor Brendon Ltd, Tiptree, Essex: London, UK, 1985; ISBN 0408013508.

24. Zhu, G.; Neises, T.; Turchi, C.; Bedilion, R. Thermodynamic evaluation of solar integration into a natural gas combined cycle power plant. *Renew. Energy* **2015**, *74*, 815–824. [CrossRef]

25. Elmohlawy, A.E.; Ochkov, V.F.; Kazandzhan, B.I. Thermal performance analysis of a concentrated solar power system (CSP) integrated with natural gas combined cycle (NGCC) power plant. *Case Stud. Therm. Eng.* **2019**, *14*, 100458. [CrossRef]

26. Rovira, A.; Abbas, R.; Sánchez, C.; Muñoz, M. Proposal and analysis of an integrated solar combined cycle with partial recuperation. *Energy* **2020**, *198*, 117379. [CrossRef]

27. Temraz, A.; Rashad, A.; Alweteedy, A.; Elshazly, K. Seasonal performance evaluation of ISCCS solar field in Kureimat, Egypt. In Proceedings of the 18th International Conference on Applied Mechanics and Mechanical Engineering (AMME18), At Military Technical College Kobry El-Kobbah, Cairo, Egypt, 3–5 April 2017; pp. 91–98.

28. Padilla, R.V.; Fontalvo, A.; Demirkaya, G.; Martinez, A.; Quiroga, A.G. Exergy analysis of parabolic trough solar receiver. *Appl. Therm. Eng.* **2014**, *67*, 579–586. [CrossRef]

29. Dunbar, W.R.; Lior, N. Sources of combustion irreversibility. *Combust. Sci. Tech.* **1994**, *103*, 41–61. [CrossRef]

30. Athari, H.; Soltani, S.; Rosen, M.A.; Seyed Mahmoudi, S.M.; Morosuk, T. Comparative exergoeconomic analyses of gas turbine steam injection cycles with and without fogging inlet cooling. *Sustainability* **2015**, *7*, 12236–12257. [CrossRef]
31. Anheden, M. *Analysis of Gas Turbine Systems for Sustainable Energy Conversion*; Royal Institute of Technology: Stockholm, Switzerland, 2000.

Article

Operational Flexibility of a CFB Furnace during Fast Load Change—Experimental Measurements and Dynamic Model

Jens Peters *, Falah Alobaid and Bernd Epple

Institute for Energy Systems and Technology, Technical University of Darmstadt, Otto-Berndt-Straße 2, 64287 Darmstadt, Germany; falah.alobaid@est.tu-darmstadt.de (F.A.); bernd.epple@est.tu-darmstadt.de (B.E.)
* Correspondence: jens.peters@est.tu-darmstadt.de; Tel.: +49-(0)-6151-16-22689; Fax: +49-(0)-6151-16-22690

Received: 24 July 2020; Accepted: 24 August 2020; Published: 28 August 2020

Abstract: The share of power from fluctuating renewable energies such as wind and solar is increasing due to the ongoing climate change. It is therefore essential to use technologies that can compensate for these fluctuations. Experiments at 1 MW_{th} scale were carried out to evaluate the operational flexibility of a circulating fluidized bed (CFB) combustor during transient operation from 60% to 100% load. A typical load following sequence for fluctuating electricity generation/demand was reproduced experimentally by performing 4 load changes. The hydrodynamic condition after a load change depends on if the load change was in positive or negative direction due to the heat stored in the refractory/bed material at high loads and released when the load decreases. A 1.5D-process simulation model was created in the software APROS (Advanced Process Simulation) with the target of showing the specific characteristics of a CFB furnace during load following operation. The model was tuned with experimental data of a steady-state test point and validated with the load cycling tests. The simulation results show the key characteristics of CFB combustion with reasonable accuracy. Detailed experimental data is presented and a core-annulus approach for the modeling of the CFB furnace is used.

Keywords: CFB combustion; operational flexibility; load transients; fluctuating electricity generation; lignite; renewables

---

1. Introduction

Climate change due to high CO_2 emissions has led to a substantial increase in the proportion of electricity generated from renewable sources around the world in the last two decades [1]. Wind and solar energy account for a large share of these renewables. These two energy sources have the disadvantage that they are not suitable to provide baseload and, depending on the conditions, the electricity generated can fluctuate strongly and rapidly. There is a demand for highly flexible technologies to ensure the energy supply with low CO_2 emissions at the same time. Circulating fluidized bed (CFB) combustion can be a key element in the future energy supply, as it can combust a high spectrum of solid fuels of different origins such as biomass and waste-derived fuels [2–5]. The technology comes with a high combustion efficiency [6] and with low emissions of SO_2 and NO_x [7,8]. CFB boilers can operate over a wide range of thermal loads and are therefore capable of compensating a fluctuating electricity demand.

The high fuel flexibility is made possible by a large mass of inert particles in the reactor, which can compensate for fluctuations in the fuel composition. However, the high thermal inertia of this material has a negative impact on the load following capability and especially the cold start-up of CFB combustors [6]. During transients from medium to high load, ramp rates of up to 7% MCR/min (MCR: maximum continuous rating) are claimed to be possible [9]. However, modern large-scale CFB boilers, such as the Polish CFBC unit at Łagisza have load following capabilities of up to 4% MCR/min, which is similar to pulverized coal power plants [6]. The need for even faster load change rates increases with the rising share of renewable energies. Therefore, there is a demand to investigate novel concepts to accelerate load ramps in CFB combustion. For example, thermal energy storage systems can be used to rapidly extract a large amount of energy from the furnace [10]. Another option is to apply smart control strategies for the water/steam side and the fuel and air mass flow [11–13]. To examine novel concepts, experimental investigations are necessary but often very expensive, especially on a large scale. Computational fluid dynamics (CFD) simulations, on the other hand, are complicated and require a lot of computing time, even though they are often the only way to study CFB combustion inside a furnace in detail [14–17]. One-dimensional dynamic process models offer the advantage of low computing time and an appropriate accuracy to evaluate new approaches before they are tested in an industrial-scale boiler [18,19]. The commercial software APROS is used extensively in industry and research for the dynamic process modeling of thermal power plants [12,20,21].

So-called 1.5-dimensional core-annulus models take into account the mass and heat flow from the core to the annulus and vice versa. Particles and particle clusters stream upwards in the core region of the furnace, while they stream downwards in the annulus, due to the low gas velocities near the wall. This internal circulation of solids in the furnace is up to 2 times higher than the external circulation via cyclone and loop seal [22]. The convective heat transfer to the walls of the furnace is mainly determined by the mass flow from the core to the annulus and from the annulus to the core, so it is important to take this mechanism into account. However, the only way to prove the suitability of process models for dynamic investigations is by validation with experimental data. Suitable models must be able to show all typical characteristics of the CFB combustion: the load change duration/behavior, the combustion chemistry, the hydrodynamic conditions (particle distribution, temperature development), and the heat transfer. The present study presents a sophisticated 1.5D-model of a CFB combustor based on the core-annulus approach for the CFB furnace. The model is validated with detailed experimental data from CFB combustion of Polish lignite in a 1 MW_{th} pilot plant under dynamic conditions.

The novelty of the paper is concluded as follows:

1. A series of load changes are carried out in a pilot-scale CFB combustor, representing a typical operation during fluctuating electricity generation by renewables. This involves rapid load cycles in positive and negative direction before stationary states are reached. The effects of ascending and descending load cycles on the temperature and pressure in the furnace are discussed in each case. To the author's best knowledge, this is done for the first time with such detailed experimental data.
2. The study presents a highly sophisticated CFB combustor model based on the core-annulus approach for the furnace. The literature is given an insight into the modeling process of this type of process model and future research in this field will be accelerated.
3. The agreement between experiment and simulation is very good and the model is suitable for the deeper investigation of fuel and load ramp flexibility in future work.
4. Weaknesses of the model are identified which can be used in future research to further increase the agreement between experiment and simulation.

2. Experimental

The following chapter describes the test facility including the pilot plant, the measurement equipment, and the cooling system. Afterward, the experimental procedure is presented with the most important boundary conditions of the experiment.

2.1. Pilot Plant Description

Experiments were carried out in a 1 MW_{th} CFB furnace for the combustion of low-rank Polish lignite. The flow diagram of the CFB reactor (CFB600) and its subsystems is shown in Figure 1.

The main components of the facility are the air-supply system (primary air, secondary air, and burner air), the reactor with a hot-loop circulation (cyclone and loop seal), the solid handling systems (solid feeding and ash extraction), the cooling system and the flue gas line (heat exchanger, bag filter, induced-draft fan).

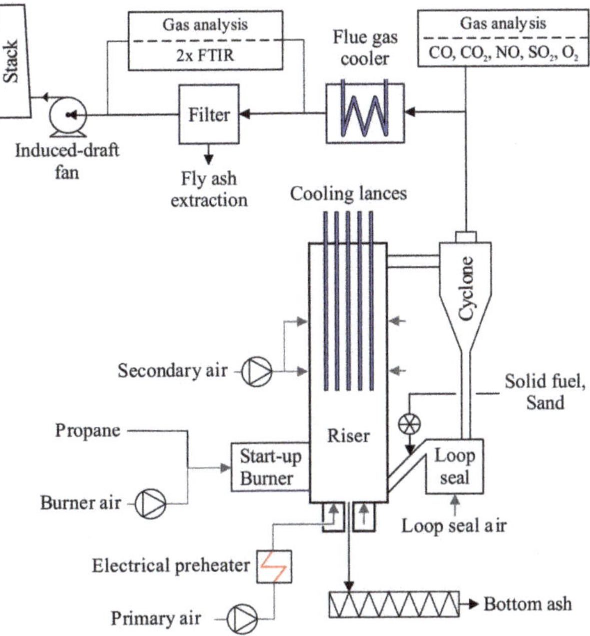

Figure 1. Simplified flow diagram of the 1 MW$_{th}$ pilot plant.

2.1.1. CFB Furnace and Auxiliary Systems

The furnace of the CFB600 reactor includes the riser and a hot-loop circulation for the recirculation of solids. The reactor itself has an inner diameter of 0.59 m and a total height of 8.6 m. It is fully refractory-lined and the outer diameter of the refractory is 1.3 m. Table 1 shows selected design parameters of the furnace. See Figure A1 in the Appendix A for a detailed geometry of the CFB combustor.

Solids that leave the riser, enter a cyclone, which separates them from the flue gas. The flue gas and fly ash enter the flue gas line with a small amount of fly ash. The main portion of the solids is recirculated to the furnace by a standpipe and a loop seal. Thereby, the residence time of char particles inside the furnace increases and the burn-out of the fuel improves. Air is injected into the loop seal at 25 °C via two nozzles to maintain fluidization of particles and to ensure a continuous recirculation. Besides the purpose of hot solid recirculation, the loop seal also provides pressure sealing between riser and standpipe/cyclone. Five water-cooled lances can be immersed vertically into the reactor at varying depth to control the combustion temperature.

Table 1. Design parameters of the furnace.

Parameter	Value
Furnace inner diameter [m]	0.59
Furnace outer diameter [m]	1.3
Furnace height [m]	8.6
Furnace free volume [m^3]	2.37
Height: Solids inlet (fuel, sand) [m]	0.481
Height: Secondary air inlet 1 [m]	2.74
Height: Secondary air inlet 2 [m]	6.0
Height: Loop seal solid recirculation to riser [m]	0.481
Height: Burner air inlet [m]	0.699
Height: Temperature sensors [m]	0.25, 1.12, 1.55, 2.38, 6.25, 8.21
Height: Pressure sensors [m]	0.11, 0.22, 0.4, 0.58, 0.91, 1.1, 2.07, 3.42, 7.31, 8.03

The combustion air is injected into the furnace at several locations and heights. The primary air is entering the furnace via a nozzle grid at the bottom of the riser. It is electrically preheated to around 300 °C before entering the reactor. A primary air fan controls the mass flow rate to a certain set point. Secondary air enters the riser at two different heights (2.74 m and 6.0 m) with two oppositely arranged nozzles at each elevation. A fan provides the desired mass flow of secondary air. The third major portion of the air is injected via the start-up burner at a height of 0.70 m. During start-up, this burner is fired with propane. During the tests, the air is supplied here to prevent backflow of particles and cool the burner components. The secondary and the burner air are not preheated and enter the furnace at an ambient temperature of around 25 °C.

A screw conveyor feeding system feeds solid fuel to a rotary valve, which is located above the return leg. The return leg connects the loop seal and riser. Thereby, the fuel is rapidly flowing to the bed of the furnace at a height of 0.48 m. The rotary valve guarantees pressure sealing between the reactor and the feeding systems. Another screw conveyor feeding system feeds sand to the same rotary valve. A water-cooled conveying screw extracts bed material via a downpipe in the middle of the nozzle grid. The particles are extracted batch-wise to keep the inventory in a suitable range. The target is to keep the bed pressure between 50 and 60 mbar.

The flue gas and fly ash leave the reactor through the cyclone. Afterward, it flows to the flue gas heat exchanger and cools down for further treatment. The heat exchanger is water-cooled and is arranged in two vertical paths. After cooling down, the flue gas is separated from the fly ash in a fabric filter. The fly ash is collected in a hopper and transported to a barrel by a rotary valve. A downstream induced-draft (ID) fan ensures a constant pressure of around 1 mbar below ambient pressure after the cyclone. After leaving the ID fan, the flue gas leaves the system through the stack at a temperature of 130–150 °C.

2.1.2. Measurement Equipment

The pilot plant is equipped with measurement equipment for temperatures, pressures, mass flow rates, and flue gas composition at several measurement locations. The temperatures and pressures are measured inside the riser, the hot loop circulation, the flue gas path, and the peripheral systems. Orifice plates and venturi nozzles measure the flow rate of the combustion air and the flue gas. The mass flow rate of solids is measured either by weight decrease (the continuous measurement of fuel and sand) or weight increase (the discontinuous measurement of bottom ash and fly ash). Solid samples are taken from these positions to analyze the chemical composition and the physical/mechanical properties of the solids. The flue gas composition is measured at three locations. The volumetric concentration (dry state) of oxygen, carbon dioxide, carbon monoxide, nitrogen oxide, and sulfur dioxide is measured after the cyclone with a paramagnetic sensor for O_2 and an NDIR (nondispersive infrared) sensor for the other gases. Before and after the fabric filter, the gas composition is measured with two FTIR (Fourier-transform infrared spectroscopy) measurement devices, which are additionally equipped with oxygen sensors. Besides the main gas components (O_2, CO_2, H_2O), the concentration of trace gases (HCl, NO_x, SO_2, NO, CH_4, CO) can be measured with these FTIRs.

2.1.3. Cooling System

Figure 2 shows a simplified scheme of the cold side of the cooling system. Heat is absorbed from the gas-side by two different subsystems. Five water-cooled lances reduce the temperature in the furnace. The lances can be moved vertically into the reactor at varying depth. The immersion depth was fixed during the experiment also at part load (two lances at 4.5 m, three lances at 6.5 m immersion depth). The cooling lances are manufactured with a double-tube design. Water is flowing down through the inner tube and flowing up through the outer tube of the lances. The second cooling subsystem is located downstream of the cyclone, where the flue gas is cooled down in a two-path heat exchanger. In the first path, the flue gas streams downwards via a tube bundle heat exchanger. It is then redirected upwards via a second tube bundle heat exchanger. The walls of both paths are designed as membrane wall heat exchangers. The mass flow rate of water is measured before and after the cooling lances and after the flue gas cooler. The water temperature is measured before and after each lance. The mixing temperature after the lances is measured and the temperatures before and after each path of the flue gas cooler are measured. The hot water is cooled down in an air re-cooling system to 110 °C before it re-enters the cooling paths. The cooling system operates at 8–16 bar and the cooling liquid is always in a liquid state (no steam generation). During the experiment, the inlet temperatures and the mass flow rates of each cooling line are fixed. Thereby, the transferred heat from the gas side to waterside can be determined by the temperature difference of the water between the inlet and outlet of the cooling subsystems. Heat losses in the subsystems are unavoidable in the pilot plant. The heat transfer calculation is therefore subject to uncertainties. The design parameters of the cooling lances and the flue gas cooler are shown in Table 2.

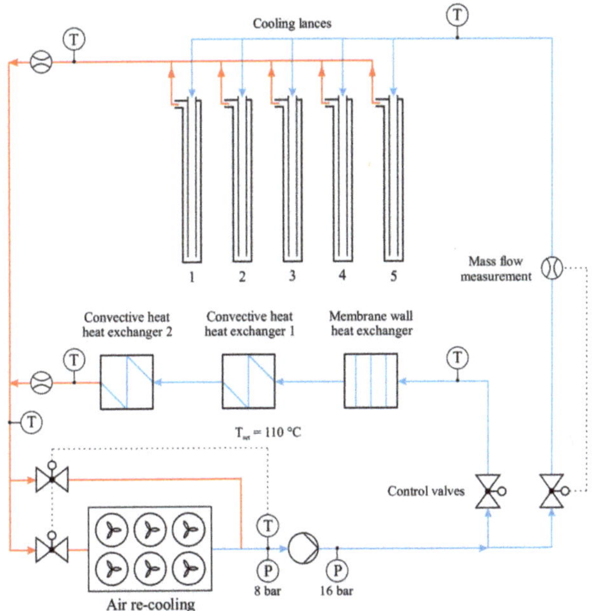

Figure 2. Simplified flow diagram of the cooling system (waterside).

Table 2. Design parameters of the cooling lances and the flue gas cooler.

Cooling Subsystem	Parameter	Value
Cooling lances	Number of lances [-]	5
	Maximum immersion depth [m]	8
	Inner diameter of inner tube [mm]	33.6
	Outer diameter of inner tube [mm]	42.4
	Inner diameter of outer tube [mm]	53.1
	Outer diameter of outer tube [mm]	60.3
Flue gas cooler	Tube bundle pipe outer diameter [mm]	31.8
	Tube bundle pipe inner diameter [mm]	25.4
	Tube bundle pipe length [m]	0.5
	Number of tube bundle pipes [-]	192

2.2. Experimental Conditions and Procedure

Low-rank polish-lignite was combusted in the experiments in the pilot CFB furnace. The lignite has a lower heating value of 11.4 MJ/kg and a moisture content of around 49%. The sand was fed to the furnace to improve the fluidization properties and keep the bed particle size in a suitable range for CFB combustion. The proximate and ultimate analyses of the raw and dried lignite are shown in Table 3, while the properties of the sand are shown in Table 4. The average particle size of the sand particles is 200 μm.

Table 3. Analyses of the polish lignite (ultimate, proximate, others).

Fuel Analysis		
Property	As Received	Dry
C [wt-%]	29.31	57.70
H [wt-%]	2.16	4.26
O [wt-%]	10.21	20.10
N [wt-%]	0.37	0.72
S [wt-%]	0.84	1.65
Moisture [wt-%]	49.20	0.00
Ash [wt-%]	7.90	15.55
Fixed C [wt-%]	18.41	36.24
Volatiles [wt-%]	24.49	48.21
Lower heating value [MJ/kg]	11.4	22.4
CaO in ash [wt-%]	16.70	16.70
D(10) [mm]	0.797	-
D(50) [mm]	8.656	-
D(90) [mm]	13.369	-
Bulk density [kg/m^3]	617	-

Table 4. Properties of sand.

Parameter	Value
D(10) [mm]	0.138
D(50) [mm]	0.204
D(90) [mm]	0.287
Particle density [kg/m^3]	2650
Specific heat capacity at 800 °C [kJ/kg K]	1.37
Heat conductivity coefficient at 800 °C [W/m K]	25

To evaluate the long-term steady-state behavior of the fuel, a 57-h test was carried out at the beginning of the test series with a thermal load of 845 kW$_{th}$. Short-term fluctuations of the fuel mass flow rate and composition are compensated in such a long period, making this test point very suitable for tuning and validation of process simulations. After the steady-state test, load change tests were carried out to evaluate the dynamic behavior and the hydrodynamic conditions at low loads. Firstly, the load was slowly reduced to 648 kW. After a stabilization time, the load was increased stepwise to 914 kW and 1032 kW. The thermal load depends on the speed of the conveying screw of the fuel, but also on changing properties of the fuel such as bulk density, moisture content, particle size, and chemical composition. To keep the excess-air factor nearly constant, the air mass flow rates were increased to the desired set point and only the fuel mass flow rate was controlled to keep an oxygen excess of around 4–6 Vol-%$_{wet}$. With the same approach, two load decreasing steps from 1032 kW to 921 kW and 699 kW were performed. After each load change, the pilot was operated in the same conditions for at least 50 min before the next load step was carried out. The large thermal inertia of the bed material acts as a buffer after a load change. Therefore, the stabilization time of ~50 min is not long enough to achieve steady-state conditions for each part-load test. However, the fluctuating behavior of wind and solar power requires quick and flexible load changes by power plants. In the energy market nowadays, it is necessary to perform load changes, when steady-state conditions are not achieved. The target of the experiments was to represent this behavior. Table 5 presents the conditions of the long-term steady-state test and the part-load tests. The experimental approach for the load changes is illustrated in Figure 3. As mentioned, the mass flow of the solid fuel fluctuates strongly, so that the figure only illustrates the average mass flow/setpoint and not the actual one. For the evaluation of the tests, mean values are calculated for the temperature, pressure, and other measured parameters for each part-load test. This evaluation period starts 15 min after each load change and ends 5 min before the next load change.

Table 5. Test conditions at steady-state and part-loads.

Property	Unit	Tuning Point 82% Load	63% Load	88% Load	100% Load	89% Load	68% Load
Primary air mass flow rate	kg/h	690.9	426.2	568.3	710.2	568.0	425.7
Secondary air mass flow rate	kg/h	615.7	347.2	490.2	633.2	490.0	346.8
Burner air mass flow rate	kg/h	84.3	84.7	85.0	84.2	84.1	84.1
Loop seal air mass flow rate	kg/h	30.8	39.1	39.1	37.2	37.3	37.3
Solid fuel mass flow rate	kg/h	266.8	202.2	286.2	326.5	289.8	219.1
Sand mass flow rate	kg/h	10.6	5.2	5.2	6.3	5.7	6.2
Excess-air factor	mol/mol	1.40	1.27	1.21	1.25	1.21	1.19

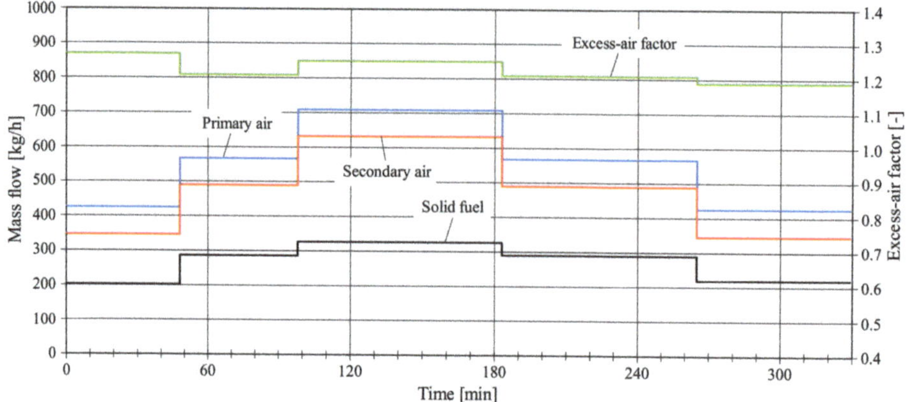

Figure 3. Schematic presentation of test conditions during part-load tests.

Part load conditions are part of the investigation in this paper. At lower loads, the velocity in the furnace decreases. To estimate the state of fluidization in the fluidized bed, it is helpful to compare the present velocity with the minimum fluidization velocity and the transition velocity. These two velocities can be calculated using various formulas. With the help of Grace et al. [23] u_{mf} is estimated to 0.015 m/s and u_{se} is estimated to 6.0 m/s. The velocity in the furnace depends on the flue gas flow rate, but it also depends on the position, as gaseous components are released along the reactor height and secondary air injection increases the gas flow rate. However, in this study, the velocity in the furnace is estimated according to the following equation, while T_{avg} is the arithmetic mean of the measured reactor temperatures, the index N means normal conditions (0 °C, 1.01325 bar), and A is the reactor cross-section:

$$v_{avg} = \frac{\dot{V}_{Flue\ gas}}{A_{Reactor}} = \frac{\dot{V}_{flue\ gas,N}}{A_{Reactor}} \cdot \frac{T_{avg} + 273.15\ K}{273.15\ K} \tag{1}$$

3. Model Description

The model of the pilot CFB furnace is created with the software APROS (Advanced Process Simulation). APROS was developed by Fortum and the Technical Research Centre of Finland (VTT) [21,24–28]. Process components such as pumps, fans, heat exchangers, or piping can be selected from a component library of the programme to model the process realistically. The components are connected via mass flow and energy equations. Data of the process components can be inserted at a high level of detail. Geometries, material properties, and characteristic curves of electrical machines can be inserted into the model. APROS is often used in literature for the modeling of thermal processes, especially thermal power plants, such as pulverized-coal fired power plants [20,29–31], municipal waste incinerators [32], circulating fluidized beds [21,33], concentrated solar power plants [34–36], and nuclear power plants [37,38].

The test facility in Darmstadt is modeled in detail with APROS, using the design data of the pilot plant. The subsystems fluidized bed combustor, air-supply, cooling system, flue gas path, and several boundary conditions for the simulation are modeled in individual nets with high detail. Where no suitable standard components are provided by APROS library (e.g., air preheater, and bag-house filter), the components are implemented with an in-house code. The homogeneous flow model describes the fluid properties and behavior in the combustion air lines, the flue gas path and the waterside of the cooling system. After the model was built, it was tuned with the long-term steady-state experiment at 82% thermal load. The test duration for this test point was 57 h, so the experimental data is considered to be very reliable. According to the experimental test schedule, the load was decreased afterward to

63%. Then, the load increasing (63–88–100%) and load decreasing steps (100–89–68%) were performed according to the test procedure described in Section 2.2. After tuning, the model was validated with the experimental data of the transient operation. Pressures, and temperatures inside and after the furnace, as well as the flue gas composition, were compared to the measurement data for validation.

3.1. APROS Model

The core of the APROS model of the pilot plant is the circulating fluidized bed reactor. Figure 4 shows the process diagram of the net of the CFB furnace including air supply, solid handling, cooling lances, insulation and the solid recirculation system. The core of the model is the fluidized bed module of the APROS simulation software. It is based on a 1.5 D core annulus approach. This allows the modelling of a typical circulating fluidized bed flow pattern, in which the particles flow upwards in the core region and downwards in the annulus near to the wall. The fluidized bed module is separated into 20 calculation nodes. Ten types of materials are considered in the riser (three fluids: air, flue gas, water and seven solids: solid fuel, sand, limestone, calcium sulfate, lime, char, ash). The solid recirculation to the riser (in the experiment by the cyclone, standpipe, and loop seal) is modelled with a solid split block (cyclone) and a heat structure module (standpipe and loop seal). In the experiments, the air is injected into the loop seal to maintain fluidization of the solids. This air mainly flows to the standpipe, heats up, mixes with the flue gas and leaves the reactor through the cyclone. To reproduce this behavior, the loop seal air is mixed with the flue gas after the cyclone at a temperature of 300 °C. The heat that is required to increase the temperature of the air from 25 °C to 300 °C is taken into account, by withdrawing exactly this amount of heat from the recirculating solids. Solid fuel is injected to the riser at 0.481 m, while the fuel is separated into three components before injection: water/moisture, lime, and the rest of the fuel (dry fuel without lime). Water is separated from the fuel to avoid numerical instabilities according to the user manual of APROS. Lime is supplied separately to take into account the desulfurization reaction ($CaO + SO_2 + 1/2O_2 \rightarrow CaSO_4$). The individual mass flow rates of the three components are calculated based on the ultimate and proximate analyses of the fuel, see Table 3.

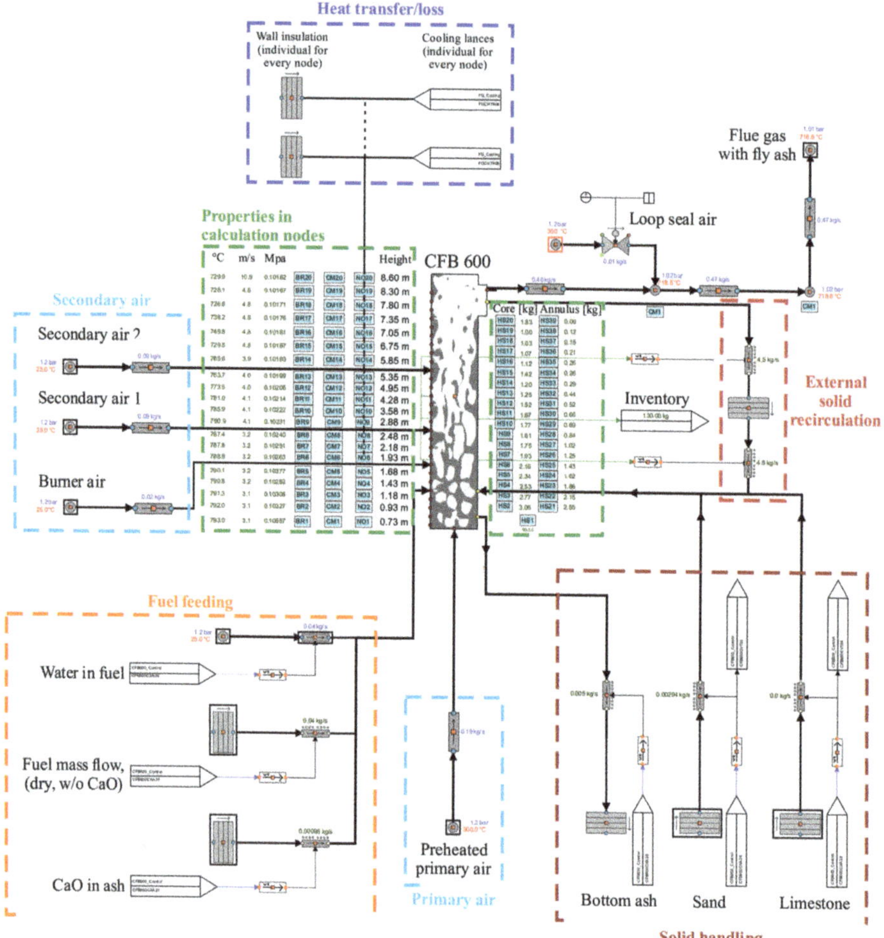

Figure 4. Circulating fluidized bed (CFB) combustor net.

In the model, the mass flow of the bottom ash is calculated by solving the solids mass balance. A major part of the incoming solid fuel is converted to gas by drying, devolatilization, and combustion reactions. Contrary, a small part of the gas is converted back to a solid state (e.g., by desulphurization reaction). After the chemical reactions, the solid materials ash, sand, lime, calcium sulfate, and unburned carbon leave the reactor either as fly ash or as bottom ash. The bottom ash mass flow is adjusted in such a way to maintain a constant inventory of a pre-defined set point of 130 kg.

The air is supplied to the reactor by three main lines: primary air, secondary air (injection at two different heights), and burner air. The air supply is modeled in a separate net, which is not shown here. The air streams enter the riser at the same height and at the same temperature as in the experiment. The primary air pre-heating is modeled by the implementation of a PI controller that either increases or decreases the heat supply to the air stream to match the temperature set point. Each of the combustion air streams is modeled by a fan, which speed is controlled by a PI controller to supply the specified mass flow rate (with the setpoint coming from the experimental data). The piping of the air supply is modeled with the APROS module "pipe". The geometrical and material data for this piping is taken from the pilot plant.

The flue gas path (from the heat exchanger to stack) is modeled in detail in another net. The flue gas flows to the two-path flue gas cooler, the filter, the ID fan, and the stack. Geometrical and material data for both paths of the heat exchanger and the piping of the flue gas lines are taken from the pilot plant design and implemented in the model. The fabric filter is not presented by a standard APROS library component. Therefore, a pre-defined pressure drop and a thermal mass are used to represent the filter in the APROS model. In the model and the experiment, the speed of the ID fan is controlled by a PI controller to maintain a pressure of around 1 mbar below ambient conditions at the cyclone outlet.

The cooling system is modeled according to the design of the cooling system in the experiment, see Figure 2 and Table 2. However, some components such as the air re-cooling unit are not implemented in the model. Instead, fresh cooling water enters the system with 110 °C and 11 bar (data taken from the experiment). The cold water is pumped to a distributor pipe before it enters the cooling lances and the flue gas cooler. The mass flow through each cooling subsystem is controlled to a certain set point by control valves, while a speed-controlled pump ensures the overall mass flow. In the model, after leaving the subsystems, the hot water is discarded. Three cooling lances are in contact with the calculation nodes 6–20 (corresponds to 6.5 m immersion depth) and two lances are in contact with the nodes 11–20 (corresponds to 4.5 m immersion depth).

3.2. Materials

In circulating fluidized beds, the high mass of particles in the bed has a high heat storage capacity. This property of the bed has a large impact during load following operation. Therefore, it is important to implement the heat capacity and the thermal conductivity of the bed materials with high accuracy, if the model is to reflect the experimental data in good agreement. Seven solid materials are present in the riser: sand, solid fuel, char, ash, lime, limestone, and calcium sulfate. Their heat capacity and thermal conductivity highly depend on the temperature, which varies significantly during part load. In the model, this temperature dependency is implemented for many of the solid materials. The thermal conductivity at 800 °C, the temperature-dependent functions for the specific heat capacity and thermal conductivity, and the density of the materials are shown in Tables 6 and 7.

Table 6. Thermal conductivity of solid particles in the riser.

Material	$\lambda_{(800\ °C)}$	Thermal Conductivity [W/m/K]				
		$\lambda_{(T)} = A + B{\cdot}T + C{\cdot}T^2 + D{\cdot}T^3 + E{\cdot}T^4$				
		A	B	C	D	E
Sand	25	1.04×10^2	-2.1×10^{-1}	1.82×10^{-4}	6.87×10^{-8}	9.17×10^{-12}
Solid fuel	0.26	0.26	0	0	0	0
Char	0.26	0.26	0	0	0	0
Ash	1.1	0.4	1×10^{-3}	0	0	0
Lime	1	1	0	0	0	0
Limestone	1.03	2.3807	-4.2×10^{-3}	4×10^{-6}	-2×10^{-9}	0
Calcium sulphate	0.17	0.17	0	0	0	0

Table 7. Density and heat capacity of solid particles in the riser.

Material	Density ρ [kg/m^3]	Heat Capacity $C_{p(T)} = A + B{\cdot}T + C{\cdot}T^2 + D{\cdot}T^3 + E{\cdot}T^4$ [kJ/kg/K]				
		A	B	C	D	E
Sand	2650	0.723	1.26×10^{-3}	-6.9×10^{-7}	2.67×10^{-10}	-3.8×10^{-14}
Solid fuel	1300	2.5	0	0	0	0
Char	1350	0.739	2.09×10^{-3}	-3.6×10^{-6}	0	0
Ash	2400	745.6	1.041	-0.6×10^{-3}	0	0
Lime	3350	0.84	0	0	0	0
Limestone	2700	0.4091	2.1×10^{-3}	-2×10^{-6}	7×10^{-10}	0
Calcium sulphate	800	1.09	0	0	0	0

Tables 8 and 9: The refractory is made of three layers, starting with concrete at the inner layer via calcium silicate in the middle layer and ending with microporous material in the outer layer.

Table 8. Thermal conductivity of furnace insulation.

Material	Thermal Conductivity [W/m/K]				
	$\lambda_{(800\,°C)}$	$\lambda_{(T)} = A + B{\cdot}T + C{\cdot}T^2 + D{\cdot}T^3$			
		A	B	C	D
Dense refractory concrete	1.657	3.1	0	-7	0
Calcium silicate	0.2874	0.09	1×10^{-4}	4×10^{-7}	-2×10^{-10}
Microporous	0.05352	0.016	2×10^{-5}	2×10^{-8}	4×10^{-11}

Table 9. Density and heat capacity of solid particles in the riser.

Material	Density ρ [kg/m^3]	Heat Capacity $C_{p(T)} = A + B{\cdot}T$ [kJ/kg/K]	
		A	B
Dense refractory concrete	2450	0.8	0
Calcium silicate	260	0.08	0.5×10^{-3}
Microporous	230	1	0

3.3. Procedure of Dynamic Simulations

The long-term steady-state test run with lignite lasted 57 h so that short-term fluctuations are compensated and the data is very reliable. Therefore, the test point is suitable to tune the APROS model to the corresponding solid fuel and the bed properties. The objective of the tuning process was to achieve a high agreement between the simulation results of the steady-state test and the experimental data, especially concerning the pressure and temperature profile. Mainly nine parameters were tuned, such as the number of calculation nodes, the heat transfer coefficient calculation method, or the global split coefficient between core and annulus and vice versa. Details of the tuning process and the tuned parameters are presented in Section 4.1.

After the model was tuned, validation of the model was done with the experimental data of the part-load tests. Therefore, no additional parameters were adjusted or tuned and only the boundary conditions were modified. The mass flow of solids (fuel and sand) and the air mass flow was modified with pre-defined setpoints according to the experimental data. Figure 3 shows a graphical visualization of the set points in the experiment and the simulation. As in the experiment, the load was increased from 63% to 88% to 100% followed by a load reduction from 100% to 89% and 68%. For details regarding the boundary conditions during the part-load simulations, see Table 5.

4. Results

Low-rank lignite was combusted in a 1 MW_{th} pilot plant to investigate the operational flexibility of CFB combustion. An APROS model was designed and adjusted to reproduce the experimental results. The underlying objective of this work is to create a model, which can predict the behavior of CFB combustion with different fuels and fuel mixtures and at varying load. The data is presented here to provide valuable input for future research in the field of the flexibility of CFB combustion. The first part of this section describes the model tuning process. Additionally, the temperature profile, the pressure profile, and the gas composition of the tuned model are presented and compared to experimental data of the long-term steady-state test. The experimental data is averaged over the complete testing time and the standard deviation is used to illustrate the fluctuations during the test. After the tuning process, the validation process with five test points at part load is presented. The development of the pressure and temperature along the furnace height is presented on average and over time to evaluate the average part-load conditions as well as the dynamic behavior of the boiler. The validation process focusses on the hydrodynamics, the flue gas composition, and the dynamic behavior during load increasing and decreasing conditions.

4.1. Steady-State Model Tuning

The model was tuned with the experimental data of the 57-h steady-state test at a load of 845 kW_{th} (82% load). Various model parameters were adjusted and chosen to meet the pressure profile, the temperature profile, and the gas composition of the experiment. Other parameters were tuned to find the best values for high numerical stability and good agreement with the experimental data. The most relevant parameters are listed in Table 10 with a short description. The riser is separated into 20 nodes. If the 20 nodes were equally distributed over the riser height, each node would have a height of 0.43 m. However, the bottom node of the riser has a height of 0.7 m. The bottom node represents the high-density bed zone. Here, no core-annulus approach is implemented due to the high inventory and the enhanced mixing of particles. For numerical stability, the bottom node must be higher than the bed height, which is calculated by APROS by dividing the mass of solid particles in the bed node by the average density of gas and solid particles in this node and the flow area of the node.

The heat transfer from the riser to the waterside depends on the temperature difference, the surface area, and the heat transfer coefficient. The heat transfer coefficient is calculated using the Mattman-Molerus-Wirth correlation. It takes into account the Archimedes number and the local pressure drop/particle fraction [39]. This correlation is recommended for the prediction of heat transfer in circulating fluidized beds [24].

The mass flow of particles from the core to the annulus and vice versa depends on the radial velocities and the mass of solids in the calculation nodes. Without further tuning, the radial velocities of the solids would be set equal to the axial velocities of the solids in the core/annulus. However, in reality, radial and axial velocities are not equal. To adjust the radial velocities and thereby the mass flow rates of the solids, so-called global split coefficients from the core to the annulus and from the annulus to the core are introduced. These coefficients have values between 0 and 1. With a value of e.g., 1, the radial velocity is equal to the axial velocity. With a value of e.g., 0.1, the radial velocity is ten times smaller than the axial velocity. Thereby, the global split coefficients have a great impact on the upflow/downflow of particles in the core/annulus and largely influence the pressure profile. The global split coefficients were adjusted to adapt the pressure profile according to the measurement data.

Table 10. Attributes/parameters of the CFB combustor model.

Parameter/Attribute	Description	Selected Value
Number of calculation nodes	The number of calculation nodes inside the riser. The parameter impacts numerical stability and model resolution.	20 [nodes] (Varied from 10–100)
Height of bottom node	The height of the bottommost node (high-density bed zone). No core-annulus approach is implemented for this node. This height must be higher than the calculated bed height for numerical stability.	0.7 [m] (Varied from 0.2–0.8)
Total mass of particles in the bed	The mass of particles in the fluidized bed. The value is estimated from the pressure drop in the experiment.	130 [kg] (Varied from 120–140)
Convection heat transfer correlation	The calculation method for the convective heat transfer coefficient between particles, gas, and tubes/walls. The parameter influences the heat transfer to the cooling lances/insulation.	Mattman-Molerus-Wirth correlation (recommended for CFB)
Heat transfer coefficient	The coefficient modifies the heat transfer coefficient calculated by the convection heat transfer correlation.	1.5 [-] (Varied from 1–5)
Global split coefficient from core to annulus/annulus to core	The coefficient reduces the calculated particle mass flows from the core to the annulus and vice versa. The parameter impacts the pressure profile.	0.03 [-] (Varied from 0–1)
Interface density	Defines the solid density on the interface between the high-density bed and the freeboard. The interface density affects the entrainment/the pressure profile.	30 [kg/m^3] (Varied from 15–50)

Another large influence on the pressure profile is given by the so-called interface density. The amount of particles that are entrained from the high-density bottom node to the first core node directly depends on the axial solid's velocity and the interface density. The interface density is a theoretical value and can be set freely by the user. It represents the suspension density in the zone between the bed and the freeboard. Therefore, it physically represents the density in the splash zone of the fluidized bed. The density can be estimated from the experimental data by using the pressure drop in this area. From the experimental data, it is reasonable to define the location of the splash zone between the two pressure sensors at 0.58 m and 2.07 m, where the density of the bed changes from high-density zone to lean zone. In the steady-state case, the calculated splash zone density is 58.2 kg/m^3. However, in the model, a value of 30 kg/m^3 was chosen, as the entrainment was highly overestimated, when using a value of 58.2 kg/m^3.

An average diameter has to be set to account for the particle size of the fluidized bed particles and no particle size distribution is implemented in the model. The average particle size is 0.211 mm, which is the average particle size of the bottom ash sample taken after the test period. To calculate the heat loss through the insulation to the environment, a boundary condition has to be set for the ambient temperature outside of the refractory lining. This temperature is set to 25 °C and is not adjusted during the simulations.

Figure 5 shows the pressure profile along with the furnace height during the long-term steady-state test with 845 kW$_{th}$ load. The average velocity according to the description in Section 2.2 is calculated to 4.2 m/s. In the model, the pressure drop in the freeboard is overestimated indicating that the entrainment is higher in the simulation compared to the experiment. The hydrodynamic conditions influence the chemical reactions, the heat transfer, and the temperature development in the furnace. Further improvements of the model are therefore useful to achieve better agreement in future studies—e.g., the interface density parameter can be set to smaller values. In this work, the interface density is already set 2 times smaller than the density in the splash zone. Therefore, it was chosen not to make the parameter even smaller, to have an appropriate compromise between the accuracy of the model and the physical foundation of the interface density. The agreement between experimental data and simulation is good both in terms of the total pressure drop and the reproduction of the typical CFB pressure profile. The figure also shows the standard errors for the experimental data calculated from the standard deviations in the measurement signals and the uncertainty of the measurement devices.

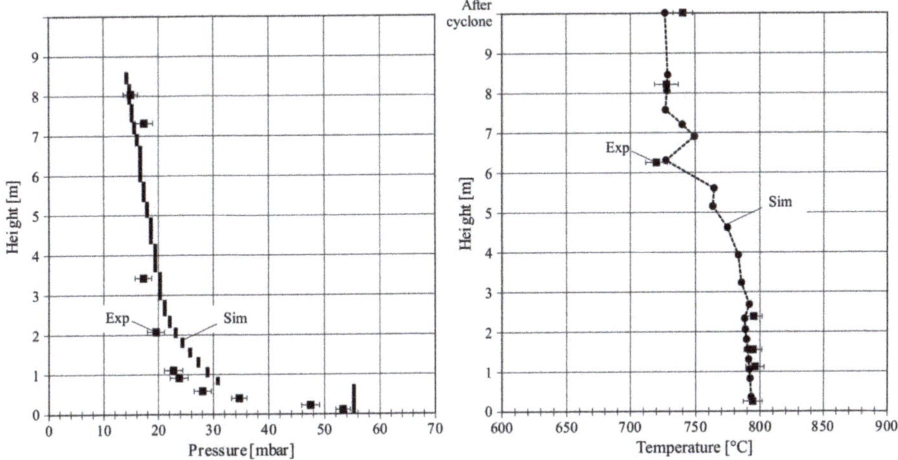

Figure 5. Pressure and temperature along with the reactor height (steady-state test).

The temperature profile is shown on the right in Figure 5. The temperatures in the simulation agree very well with the experiment. In the bed zone, the temperatures are almost constant due to the high level of mixing in the bottom region. At 2.7 m, secondary air is injected. However, the measurement shows decreasing temperatures from 2.4 to 6.3 m potentially for two reasons. Firstly, the cooling lances and the non-preheated secondary air might compensate for the combustion reactions. Secondly, no measurement is installed between these two locations. Therefore, the temperature might rise above the secondary air injection and it is just not shown by the available experimental data. However, the simulation does not show a temperature increase above 2.7 m either. In the experiment, the temperature increases above the secondary air injection at 6.0 m to the last measurement at 8.2 m. In the model, a temperature peak at 7.0 m is observed, before the temperature decreases again due to the impact of the cooling lances. At 8.2 m, experiment and simulation correspond. In addition to the riser temperatures, the temperature measurement after the cyclone is shown in Figure 5. In the experiment, the temperature increases between riser and cyclone outlet. It is reasonable, that the mixing of secondary air, volatiles, and char is considerably smaller in the experiment, as the penetration depth of the secondary air is not optimized in the pilot. Thereby, the combustion reactions are shifted to the top of the reactor and the cyclone. In the cyclone, the high velocity and the turbulent flow ensure a strong mixing leading to the post-combustion of char. In the model, the mixing is very efficient, due to its 1-dimensional nature. Therefore, the combustion reactions are taking place only in the riser and the post-combustion is not reproduced sufficiently.

The flue gas composition is presented in Figure 6. On the left-hand side, the main gas components O_2, CO_2, and H_2O are shown. The standard errors are presented, taking into account the measurement device uncertainty and the standard deviation of the fluctuating measurement signal. The simulation agrees very well with the experiment. Simulation results are always within the standard error of the experiment and the relative deviations are very small. Considering that the composition of the fuel is subject to fluctuations during the experiment, the agreement is considered very high.

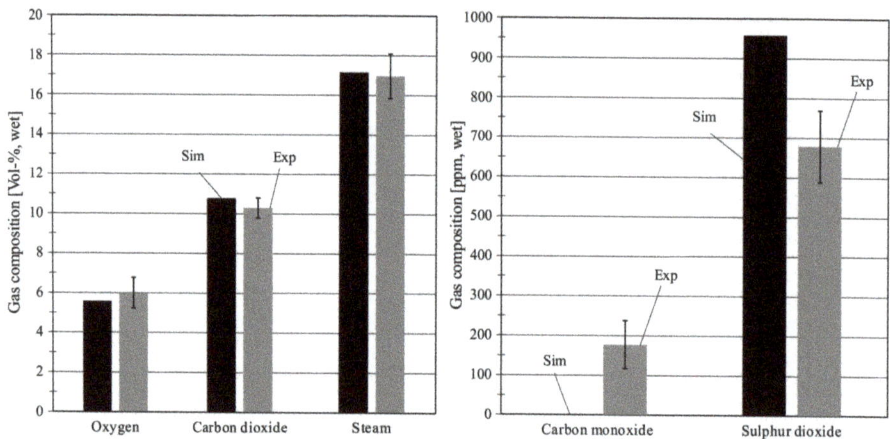

Figure 6. Flue gas composition (steady-state test).

On the right-hand side of Figure 6, the content of the trace gases CO and SO_2 is presented. There is no CO in the flue gas in the simulation. This supports the findings from the temperature profile of the simulation, where all combustion reactions are completed in the riser. In the experiment, the CO content is around 200 ppm$_{wet}$, so the combustion reactions are not completed even after the post-combustion in the cyclone. The SO_2 content depends on the varying sulfur content in the fuel and on the efficiency of the sulfur capture with calcium oxide from the fuel ash. The sulfur capture in a CFB combustor is a complex process, which depends on many parameters, such as the flue gas composition (locally reducing conditions, water vapour content), the temperature, the residence times and the properties of the CaO-particles (porosity, size, shape) [40–43]. Taking into account all these uncertainties and the fact, that especially the particle properties cannot be defined to this extent in the APROS model, the relative deviation of 41% between model and experiment is in a comprehensible range. It should be mentioned at this point that no tuning was carried out concerning the desulphurization reactions in the model, besides the definition of the S-content in the fuel and the CaO content in the fuel ash.

4.2. Model Validation

After tuning with the steady-state test, no further tuning or modification of the model took place during the validation process. The model is validated with the measurement data of five test points, with a thermal load from 63% to 100%. The model validation focusses on the hydrodynamic conditions, the flue gas composition, and the dynamic behavior of the fluidized bed.

4.2.1. Hydrodynamic Condition in the Furnace

The hydrodynamic conditions in a CFB furnace depend on the load, the bed material properties, the fuel, the combustion air, the heat exchangers inside the furnace, and other parameters, such as the insulation of the furnace. All these parameters are considered in the presented APROS model. The temperature and pressure profiles are a suitable way to validate a process model in terms of the hydrodynamics. Figure 7 shows the pressure and temperature profile of the 100% load case (1032 kW) with a calculated average velocity of 4.60 m/s. The agreement between experiment and simulation is very good for the temperature profile. The bed temperature, as well as the temperature development along the reactor and the temperature at the riser top in the simulation, are within the standard errors of the experimental data. As in the steady-state case, the post-combustion in the cyclone does not happen in the model. Therefore, there is a temperature deviation of about 20 K between simulation and experiment for the temperature after the cyclone. The pressure profile shows greater deviations than the temperature profile. The overestimation of the pressure drop in the lean zone is similar to the steady-state test. However, the total pressure drop and the slope of the pressure profile in the upper reactor is still in good agreement.

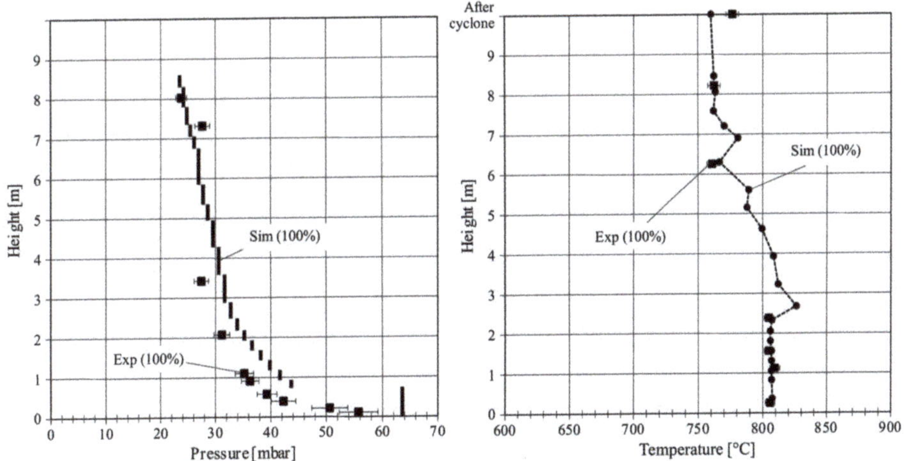

Figure 7. Pressure and temperature along the reactor at high load.

Figure 8 presents the hydrodynamics at medium load (88%, 89%). The average calculated velocities are 3.70 m/s (88%) and 3.77 m/s (89%). The pressure profile is similar for ascending and descending load. The load is slightly higher than in the steady-state tuning test point. However, the velocities are smaller, as the excess air factor is lower in these cases. The pressure drop in the freeboard is overestimated by the model, which could potentially be improved by adjusting the interface density in the model. However, the numerical simulation is capable of predicting the reduced entrainment at decreasing velocities/load cases on the same scale as in the experiment. The agreement between experiment and simulation is good regarding the total pressure drop and the bed density.

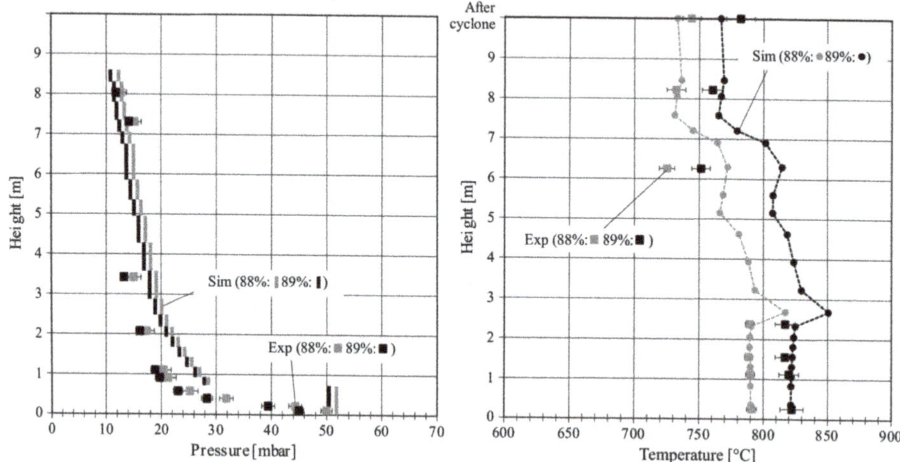

Figure 8. Pressure and temperature along the reactor at medium load.

It is noticeable that the average temperature differs significantly at descending and ascending load changes, despite similar loads. The refractory and the inert bed material play a major role here. A large amount of energy is stored and is then slowly released when the load decreases. Therefore, the temperature level is higher in the case of descending load (89% load case), as long as steady-state conditions are not achieved. This mechanism is well shown also by the simulation, where this temperature difference is also present. It is necessary to mention here, that the walls of commercial CFB boilers are made of water walls. Therefore, the effect observed here will hardly affect industrial scale boilers.

Despite the different average temperatures, the temperature development along the riser is similar for both medium load tests. Partly, bed temperatures are higher at medium loads compared to high loads. Additionally, the temperature drop along the reactor increases at lower loads (see Figure 7), due to several reasons. Less particles (e.g., char) are entrained from the bed at lower loads/velocities. Thereby the combustion reactions are shifted to the bed zone, which increases the bed temperatures and decreases the temperatures in the upper reactor. When the circulation of particles decreases, fewer particles are cooled by contact with the cooling lances and the surrounding flue gas. This lack of particle-cooling can result in an increased temperature of the solids and thereby an increasing bed temperature. These mechanisms, in combination with the heat stored in the insulation, can lead to an increase in the bed temperature, although the load is reduced. The simulated temperatures match the experiment very well in the high-density bed region and at the reactor top. At 6.3 m height, there is a larger deviation between model and experiment. Potentially, the heat transfer to the cooling lances is not correctly reproduced at this low load. As already discussed, also the combustion reactions with the secondary air, which is injected at a height of 6.0 m, are completed much faster in the model due to the idealized mixing. This leads to a higher temperature in this area.

The hydrodynamic profiles of the tests with 63% and 68% load are illustrated in Figure 9. The calculated velocities are 2.78 m/s (63%) and 2.88 m/s (68%). In the experiment, there is only a minor pressure drop along the lean zone of the riser, while the simulation results still show significant entrainment. As discussed before, the model parameter interface density, which largely influences the entrainment, is kept constant for validation purposes. In the experiment, on the other hand, the density in the splash zone is slightly decreasing at lower loads (58.2 mbar at 82% load and 48.2 mbar at 63% load). It is therefore reasonable, that the deviation between the pressure profile of the simulation and the experiment may get worse at lower loads. The model gives a good indication of the total pressure loss of the experiment, even if the deviations are greater than at higher load.

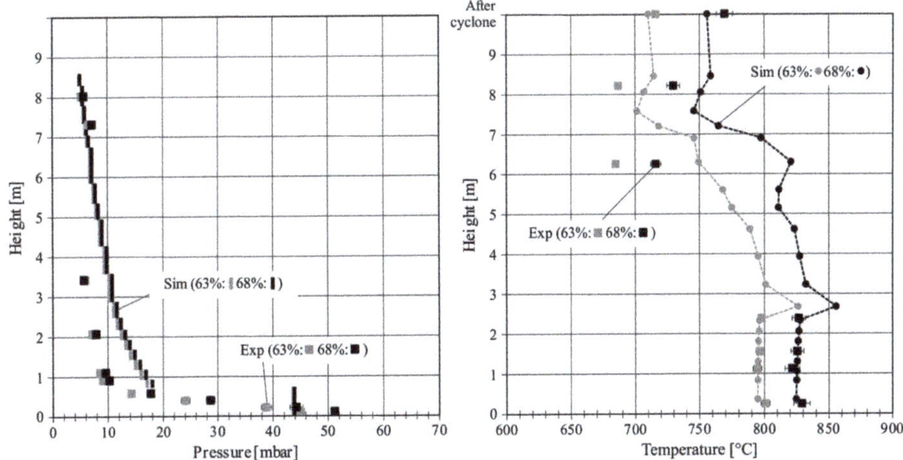

Figure 9. Pressure and temperature along the reactor at low load.

Again, the temperature level is higher for the load reduction test (68% load) compared to the load elevation test (63%), which is partly due to the energy stored in the refractory and bed material. However, the load in the latter is also significantly lower, which is also responsible for the higher temperatures. There is an excellent agreement between experiment and simulation for the bed temperatures. The model also shows that the temperature along the reactor decreases much more than it did at higher loads, which matches the experimental data. At the top of the reactor and around 6 m, the deviation between experiment and simulation is higher than at high and medium load, whereby the agreement between experiment and model is much better for the temperature after the cyclone. In the experiment, there is a major level of post-combustion in the cyclone, which is not present in the model due to the reasons discussed before. As the combustion takes place entirely in the riser, the temperatures in the simulation are higher at 6.3 and 8.2 m.

4.2.2. Flue Gas Composition

The flue gas composition is measured at several positions in the pilot plant. The FTIR measurement after the flue gas cooler includes all main species including water vapor and many trace gases. This measurement data is compared to the simulation results for the validation of the model. Figure 10 shows the flue gas composition of the part-load tests. In the simulation, the content of carbon dioxide and water vapor is within the standard error of the experiment for all tests. The oxygen content is always at or slightly below the lower limit of the standard error in the experiment. Regarding the fluctuating fuel composition, the model shows an excellent agreement with the experiment.

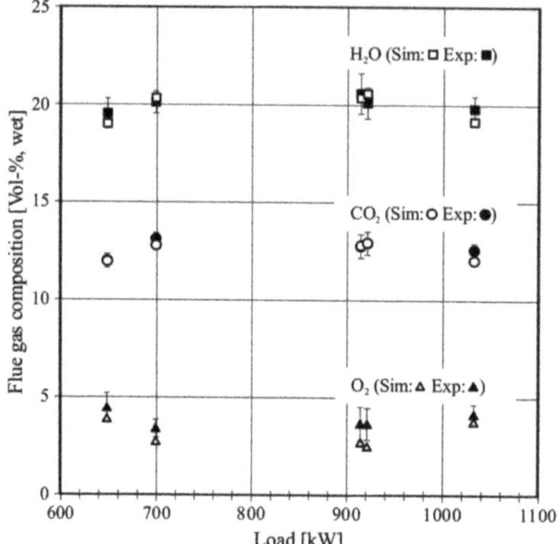

Figure 10. Flue gas composition at part load.

The CO content in dependency of the average riser temperature is shown in Figure 11. Large fluctuations are present for the CO content in the experiment. The CO content tends to decrease with increasing temperature, as high temperatures accelerate the combustion reactions. However, more parameters influence the CO content, such as the oxygen content and the efficiency of the post-combustion in the cyclone for instance. In the simulation, no carbon monoxide remains in the flue gas and the combustion reactions are completed inside the riser for all tests. As discussed before, the mixing of oxygen and burnable components (CO, char) is modeled as ideal in the 1.5 D APROS model. In the experiment, the mixing depends on several parameters, such as the injection speed of the secondary air and the conditions in the riser.

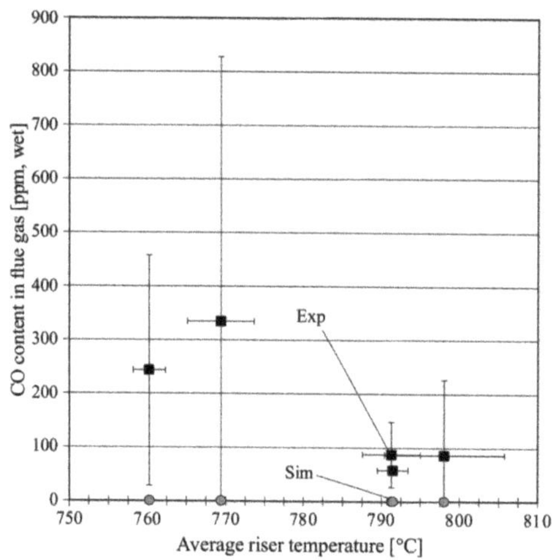

Figure 11. CO content in flue gas for varying riser temperature.

4.2.3. Dynamic Evaluation

Besides the evaluation of the part-load tests with mean values for the temperature and pressure, also the dynamic behavior during load change is relevant to validate the process model. In this chapter, the dynamic response of the furnace is evaluated and the model prediction of this response is assessed. Figure 12 shows the development of the temperature in the bed and at 8.2 m, the pressure at several heights, and the flue gas composition. The fuel mass flow is varying, due to differences in the fuel properties such as particle size, shape, density, or moisture. In the simulation, these fluctuations are not modeled. This results in fluctuating values for the flue gas composition and the temperatures.

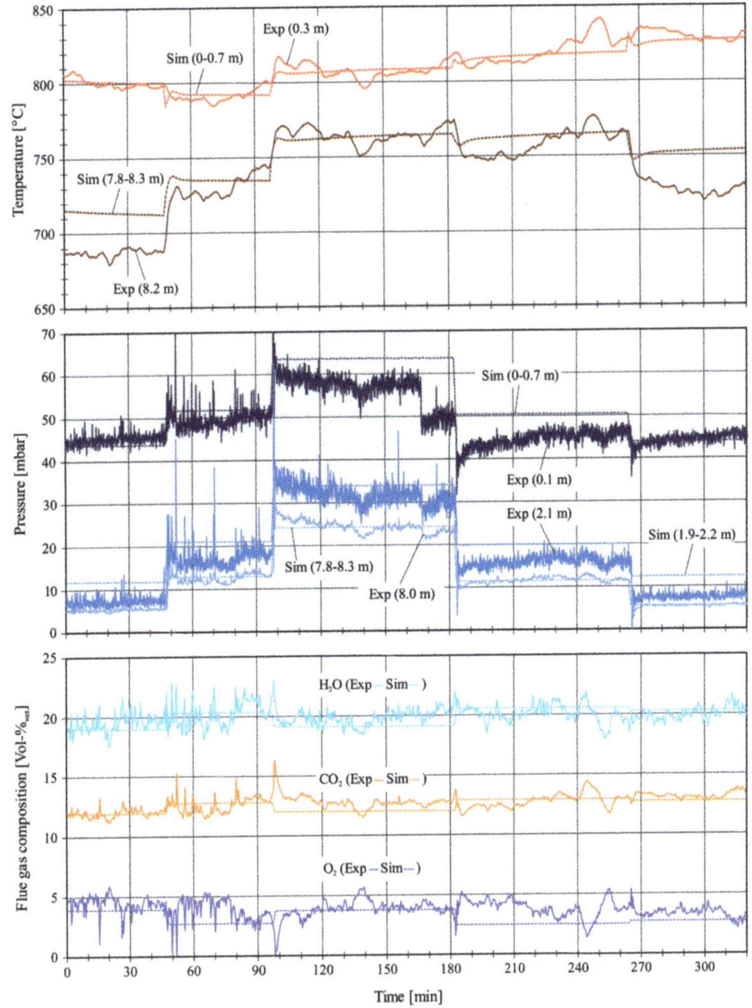

Figure 12. Temperature, pressure, and fuel mass flow during load change tests.

At the starting point at 63% load, the bed temperature of the numerical analysis and the experiment agree very well, while the temperature at the top of the reactor is significantly higher in the simulation. It is also apparent that the pressure above the bed (2.1 m) is overestimated, while the pressure in the bed and at the top, as well as the flue gas composition agree well with the experiment. The reasons for the deviations are discussed in detail in Section 4.2.1. After 47 min, the load is increased to 88%. In the simulation, the temperatures oscillate shortly after the load changes, especially in the dense bed region. PI-controllers control the mass flows of combustion air in the model. These controllers cause a temporary excess/shortage of air and thus they are responsible for the temperature oscillation. Further optimization of the controller is necessary to match the air-flow of the experiment at all times. There is a very good agreement for the overall duration of the load change. In the experiment and the simulation, 75% of the absolute change in temperature is completed after 3–4 min. After 97 min, the load is increased to 100%. The load change duration is similar to the first load transient. Despite the oscillation after the load change in the numerical model, the furnace response is predicted well by the model and 75% of the new temperature level is reached after 2–3 min. As already discussed in Section 4.2.1, the model is more accurate at higher loads with regard to the average pressure and temperature in the furnace.

The first load decreasing step (100–89%) is performed after 182 min. Due to the fuel mass flow fluctuations, shortly after the load change, there is a higher oxygen excess in the flue gas in the experiment compared to the simulation. When the fuel mass flow decreases slightly over time, the average oxygen excess decreases and is closer to the model prediction. This leads to a lower temperature level at the beginning of the 88% load test and an increasing temperature to the end of this test. For both, experiment and process model, the temperature difference between the 100% and 89% load test is low compared to the other tests. Therefore, and due to the fuel mass flow fluctuations, it is difficult to assess the duration of the temperature change in detail. After 264 min, the load changes from 89% to 68% load. The temperature response directly after the load change is reproduced well by the model. However, in the experiment, there is a long period (20–25 min), where the temperature converges to the final value, which is not observed in the simulation.

Compared to the partly slowly changing temperatures, the pressures are changing very fast (a few minutes) in all load transient tests—for the experiment and the numerical model. This is because the hydrodynamic condition mainly depends on the gas velocity and the solid particle properties. It depends only indirectly on the temperature—mainly by changing gas density and viscosity. On the other hand, the temperatures do of course depend strongly on the hydrodynamic conditions. However, the new temperature levels do not adjust as quickly as the pressure, due to the heat storage capacity of the bed material and the refractory. The large fluctuations of the pressure in the experiment in areas with high solid content (bed and splash zone) are typical for a CFB. The fluctuations decrease with the height of the reactor. The fluctuations are not given in the simulation, due to its one-dimensional nature.

Concluding this, the response of pressure and temperature on positive load changes can be predicted well by the model. It is not clear if this is also valid for negative load transients, but the first response after the load change seems to be predicted with good accuracy. In future studies, the fuel mass flow including the fluctuations should be modeled with higher details to further assess the dynamic properties of the model. However, with some improvements, the model seems to be suitable to predict the load transient response of a boiler very well. Therefore, core annulus models could be used to test novel concepts to increase the maximum load ramps of CFB combustors.

5. Conclusions

The conclusions are as follows:

1. The average temperature in the riser is significantly higher at descending load compared to ascending loads, which is due to the heat that is stored in the refractory/bed material at high loads and released when the load decreases. The process model shows this specific characteristic of CFBs with high accuracy, especially for the bed temperature.
2. The model predicts the duration of the positive load changes correctly. A statement for negative load changes cannot be given with certainty, but the response of the furnace directly after the load change seems to be reproduced well also at negative load transients. A high level of detail is necessary when modeling the material properties of the refractory and the bed material to achieve a high agreement between experiment and simulation for the dynamic behavior.

3. In the pilot, the freeboard temperature decreases sharply at lower loads, because the combustion reactions shift to the high-density bed. The model can reproduce this behavior well, although it overestimates the temperature in the freeboard at low loads.

4. The CO concentration increases at lower loads due to decreasing temperatures in the freeboard and an imperfect mixing of air and burnable components in the riser. The model neglected this imperfect mixing and was not able to show the incomplete combustion at low loads/temperatures.

5. As expected, the entrainment increases with increasing load, while the overall pressure difference from the bottom to the top only slightly changes. The APROS model with the annulus core approach overestimates the entrainment but reproduces the total pressure drop well. Improvements are necessary for future studies to match the pressure profile of the experiment with higher accuracy.

This study is relevant for future research on the flexible operation of CFB combustion regarding load change operation. The presented CFB model is highly suitable to examine concepts for accelerated load ramps. The work gives valuable concepts for, and insight into the process modeling of the CFB combustion technology. In ongoing work, the developed model is further optimized for different kinds of fuels such as refuse-derived fuel (RDF) and biomass and will be validated with experimental data of co-combustion tests with lignite and straw/RDF.

Author Contributions: Conceptualization, J.P. and F.A.; Methodology, J.P. and F.A.; Validation, J.P. and F.A.; Investigation, J.P.; Resources, B.E.; Writing—Original Draft Preparation, J.P.; Writing—Review & Editing, J.P. and F.A.; Visualization, J.P.; Supervision, B.E. and F.A.; Project Administration, B.E.; Funding Acquisition, B.E. All authors have read and agreed to the published version of the manuscript.

Funding: This research was funded by the Research Fund for Coal and Steel (RFCS) grant number 754032 and the APC was funded by the German Research Foundation and the Open Access Publishing Fund of Technical University of Darmstadt.

Acknowledgments: Financial support is acknowledged from the RFCS project of the European Commission under grant agreement n. 754032 (FLEXible operation of FB plants co-Firing LOw rank coal with renewable fuels compensating vRES—FLEX FLORES). We acknowledge support by the German Research Foundation and the Open Access Publishing Fund of Technical University of Darmstadt.

Conflicts of Interest: The authors declare no conflict of interest.

Appendix A

Table A1. Test results at steady-state and part-loads.

Property	Unit	Tuning Point 82% Load	63% Load	88% Load	100% Load	89% Load	68% Load
Temp. at 0.25 m	°C	794.3	800.8	790.9	805.6	822.3	829.1
Temp. at 1.12 m	°C	796.1	794.4	789.9	809.3	820.0	821.8
Temp. at 1.55 m	°C	794.6	796.6	788.6	804.8	817.0	825.6
Temp. at 2.38 m	°C	795.1	797.8	789.5	804.9	817.1	826.8
Temp. at 6.25 m	°C	720.3	685.0	725.2	760.8	751.5	715.9
Temp. at 8.21 m	°C	728.2	686.9	732.2	762.3	760.5	729.6
Temp. after cyclone	°C	740.9	715.9	744.1	776.5	782.2	769.6
Pressure at 0.11 m	mbar	53.3	45.1	50.0	55.7	45.0	51.2
Pressure at 0.22 m	mbar	47.5	38.8	44.3	50.6	39.4	44.3
Pressure at 0.4 m	mbar	34.7	24.1	31.9	42.3	28.4	28.7
Pressure at 0.58 m	mbar	28.1	14.2	25.3	39.3	23.1	17.7
Pressure at 0.91 m	mbar	23.8	9.1	21.3	36.3	19.8	10.3
Pressure at 1.1 m	mbar	22.8	8.6	20.4	35.2	18.9	9.6
Pressure at 2.07 m	mbar	19.5	7.2	17.3	31.2	16.1	7.9
Pressure at 3.42 m	mbar	17.2	5.9	15.0	27.4	13.2	5.7
Pressure at 7.31 m	mbar	17.4	6.6	15.2	27.6	14.1	7.2
Pressure at 8.03 m	mbar	14.9	5.3	12.7	23.7	11.8	5.7

Table A1. *Cont.*

Property	Unit	Tuning Point 82% Load	63% Load	88% Load	100% Load	89% Load	68% Load
CO_2 in flue gas	Vol-%$_{wet}$	10.3	12.0	12.8	12.6	12.9	13.1
O_2 in flue gas	Vol-%$_{wet}$	6.0	4.5	3.7	4.2	3.7	3.4
H_2O in flue gas	Vol-%$_{wet}$	17.0	19.5	20.6	19.8	20.1	20.1
CO in flue gas	ppm$_{wet}$	177.5	243.3	335.0	87.8	86.6	58.6
SO_2 in flue gas	ppm$_{wet}$	678.7	944.4	739.8	849.7	1624.8	1300.1
Flue gas flow rate	Nm^3/h	1364.5	921.2	1214.1	1477.7	1205.9	926.9
v_{avg}	m/s	4.16	2.78	3.70	4.60	3.77	2.88

Table A2. Bottom ash analysis.

Property	Unit	Value
Bulk density	kg/m^3	1350
Total carbon	wt-%$_{dry}$	0.4
D (10)	mm	0.132
D (50)	mm	0.211
D (90)	mm	1.100

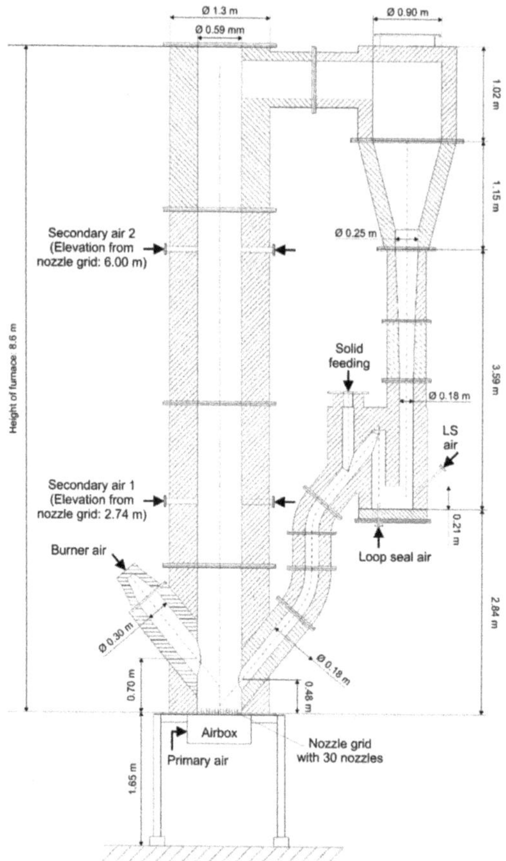

Figure A1. Detailed geometry of CFB combustor.

References

1. REN21. *Renewables 2015 Global Status Report*; REN21 Secretariat: Paris, France, 2015.
2. Walter, H.; Epple, B. *Numerical Simulation of Power Plants and Firing Systems*; Springer: Vienna, Austria, 2017.
3. Yates, J.G.; Lettieri, P. *Fluidized-Bed Reactors: Processes and Operating Conditions*; Springer International Publishing: Basel, Switzerland, 2016.
4. Oka, S. *Fluidized Bed Combustion*; Marcel Dekker, Inc.: New York, NY, USA, 2003.
5. Nuortimo, K.; Eriksson, T.; Nevalainen, T. *Large Scale Utility CFB Technology in Worlds Largest Greenfield 100% Biomass Power Plant*; Sumitomo Heavy Industries, Ltd.: Espoo, Finland, 2017.
6. Lockwood, T. *Techno-Economic Analysis of PC Versus CFB Combustion Technology, Ccc/226*; IEA Clean Coal Centre: London, UK, 2013.
7. Miller, B.G. *Clean Coal Engineering Technology*; Elsevier: Oxford, UK, 2010.
8. Basu, P. Combustion of coal in circulating fluidized-bed boilers: A review. *Chem. Eng. Sci.* **1999**, *54*, 5547–5557. [CrossRef]
9. Mills, S.J. *Integrating Intermittent Renewable Energy Technologies with Coal-Fired Power Plant, Ccc/189*; IEA Clean Coal Centre: London, UK, 2011.
10. Stefanitsis, D.; Nesiadis, A.; Koutita, K.; Nikolopoulos, A.; Nikolopoulos, N.; Peters, J.; Ströhle, J.; Epple, B. Simulation of a CFB boiler integrated with a thermal energy storage system during transient operation. *Front. Energy Res.* **2020**, *8*, 169.
11. Gao, M.; Hong, F.; Liu, J. Investigation on energy storage and quick load change control of subcritical circulating fluidized bed boiler units. *Appl. Energy* **2017**, *185*, 463–471. [CrossRef]
12. Alobaid, F.; Mertens, N.; Starkloff, R.; Lanz, T.; Heinze, C.; Epple, B. Progress in dynamic simulation of thermal power plants. *Prog. Energy Combust. Sci.* **2017**, *59*, 79–162. [CrossRef]
13. Henderson, C. *Increasing the Flexibility of Coal-Fired Power Plants*; IEA Clean Coal Centre: London, UK, 2014; p. 15.
14. Huttunen, M.; Peltola, J.; Kallio, S.; Karvonen, L.; Niemi, T.; Ylä-Outinen, V. Analysis of the processes in fluidized bed boiler furnaces during load changes. *Energy Procedia* **2017**, *120*, 580–587. [CrossRef]
15. Wu, Y.; Liu, D.; Ma, J.; Chen, X. Three-Dimensional Eulerian–Eulerian Simulation of Coal Combustion under Air Atmosphere in a Circulating Fluidized Bed Combustor. *Energy Fuels* **2017**, *31*, 7952–7966. [CrossRef]
16. Xie, J.; Zhong, W.; Shao, Y.; Liu, Q.; Liu, L.; Liu, G. Simulation of combustion of municipal solid waste and coal in an industrial-scale circulating fluidized bed boiler. *Energy Fuels* **2017**, *31*, 14248–14261. [CrossRef]
17. Farid, M.M.; Jeong, H.J.; Kim, K.H.; Lee, J.; Kim, D.; Hwang, J. Numerical investigation of particle transport hydrodynamics and coal combustion in an industrial-scale circulating fluidized bed combustor: Effects of coal feeder positions and coal feeding rates. *Fuel* **2017**, *192*, 187–200. [CrossRef]
18. Kettunen, A.; Hyppanen, T.; Kirkinen, A.-P.; Maikkola, E. Model-Based Analysis of Transient Behavior of Large Scale CFB Boilers. In Proceedings of the International Conference on Fluidized Bed Combustion, Jacksonville, FL, USA, 18–21 May 2003; pp. 523–531.
19. Selcuk, N.; Ozkan, M. Simulation of circulating fluidized bed combustors firing indigenous lignite. *Int. J. Therm. Sci.* **2011**, *50*, 1109–1115. [CrossRef]
20. Starkloff, R.; Alobaid, F.; Karner, K.; Epple, B.; Schmitz, M.; Boehm, F. Development and validation of a dynamic simulation model for a large coal-fired power plant. *Appl. Therm. Eng.* **2015**, *91*, 496–506. [CrossRef]
21. Lappalainen, J.; Tourunen, A.; Mikkonen, H.; Hänninen, M.; Kovács, J. Modelling and dynamic simulation of a supercritical, oxy combustion circulating fluidized bed power plant concept—Firing mode switching case. *Int. J. Greenh. Gas Control* **2014**, *28*, 11–24. [CrossRef]
22. Kaikko, J.; Mankonen, A.; Vakkilainen, E.; Sergeev, V. Core-annulus model development and simulation of a CFB boiler furnace. *Energy Procedia* **2017**, *120*, 572–579. [CrossRef]
23. Grace, J.R.; Knowlton, T.; Avidan, A. *Circulating Fluidized Beds*; Blackie Academic and Professional: London, UK, 1997.
24. APROS. *Apros 6 Feature Tutorial*; Apros: Seoul, Korea, 2018; pp. 1–520.
25. Lappalainen, J.; Lalam, V.; Charreire, R.; Ylijoki, J. Dynamic modelling of a CFB boiler including the solids, gas and water-steam systems. In Proceedings of the 12th International Conference on Fluidized Bed Technology (CFB), Krakow, Poland, 23–27 May 2017; pp. 321–328.

26. Lappalainen, J.T.; Hiidenkari, H.; Tuuri, S.; Ritvanen, J. Dynamic Core-Annulus Model for Circulating Fluidized Bed Boilers. In Proceedings of the Nordic Flame Days 2019, Turku, Finland, 28–29 August 2019.

27. Hänninen, M. *Phenomenological Extensions to APROS Six-Equation Model. Non-Condensable Gas, Supercritical Pressure, Improved CCFL and Reduced Numerical Diffusion for Scalar Transport Calculation*; VTT: Espoo, Finland, 2009.

28. Hänninen, M.; Ylijoki, J. *The One-Dimensional Separate Two-Phase Flow Model of APROS*; VTT: Espoo, Finland, 2008.

29. Starkloff, R.; Postler, R.; Al-Maliki, W.A.K.; Alobaid, F.; Epple, B. Investigation into Gas Dynamics in an Oxyfuel Coal Fired Boiler during Master Fuel Trip and Blackout. *J. Process Control* **2016**, *41*. [CrossRef]

30. Schuhbauer, C.; Angerer, M.; Spliethoff, H.; Kluger, F.; Tschaffon, H. Coupled simulation of a tangentially hard coal fired 700 °C boiler. *Fuel* **2014**, *122*, 149–163. [CrossRef]

31. Hentschel, J.; Zindler, H.; Spliethoff, H. Modelling and transient simulation of a supercritical coal-fired power plant: Dynamic response to extended secondary control power output. *Energy* **2017**, *137*, 927–940. [CrossRef]

32. Alobaid, F.; Al-Maliki, W.A.K.; Lanz, T.; Haaf, M.; Brachthäuser, A.; Epple, B.; Zorbach, I. Dynamic simulation of a municipal solid waste incinerator. *Energy* **2018**, *149*, 230–249. [CrossRef]

33. Mikkonen, H.; Lappalainen, J.; Kuivalainen, R. Modelling and dynamic studies of the second generation oxy-fired CFB boiler. In Proceedings of the 22nd International Conference on Fluidized Bed Conversion, Turku, Finland, 14–17 June 2015.

34. Al-Maliki, W.A.K.; Alobaid, F.; Kez, V.; Epple, B. Modelling and dynamic simulation of a parabolic trough power plant. *J. Proces Control* **2016**, *39*, 123–138. [CrossRef]

35. Al-Maliki, W.A.K.; Alobaid, F.; Starkloff, R.; Kez, V.; Epple, B. Investigation on the dynamic behaviour of a parabolic trough power plant during strongly cloudy days. *Appl. Therm. Eng.* **2016**, *99*, 114–132. [CrossRef]

36. Henrion, T.; Ponweiser, K.; Band, D.; Telgen, T. Dynamic simulation of a solar power plant steam generation system. *Simul. Model. Pract. Theory* **2013**, *33*, 2–17. [CrossRef]

37. Kurki, J. Simulation of thermal hydraulics at supercritical pressures with APROS. In Proceedings of the International Youth Nuclear Congress (IYNC 2008), Interlaken, Switzerland, 20–26 September 2008.

38. Arkoma, A.; Hänninen, M.; Rantamäki, K.; Kurki, J.; Hämäläinen, A. Statistical analysis of fuel failures in large break loss-of-coolant accident (LBLOCA) in EPR type nuclear power plant. *Nucl. Eng. Des.* **2015**, *285*, 1–14. [CrossRef]

39. Molerus, O.; Wirth, K.-E. *Heat Transfer in Fluidized Beds*; Chapman & Hall: London, UK, 1997.

40. Lyngfelt, A.; Leckner, B. Sulphur capture in circulating fluidized-bed boilers: Can the efficiency be predicted? *Chem. Eng. Sci.* **1999**, *54*, 5573–5584. [CrossRef]

41. Makarytchev, S.; Cen, K.; Luo, Z.; Li, X. High-temperature sulphur removal under fluidized bed combustion conditions—A chemical interpretation. *Chem. Eng. Sci.* **1995**, *50*, 1401–1407. [CrossRef]

42. Mattisson, T.; Lyngfelt, A. A sulphur capture model for circulating fluidized-bed boilers. *Chem. Eng. Sci.* **1998**, *53*, 1163–1173. [CrossRef]

43. Wang, C.; Jia, L.; Tan, Y.; Anthony, E. The effect of water on the sulphation of limestone. *Fuel* **2010**, *89*, 2628–2632. [CrossRef]

Article

A Detailed One-Dimensional Hydrodynamic and Kinetic Model for Sorption Enhanced Gasification

Marcel Beirow *, Ashak Mahmud Parvez, Max Schmid and Günter Scheffknecht

University of Stuttgart, Institute of Combustion and Power Plant Technology (IFK), Pfaffenwaldring 23, 70569 Stuttgart, Germany; ashak.parvez@ifk.uni-stuttgart.de (A.M.P.); max.schmid@ifk.uni-stuttgart.de (M.S.); guenter.scheffknecht@ifk.uni-stuttgart.de (G.S.)
* Correspondence: marcel.beirow@ifk.uni-stuttgart.de; Tel.: +49-0711-6856-8938

Received: 21 July 2020; Accepted: 27 August 2020; Published: 3 September 2020

Abstract: Increased installation of renewable electricity generators requires different technologies to compensate for the associated fast and high load gradients. In this work, sorption enhanced gasification (SEG) in a dual fluidized bed gasification system is considered as a promising and flexible technology for the tailored syngas production for use in chemical manufacturing or electricity generation. To study different operational strategies, as defined by gasification temperature or fuel input, a simulation model is developed. This model considers the hydrodynamics in a bubbling fluidized bed gasifier and the kinetics of gasification reactions and CO_2 capture. The CO_2 capture rate is defined by the number of carbonation/calcination cycles and the make-up of fresh limestone. A parametric study of the make-up flow rate (0.2, 6.6, and 15 kg/h) reveals its strong influence on the syngas composition, especially at low gasification temperatures (600–650 °C). Our results show good agreement with the experimental data of a 200 kW pilot plant, as demonstrated by deviations of syngas composition (5–34%), lower heating value (*LHV*) (5–7%), and *M* module (23–32%). Studying the fuel feeding rate (22–40 kg/h), an operational range with a good mixing of solids in the fluidized bed is identified. The achieved results are summarized in a reactor performance diagram, which gives the syngas power depending on the gasification temperature and the fuel feeding rate.

Keywords: one-dimensional SEG model; dual fluidized bed; sorbent deactivation; hydrodynamics; kinetics; fuel feeding rate; biomass

1. Introduction

Global greenhouse gas (GHG) emissions have been increasing exponentially for the past 60 years, mostly due to the use of oil, coal, and natural gas [1,2]. Hereby, CO_2 is the major GHG, accounting for 65% of the total amount. Replacing the usage of fossil fuels in combustion or gasification processes with biomass enables the reduction of CO_2 emissions.

Sorption enhanced gasification (SEG) has been considered a promising technology for tailored syngas production with in situ CO_2 capture. The plant configuration of a SEG process is based on the conventional steam gasification process carried out in indirect heated (or allothermal) dual fluidized bed systems. This conventional steam gasification process uses an inert bed material as a heat carrier to deliver the heat for the endothermic gasification, enabling N_2-free syngas with typical hydrogen content of 20–40 vol% (on a dry basis). The syngas of this process can be used for power and heat generation, and the technology has already been applied in a number of industrial-scale facilities with thermal power ranging from 2–20 MW [3–5]. Accordingly, a number of different models describing the steam gasification of biomass have been developed and published [6–8].

Gordillo and Belghit [9] developed a two-phase model for steam gasification of biomass char which considers hydrodynamic phenomena; however, pyrolysis was not included, which is an important step in the biomass gasification process. Agu et al. [10] proposed a detailed one-dimensional model for steam gasification in a bubbling fluidized bed, which considers a Lagrangian approach for solid fuel particles extending the often used assumption of a uniform distribution of fuel particles. In contrast to conventional steam gasification, the SEG uses a reactive bed material (e.g., limestone) as the heat carrier, which enables the in situ capture of CO_2 in the gasifier. This CO_2 capture affects gasification reactions and shifts hydrogen concentrations in the syngas up to 75 vol% (on a dry basis) [11]. Thus, SEG has been considered to be a suitable process for the synthesis of hydrogen, transport fuels, and chemicals [12,13]. Hereby, models are important in finding the best operation strategies for the gasifier, especially if the gasifier is part of a complex production path, including mass and energy integration. However, few models exist which have considered the CO_2 sorption characteristics in fluidized bed reactors. Inayat et al. [14] and Sreejith et al. [15] used a detailed approach based on reaction kinetics, but hydrodynamics were not in focus. In the work of Hejazi et al. [16,17], a model to describe sorption enhanced gasification in a bubbling fluidized bed was presented. This model considers hydrodynamics in the fluidized bed, but evaluations of process parameter over reactor height were not included due to the assumption of a uniform temperature throughout the dense bed. Another comprehensive SEG model has been published by Pitkäoja et al. [18]. This model considers a transport disengaging height to describe the amount of reacting particles in the freeboard, which is important for the overall syngas composition. However, minor focus is placed on the bed activity, which depends (in a dual fluidized system) on the mean particle residence time, as well as the carbonation/calcination cycles. Considering the issues discussed above, we developed a detailed SEG model focusing on various key aspects, including (i) hydrodynamic features of the fluidized bed reactor and evaluation of the process parameters over reactor height; (ii) kinetic aspects of the steam gasification process; and (iii) reaction kinetics of CO_2 capture through carbonation in the fluidized bed, as well as by elutriated particles in the freeboard. As deactivation of the sorbent is an important issue in a SEG process, a deactivation model is included in this work.

2. SEG Process Description and Model Development

2.1. Description of the SEG Process

Based on availability of experimental data, atmospheric conditions were considered for sorption enhanced gasification (SEG). In this work, calcined limestone (CaO) was used as reactive bed material to capture CO_2 from the syngas through the carbonation reaction ($CaO + CO_2 \rightarrow CaCO_3$). This shifts gasification and water–gas shift reactions towards hydrogen production. The carbonation reaction proceeds until the equilibrium condition is reached, where the equilibrium depends on the temperature and CO_2 partial pressure. Hence, the grade of CO_2 capturing can be set by different operating conditions. Adjustment of syngas composition by selecting gasification temperature is a great advantage of the SEG process. Poboss [19] reported that there is a relatively high CO_2 capture efficiency below 650 °C. For higher temperatures, the capture efficiency decreases rapidly. Based on the CO_2 volume concentration in the wet syngas, the CO_2 capture ceases at a temperature of 750–770 °C [19]. To a certain extent, operation above 750 °C and without CO_2 separation can also be achieved with a dual fluidized bed system by increasing the solids circulation rate. The concept of SEG, with its two fluidized bed reactors, is illustrated in Figure 1. The energy needed for the endothermic gasification reactions in gasifier is supplied by the highly exothermic CaO carbonation reaction and by the sensible heat of circulating solids flowing from the regenerator (combustion reactor). This means if the temperature in the gasifier exceeds the range for carbonation reaction, the process is only driven by the temperature difference between regenerator and gasifier. However, the temperature of the regenerator cannot be set arbitrarily high since otherwise sintering of CaO will increasingly occur, resulting in a loss of activity [20]. To avoid sintering significantly, the maximum regenerator temperature should be below 950 °C. Thus, in a limestone-based process, the maximum gasification temperature is fixed at about 850 °C; if gasification is to take place at higher temperatures, other processes must be considered. The temperatures in the regenerator of 850–950 °C are obtained by the combustion of unreacted char contained in the solids leaving the gasifier and flowing back to the regenerator. At these temperatures, the sorbent is regenerated through the endothermic calcination reaction ($CaCO_3 \rightarrow CaO + CO_2$). If required, extra fuel can be added into the regenerator.

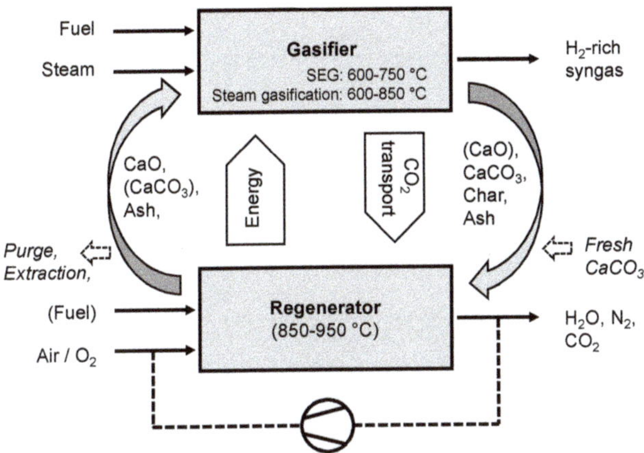

Figure 1. Schematic diagram of sorption enhanced gasification (SEG) process (up to 750 °C) and extended steam gasification mode (up to 850 °C); option of oxy-fuel operation.

The fluidized bed gasifier consists of a fluidized bed and a freeboard. Since main gasification reactions are located close proximity to solids, the gasification temperature is referred to as the average temperature in the fluidized bed. In this work, a gasifier temperature between 600–850 °C was investigated by means of simulations.

The syngas produced from the SEG process is N_2-free and lean in CO_2 and, therefore, has a high calorific value. Experimental results on SEG have shown that, at low gasification temperatures (around 600 °C), the CO_2 absorption is high and a H_2 fraction on the order of 70–75 vol% (on a dry basis) can be reliably reached [11,21,22]. There is also an additional advantage associated with the use of CaO as a bed material: CaO-based bed materials are known to be catalytically active towards tar cracking. Therefore, despite the lower gasification temperatures used in the SEG process, it has been experimentally demonstrated that tar production can be up to 5 times lower than that in classical (steam) fluidized bed gasification processes without CaO [23,24]. Apart from combustion with air, the regenerator can also be operated with pure oxygen and recirculated flue gas, known as the Oxy-SEG process. In this case, the flue gas contains no nitrogen but mostly consists of CO_2, which can be stored or utilized [25].

2.2. Development of a Dual Fluidized Bed Gasifier Model

In Figure 2, the SEG modeling concept considered in this work is illustrated. The proposed model is based on details of the 200 kW dual fluidized bed (DFB) facility at the IFK. It consists of a bubbling fluidized bed (BFB) gasification reactor and a circulating fluidized bed (CFB) regenerator reactor. The gasifier model was discretized along the reactor length, in order to calculate all relevant transport values and for model validation purposes, to compare with the underlying experimental data. The regenerator is modeled in a simplified way, considering mass and energy balances as well as combustion and calcination reactions, in order to enable the coupling with the gasifier through the circulating solid.

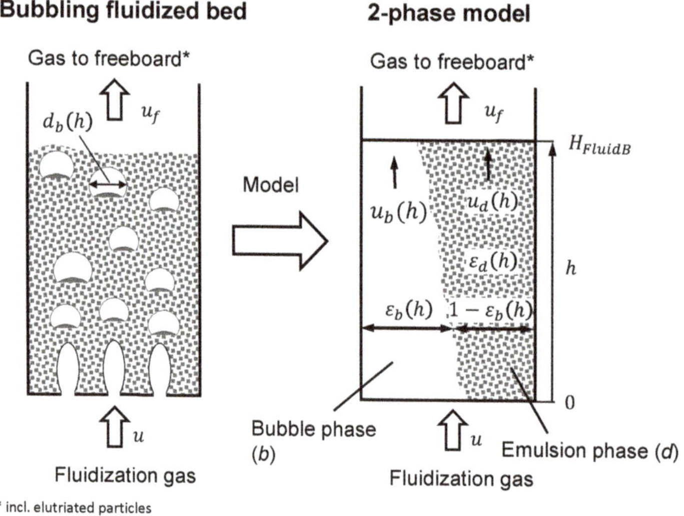

Figure 2. Modeling concept of SEG in a dual fluidized bed system.

2.3. Hydrodynamics

For modeling purposes, the fluidized bed was split into two phases, a solid-free bubble phase (fraction: ε_b) and a solid-loaded dense phase (fraction: $1\text{-}\varepsilon_b$), as shown in Figure 3.

Figure 3. Simplified reactor scheme of the modeled process.

2.3.1. Dense Phase

In the model, the dense phase is considered to be a perfused pack with porosity ε_d and gas velocity u_d. The calculation of u_d and ε_d is performed following the methods presented by Hilligardt [26]. The minimum fluidization gas velocity, u_{mf}, is determined using the Ergun equation (Equation (1)), which is formulated based on the definitions of the Archimedes and Reynolds numbers, as well as the

Sauter diameter (Equations (2)–(4)). In the present work, a sphericity of $\psi = 0.75$ [27] and a porosity at minimum fluidization $\varepsilon_{mf} = 0.45$ [27] are assumed.

$$Ar = 150\frac{1-\varepsilon_{mf}}{\psi^2\cdot\varepsilon_{mf}^3}Re_{mf} + \frac{1.75}{\psi\cdot\varepsilon_{mf}^3}Re_{mf}^2 \tag{1}$$

$$Ar = \frac{g\cdot d_{sv}^3}{v_g^2}\frac{\rho_p - \rho_g}{\rho_g} \tag{2}$$

$$Re_{mf} = \frac{d_{sv}\cdot u_{mf}}{v_g} \tag{3}$$

$$d_{sv} = \sqrt{\varphi\cdot\overline{d_p}} \tag{4}$$

Analysis of the fluidized bed inventory showed a mean particle size, $\overline{d_p}$, of 350 μm [11]. According to Hilligardt [26], the real velocity in the dense phase is higher than the calculated minimum fluidization gas velocity, u_{mf}, and it can be estimated with the following empirical equation:

$$u_d|_{h=0} = u_{mf} + \frac{1}{4}\left(u_{empty}|_{h=0} - u_{mf}\right) \tag{5}$$

For the remaining reactor heights $u_d(h)$ is determined by the continuity equation.
The porosity ε_d is calculated, as proposed by Richardson and Zaki [28], as

$$\varepsilon_d(h) = \varepsilon_{mf}\left(\frac{u_d(h)}{u_{mf}}\right)^{\frac{1}{n_{Rz}}}. \tag{6}$$

For the parameter n_{Rz}, Richardson and Zaki [28] provided the empirical equation:

$$n_{Rz} = \begin{cases} 4.65 & \text{if } Re_s \leq 0.2 \\ 4.4\cdot Re_s^{-0.03} & \text{if } 0.2 \leq Re_s \leq 1 \\ 4.4\cdot Re_s^{-0.1} & \text{if } 1 \leq Re_s \leq 500 \\ 2.4 & \text{if } Re_s > 500 \end{cases} \tag{7}$$

where Re_s is the Reynolds number from the rate of descent of a single particle. In this work, the parameter n_{Rz} is used to adjust the calculated bed height to the experimental values. With $n_{Rz} = 5.5$, the model could be fitted to the real bed height determined experimentally.

2.3.2. Bubble Phase

Werther [29] developed a model (Equation (8)) to determine the bubble diameter, d_b, depending on the height of the fluidized bed, taking into account the coalescence and separation of the bubbles:

$$\frac{d(d_b)}{dh} = \left(\frac{2\varepsilon_b}{9\pi}\right)^{1/3} - \frac{d_b}{3\cdot280\frac{u_{mf}}{g}\cdot u_b^*(h)} \tag{8}$$

At the position $h = 0$, the initial bubble diameter is calculated using the correlation $d_b = 1.3\left(\dot{V}_{steam}^2/g\right)^{0.2}$, according to Tepper [27] and Davidson [30]. The initial bubble diameter depends on the volume flow of steam, $\dot{V}_{steam}$, through an orifice of a gas distributor. The ascent velocity, u_b^*, of a bubble can be determined by Equations (9)–(11) from Hilligardt [26] and Tepper [27]:

$$u_b^* = \sqrt{\frac{4gd_b}{3C_{D,b}}} \tag{9}$$

$$C_{D,b} = \frac{16}{Re_b} + 2.64 \tag{10}$$

$$Re_b = \frac{d_b \cdot u_b^*}{v_d} \text{ with } v_d = 60 \cdot d_{sv}^{3/2} \cdot g^{1/2}. \tag{11}$$

The parameter $C_{D,b}$ is the drag coefficient of a single bubble and v_d is the viscosity of the suspension phase. However, the ascent velocity u_b^* does not equal the gas velocity in the bubble phase u_b, as the bubbles are additionally perfused by gas streams coming from the suspension phase [26,27]. Following the method proposed by Hilligardt [26], the gas velocity u_b can be determined using the following empirical correlation:

$$u_b = u_b^* + 2.7u_d \tag{12}$$

2.3.3. Fluidized Bed Height

With a defined inventory of the fluidized bed, M_{FluidB}, the height of the fluidized bed, H_{FluidB}, can be calculated by integrating the solid mass along the axial co-ordinate h [27]:

$$M_{FluidB} = \int_0^{H_{FluidB}} (1 - \varepsilon_b)(1 - \varepsilon_d)\rho_p A_c dh \tag{13}$$

2.3.4. Elutriation Rate

When gas bubbles rise to the surface of the fluidized bed and break, solid particles are thrown into the freeboard and entrained by the upward gas volume flow [31]. While the major fraction of these particles fall back into the fluidized bed, small particles whose terminal velocity is lower than the gas velocity are elutriated from the freeboard [31].

In the 200 kW facility, a cyclone is installed at the exit of the freeboard to reduce the extraction of bed inventory by leading the particles back into the fluidized bed [11]. It is assumed that there is an additional CO_2 capture effect in the freeboard: Due to lower freeboard temperatures, the position of the chemical equilibrium can be changed, which enables a further carbonation reaction. However, as bigger particles fall or are transferred back into the fluidized bed by the cyclone, it is assumed that only fine particles have a contribution to the additional carbonation reaction. According to [31], the elutriation rate of particles (in g/s) is described by

$$\dot{M}_{elut} = k_{elut} x_{fine} \tag{14}$$

Herein, k_{elut} is the elutriation rate constant and the weight fraction of fine particles, x_{fine}, present in the bed was identified by measurements at the 200 kW facility. With a secondary cyclone, a mean particle size $\overline{d_p}$ of 25 μm of elutriated particles was found. Considering the particle size distribution of the raw limestone, it can be derived that fine particles do not originate from the make-up of the raw material. Thus, the source of these particles must be attrition or fragmentation effects. From experiments, a correlation was derived to calculate the weight fraction of the fine particles:

$$x_{fine} = a_1 \tanh((u - a_2)/a_3) + a_4, \tag{15}$$

with the parameters $a_1 = -0.12697$, $a_2 = 0.71214$, $a_3 = -0.01191$, and $a_4 = 0.12807$.

2.4. Gasifier Dimensions

In Figure 4a, a schematic diagram of the bubbling fluidized bed gasifier facility is shown, including the inlet/outlet gas and solid flows, as well as details of their axial position (in mm). These data were used to parametrize the simulation model. Here, $\dot{M}_{RegOut}$ is the mass flow at the outlet of the regenerator to the gasifier. By tuning this mass flow rate, the desired gasification temperature can be achieved. The height of the fluidized bed, H_{FluidB}, depends on the fluidization velocity and, hence, the distance between the inlet of $\dot{M}_{RegOut}$ and H_{FluidB} is variable.

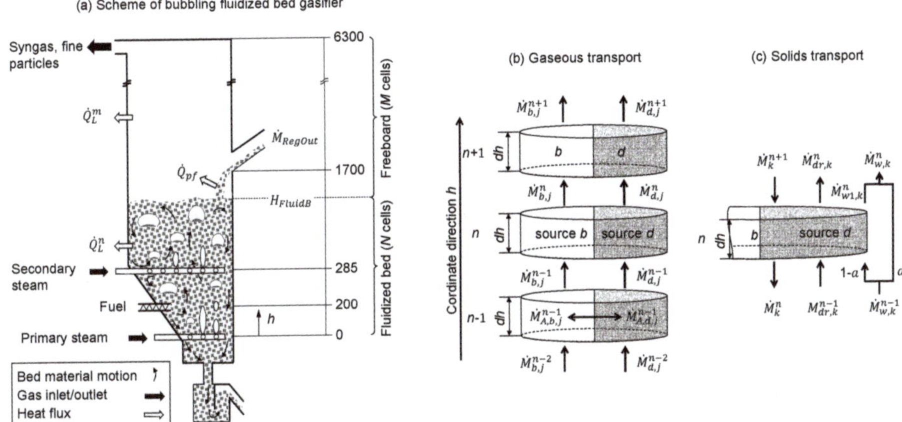

Figure 4. (a) Scheme of bubbling fluidized bed gasifier with levels (in mm) of inlet/outlet flows; Cell model of the fluidized bed, (b) Gaseous flows, and (c) Solid flows.

2.5. Mass Balance

The gasifier is discretized along the reactor height. This includes the bubbling bed as well as the freeboard area. Figure 4b,c show how the calculation with axial discretization in cells proceeds through the fluidized bed for both gaseous and solid components. In each calculation cell of this 1d model, solid and gas components are considered to be fully mixed.

2.5.1. Fluidized Bed

According to the discretization shown in Figure 4b, the mass balance for each gaseous component in both dense phase ($P = d$) and bubble phase ($P = b$) is stated in Equation (16):

$$\frac{dM_{P,j}^n}{dt} = 0 = \dot{M}_{P,j}^{n-1} - \dot{M}_{P,j}^n + \dot{M}_{A,P,j}^n + MW_j \cdot \sum_{I_P} v_{i,j} R_{P,i}^n + \dot{M}_{in,P,j}^n \tag{16}$$

where $\dot{M}_{P,j}$ is the convective gas mass flow and $\dot{M}_{A,P,j}$ is the exchange mass between bubble and suspension phases of component j in phase P; R_i is the reaction rate of reaction i and $v_{i,j}$ is the stoichiometric coefficient of component j in reaction i; I_P describes the maximum number of reactions taking place in the phase and, with the term $\dot{M}_{in,P,j}$, external inflows (e.g., from a secondary steam inlet) can be considered. Convective mass flows between adjacent calculation cells are defined by [27]

$$\dot{M}_{d,j}^n = \rho_d^n u_d^n x_{d,j}^n \varepsilon_d^n A_c^n \tag{17}$$

$$\dot{M}_{b,j}^n = \rho_b^n u_b^n x_{b,j}^n \left(1 - \varepsilon_d^n\right) A_c^n \tag{18}$$

The exchange between bubble and suspension phases inside the cell n is defined by [27]

$$\dot{M}_{A,d,j}^n = K_{db}^n A_{db}^n \left(\rho_b^n x_{b,j}^n - \rho_d^n x_{d,j}^n\right) \tag{19}$$

$$\dot{M}^n_{A,b,j} = K^n_{db} A^n_{db} \left(\rho^n_d x^n_{d,j} - \rho^n_b x^n_{b,j} \right) \tag{20}$$

According to Hilligardt [26], the mass transfer coefficient between the bubble and suspension phase is calculated as $K^n_{db} = \frac{2.7 u^n_d}{4}$, and the mass exchange area over all bubbles in the cell n is $A^n_{db} = 6\varepsilon^n_b \cdot A_c \cdot dh / d^n_b$. Additionally, overall mass balances for the suspension ($P = d$) and bubble ($P = b$) phases are developed in Equation (21), including the molar weight MW_j of each component j:

$$\frac{dM^n_P}{dt} = 0 = \dot{M}^{n-1}_P - \dot{M}^n_P + \sum_J \dot{M}^n_{A,P,j} + \sum_J \sum_{I_P} MW_j v_{i,j} R_i + \dot{M}^n_{in,P}. \tag{21}$$

Beside a balance for the gaseous components, a separate balance equation (Equation (22)) exists for the solids. According to Figure 4c, the amount of each solid component k in a calculation cell n is considered:

$$\frac{dM^n_k}{dt} = 0 = \dot{M}^{n-1}_{dr,k} + (1-\alpha)\dot{M}^{n-1}_{w,k} + \dot{M}^{n+1}_k - \dot{M}^n_{dr,k} - \dot{M}^n_{w1,k} - \dot{M}^n_k + MW_j \cdot \sum_{I_d} v_{i,k} R^n_{d,k} + \dot{M}^n_{in,k} \tag{22}$$

The cell adjacent to the freeboard ($n = N$) additionally includes a sink term for elutriated fine particles $\dot{M}_{elut}$. It is important to split the mass balance into a description of gaseous and solid components as, in real plant operations, there exists a downward flow from the surface of the fluid bed (H_{FluidB}) to the bottom (leaving the gasifier through the loop seal) and an upward flow due to the wake and drift of each rising gas bubble. Here, the term wake ($\dot{M}_w$) describes solids that fasten to the bubbles on their way upwards and drift ($\dot{M}_{dr}$) refers to solids that are loosely drawn upwards through the bubble movement. The solid transport by bubbles, together with the conical asymmetric cross section, see Figure 4a, in the lower part of the fluidized bed, leads to the strong exchange and motion of solids. To consider these effects, a model was developed in which the drift part of a rising bubble is fully mixed in each cell, whereas the proportion of wake that is mixed with or bypasses each cell can be chosen using the parameter α. The Term $\left(MW_j \cdot \sum_{I_d} v_{i,k} R^n_{d,k} \right)$ considers chemical reactions and with the term ($\dot{M}^n_{in,k}$), inflows (e.g., from solid circulation) are included in the equation. The amount of solids which are transported with each bubble can be described by the empirical Equations (23) and (24) mentioned in [27]:

$$\dot{M}_w = A_c \rho_P (1 - \varepsilon_d) \varepsilon_b u^*_b \cdot [0.59 - 0.046 \ln(Ar)] \tag{23}$$

$$\dot{M}_{dr} = A_c \rho_P (1 - \varepsilon_d) \varepsilon_b \cdot 0.38 u^*_b \cdot [1.5 - 0.135 \ln(Ar)] \tag{24}$$

In Figure 4c, the concept is illustrated to adapt the 1d model for gasification experiments with this highly three-dimensional behavior of the real system, through the adjustable parameter α.

2.5.2. Freeboard

The freeboard consists of a gas phase with a small amount of very fine particles with almost the same velocity. The gas component balance is

$$\frac{dM^m_{f,j}}{dt} = 0 = \dot{M}^{m-1}_{f,j} - \dot{M}^m_{f,j} + MW_j \cdot \sum_{I_f} v_{i,j} R^m_{f,i} \tag{25}$$

and the solid component balance is

$$\frac{dM_k^m}{dt} = 0 = \dot{M}_k^{m-1} - \dot{M}_k^m + MW_k \cdot \sum_{I_k} v_{i,k} R_{k,i}^m \tag{26}$$

Therefore, the overall mass balance expression is written as

$$\frac{dM^m}{dt} = 0 = \dot{M}_f^{m-1} + \dot{M}_k^{m-1} - \dot{M}_f^m - \dot{M}_k^m. \tag{27}$$

The linkage of the freeboard with the fluidized bed is carried out by applying Equation (28) for gases and Equation (29) for solids:

$$\dot{M}_f^{m=0} = \dot{M}_d^{n=N} + \dot{M}_b^{n=N} \tag{28}$$

$$\dot{M}_k^{m=0} = \dot{M}_{elut} \tag{29}$$

2.6. Energy Balances

2.6.1. Fluidized Bed

For each cell, a thermally fully developed mixture with a constant temperature is assumed. Therein, temperatures of solids and gases are equal; however, the temperature of the wake can differ, applying a vertical heat transfer between the cells.

Enthalpy fluxes for solids and gases are generally defined by mass flows: $\dot{H} = \dot{M} \cdot h$. At the boundary of a cell, the enthalpy h of a mass flow of mixtures (e.g., gas inlet/outlet) is calculated as

$$h = \sum_J x_j \cdot h_j(T). \tag{30}$$

For solid mixtures, Equation (30) is written with index k. The temperature dependency of the enthalpy of a gas component j or a solid component k is calculated using polynomials from the software package FactSage®. In this approach, the enthalpies of the chemical reactions need not be considered additionally. The energy balance for a certain cell n is defined as

$$\frac{dH^n}{dt} = 0 = \dot{H}_d^{n-1} - \dot{H}_d^n + \dot{H}_b^{n-1} - \dot{H}_b^n + \dot{H}_k^{n-1} - \dot{H}_k^n + \dot{H}_{dr}^{n-1} - \dot{H}_{dr}^n + (1-\alpha)\dot{H}_{w,k}^n - \dot{H}_{w1,k}^n + \dot{H}_{in}^n - k_{FluidB}\pi d_r dh(T^n - T_c) \tag{31}$$

The last term of Equation (31) describes the heat loss through the reactor wall $\dot{Q}_L^n$. It is calculated by using the temperature of the cooling jacket $T_c = 40\,°C$ and, by adapting to experimental pilot plant data, a heat transfer coefficient of $k_{FluidB} = 12.9\ \mathrm{Wm^{-2}K^{-1}}$ was found.

2.6.2. Freeboard

The enthalpy balance for the freeboard section of the reactor can be expressed by

$$\frac{dH_f^m}{dt} = 0 = \dot{H}_f^{m-1} - \dot{H}_f^m - k_f \pi d_r dh(T^m - T_c) + k_p \frac{6}{d_p \rho_p} \dot{M}_p^m (T_p^m - T^m) \cdot v_p, \tag{32}$$

where the term with the heat transfer coefficient, k_f, describes the heat loss $\dot{Q}_{L,f}^m$ through the reactor wall in the freeboard section. A value of $3.4\ \mathrm{Wm^{-2}K^{-1}}$ was selected by fitting the simulated temperature of the experimental temperature profile. According to the reactor design (see Figure 4a), hot solids from the regenerator ($\dot{M}_{RegOut}$) flow into the gasifier. However, the level of the inflow is located in the freeboard and above the surface of the fluidized bed (i.e., at position H_{FluidB}). Hence, the particles pass the lower region of the freeboard before they dip in the fluidized bed. In this region, heat transfer

$\dot{Q}_{p,f}^{m}$ between solid particles (p) and the gas phase (f) of the freeboard occurs. For this, a heat transfer coefficient, k_p, with a value of 160.7 Wm^{-2}K^{-1} was determined, which is well-aligned with values reported in the literature [32,33]. The temperature, T_p^n, of the bed material in the region between the inlet and fluidized bed is calculated by solving the energy balance:

$$\frac{dH_p^n}{dt} = 0 = \dot{M}_{RegOut}c_{p,p}^{n+1}T_p^{n+1} - \dot{M}_p c_{p,p}^n T_p^n - k_{pf}\frac{6}{d_p \rho_p}\dot{M}_p\left(T_p^n - T^n\right)\cdot v_p \tag{33}$$

In this case, $T_p^{m=m_{in}} = T_{RegOut}$ and v_p describes the velocity of a falling particle.

2.7. Chemical Reactions

Table 1 lists all the chemical reactions considered in the model, including equations to calculate the reaction rates. The pyrolysis step (reaction 1) is modelled with a one-step reaction kinetic considering the products: Ash, char, H_2O, gases (CO_2, CO, CH_4, and H_2), non-condensable hydrocarbons (simplified as C_2H_4), and tars (simplified as Naphthalene: $C_{10}H_8$). For the mass fractions ω_j, experimental data from Fagbemi et al. [34] were used and interpolated to consider a temperature-dependent pyrolysis product composition. The values for the amounts of tar, however, originate from experiments with the 200 kW DFB system [19] and, hence, secondary pyrolysis reaction modelling was not required. With this assumption, catalytic effects of CaO on tar conversion are indirectly considered by measured concentrations. Residual char was handled as a mixture of C, H, and O and, according to the char analysis from Fagbemi et al. [34], the composition was also interpolated for different gasification temperatures. It is worth noting that the used elemental analysis of wood pellets (C: 48.99 wt.-%$_{waf}$, H: 6.97 wt.-%$_{waf}$, O: 44.04 wt.-%$_{waf}$) differed from the analysis of biomass in Fagbemi et al. [34]. Thus, yields from pyrolysis ω_j needed to be adapted, in order to satisfy the elemental balance. Therefore, a linear equation system for the elements C, H, and O had to be solved. While coefficients from $C_{10}H_8$, CH_4, C_2H_4, H_2O, and char were fixed, the coefficients of CO, CO_2, and H_2 were fitted to close the elemental balance of wood pellets.

Results for the mass fractions ω_j are listed in Table A1. The reaction kinetics of ethene reformation (reaction 6) were adapted to fit the simulated H_2 and CO concentrations to experimental values from the 200 kW DFB pilot plant. For the carbonation reaction, sorbent deactivation, which is dependent on the number of calcination–carbonation cycles, was taken into account through the parameter X_{ave}. Details of the calculation of X_{ave} are described in the subsequent section. For all gasification reactions, it is assumed that they occur only in the emulsion phase, due to the catalytic behavior of CaO and char, which enhance reaction rates compared to those in the gas phases [42].

Table 1. List of chemical reactions and reaction rates.

Reaction	Chemical Reaction and Kinetics	Reference
1. Pyrolysis	$BM_{wf} \rightarrow \omega_{Ash}\,Ash + \omega_{Char}\,Char + \omega_{H2O}\,H_2O + \omega_{CO2}\,CO_2 + \omega_{CO}\,CO + \omega_{CH4}\,CH_4 + \omega_{H2}\,H_2 + \omega_{C2H4}\,C_2H_4 + \omega_{C10H8}\,C_{10}H_8$ $r = 1.516{\cdot}10^3 s^{-1}{\cdot} \exp(-6043\ K/T){\cdot}c_{BM_{wf}}^{mass}$ in $kgs^{-1}m^{-3}$	[35]
2. Heterogeneous water–gas	$C + H_2O \rightarrow CO + H_2$ $r = 1.23{\cdot}10^7 bar^{-0.75} s^{-1}{\cdot} \exp(-23815\ K/T){\cdot}c_{H2O}^{0.75}{\cdot}c_{Char}$ in mol $s^{-1}\ m^{-3}$	[36]
3. Boudouard	$C + CO_2 \rightarrow 2CO$ $r = \dfrac{k_1 p_{CO2}}{1 + \frac{k_1}{k_3}p_{CO2} + \frac{k_2}{k_3}p_{CO}} c_{Char}$ in mol $s^{-1}m^{-3}$ $k_1 = 1.2{\cdot}10^{11} bar^{-1}s^{-1}{\cdot}\exp(-19245\ K/T)$ $k_2 = 5.9{\cdot}10^{08} bar^{-1}s^{-1}{\cdot}\exp(-20447\ K/T)$ $k_3 = 2.2{\cdot}10^{10} bar^{-1}s^{-1}{\cdot}\exp(-33678\ K/T)$	[37]
4. Carbonation	$CO_2 + CaO \rightarrow CaCO_3$ $r = 0.26 s^{-1}{\cdot}X_{ave}c_{CaO}\dfrac{p_{CO2}-p_{CO2,eq}}{p}$ $p_{CO2,eq} = 4.192{\cdot}10^7 bar{\cdot}\exp(-20474\ K/T)$ in mol $s^{-1}m^{-3}$	[38] [39]
5. Water–gas shift	$CO + H_2O \longleftrightarrow CO_2 + H_2$ $r = 2.78{\cdot}\exp(-1513\ K/T){\cdot}\left(c_{CO}c_{H2O} - \dfrac{c_{CO2}c_{H2}}{0.0265{\cdot}\exp(-3966\ K/T)}\right)$ in mol $s^{-1}m^{-3}$	[40]
6. Ethene reformation	$C_2H_4 + 2H_2O \rightarrow 2CO + 4H_2$ $r = 230{\cdot}\exp(-3789\ K/T){\cdot}c_{C2H4}$ in mol $s^{-1}m^{-3}$	[41] [1]

[1] Kinetics adapted according to experimental data.

2.8. Sorbent Deactivation

If limestone is subjected to several calcination–carbonation cycles, its CO_2 sorption capacity is reduced, due to sintering phenomena on the particle surface [43]. For limestone, Grasa and Abanades [44] described the decay of the CO_2 carrying capacity X_N as

$$X_N = \frac{1}{\frac{1}{1-X_r} + kN} + X_r. \tag{34}$$

Equation (34) depends on the number of calcination–carbonation cycles N and uses the empirical constants $k = 0.52$ and $X_r = 0.075$. However, in a fluidized bed system, particles have different residence times, which leads to a distribution of the average CO_2 carrying capacity. According to the references [45–47], an average carrying capacity X_{ave} is calculated by a population balance:

$$X_{ave} = \sum_{N=1}^{\infty} \frac{F_0}{F_R}\left(1 - \frac{F_0}{F_R}\right)^{N-1} \cdot X_N \tag{35}$$

In the current study, Equation (35) was integrated into the model to calculate the average CO_2 carrying capacity of the particle system with regard to the carbonation reaction (Table 1, Reaction 4). Therein, F_R is the molar-based flow of $\dot{M}_{RegOut}$, describing the circulation flow of CaO from the regenerator into the gasifier. To compensate mass losses, mostly due to attrition, the reactor inventory was maintained by an input flow of raw limestone. In Equation (35), this input flow F_0 was considered on a molar basis. As, in practice, the measured material flow of fresh limestone (by dosing units) contains particles which are small enough to be directly discharged, especially when feeding into the regenerator with fluidization velocities up to 5.5 m/s [11], it was assumed that F_0 only represents the effective material flow that remains longer in the system. Fresh limestone and purge material are not considered in the mass and energy balance calculations, as their flow rates are very low compared to the circulating CaO mass flow. If this should be taken into account, detailed information is needed regarding the size of the particles and their degree of calcination when they are directly discharged from the regenerator due to their hydrodynamic properties.

2.9. Simulation Algorithm

In the SEG system, internal solid circulation (mass flow from the Regenerator $\dot{M}_{RegOut}$), the biomass feed stream, and the steam input are important parameters defining the gasification temperature. Therefore, in the simulation model, the molar calcium looping ratio (F_{CaO}/F_C) is used to set the gasification temperature. In this ratio, F_C considers the molar flow of carbon contained in the biomass. The corresponding flow chart of the model is shown in Figure 5.

After setting boundary conditions and model parameters, the average CO_2 carrying capacity is calculated using the make-up flow of fresh limestone and the looping ratio. Starting from an initial temperature, first the pyrolysis products and then the fluid dynamics of the gasifier are calculated. After solving the mass and energy balances, a new gasification temperature is calculated, and the pyrolysis step is updated for a certain looping ratio. This is iterated until the change of the empty reactor velocity, u_{empty}, is smaller than a defined value δ; then, the results of the operation point are saved in a file.

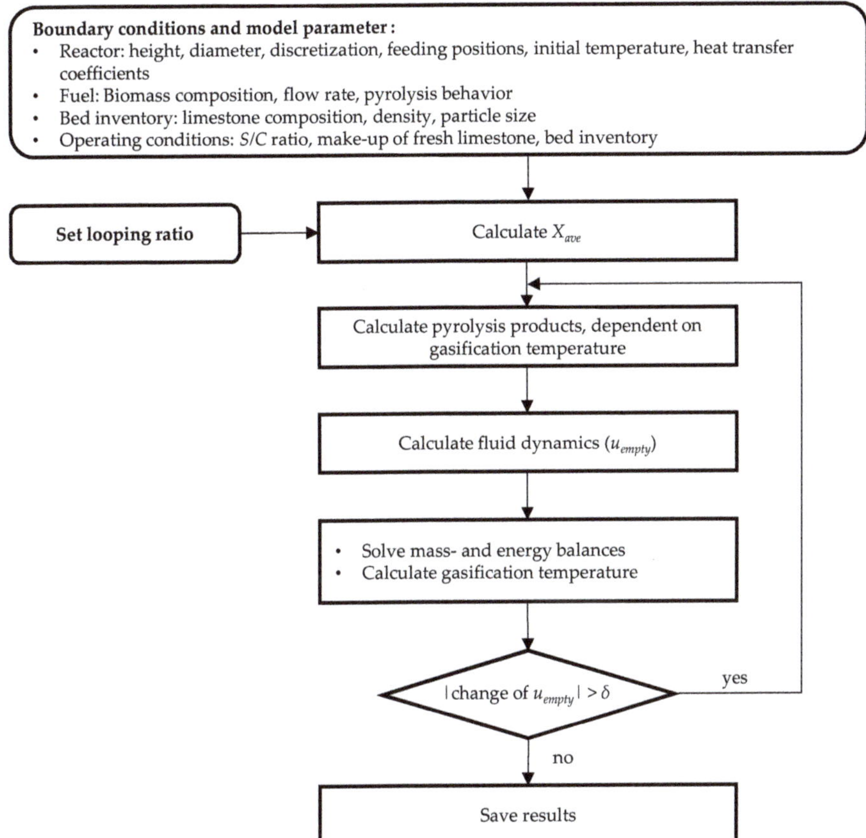

Figure 5. Simulation flow chart for SEG fluidized bed gasifier model.

3. Results and Discussion

In Section 3.1, the model parameters (e.g., heat transfer coefficients) are verified against experimental temperature data measured in the 200 kW fluidized bed gasifier [11]. Section 3.2 contains a model validation with measured gas compositions (H_2, CO, CO_2, CH_4, C_2H_4), the lower heating value (*LHV*) of the syngas, and the *M* module from the 200 kW fluidized bed gasifier [11,19] in a temperature range of 600–850 °C. According to [48,49], the accuracy of the parameter verification and model validation was quantified by the sum squared deviation method:

$$\text{MeanError}: \overline{E} = \sqrt{\frac{\sum_{n=1}^{N}\left(\frac{\varphi_{exp}-\varphi_{model}}{\varphi_{exp}}\right)^2}{N}} \tag{36}$$

For parameter verification, a low mean error of the temperature distribution along the reactor height of 10.6% was found. The model validation was carried out with a limestone make-up flow rate (*MU*) of 6.6 kg/h, according to experiments [11]. To characterize the effect of limestone make-up flow rate, a parametric study with 0.2 kg/h, 6.6 kg/h, and 15 kg/h is also included in the result diagrams. Mean errors to quantify the model prediction accuracy are listed in Table A2.

3.1. Verification of Model Parameter

In Figure 6, simulated temperature profiles from variations of the parameter α are compared with temperatures measured on different positions in the fluidized bed and the freeboard (available in [11]).

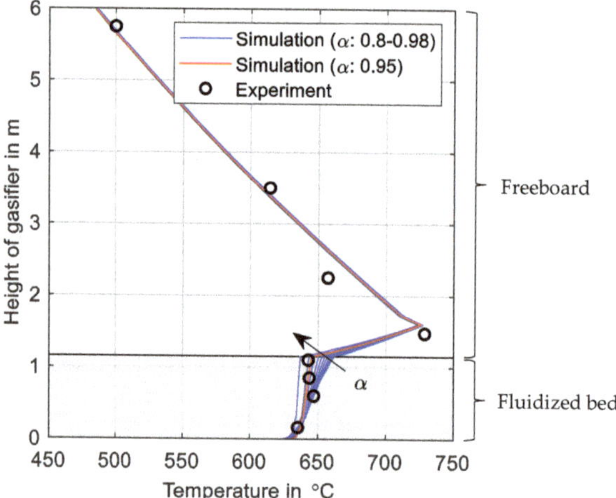

Figure 6. Simulated temperature profiles of the gasifier for various α (ranging from 0.8 to 0.98); $\dot{M}_{BM,wf}$ = 29.7 kg/h, S/C = 2.2 mol/mol, *WHSV* (weight hourly space velocity) = 0.68 1/h, limestone make-up 6.6 kg/h, and comparison with experimental data from 200 kW dual fluidized bed (DFB) pilot plant [11].

The profile of the measured temperatures (circles in Figure 6) over the height of the gasifier can be explained as follows: due to a good mixing of solids in the fluidized bed (0 m to 1.15 m) the temperatures were close to 640 °C, followed by an inflection, which was caused by the inlet of hot solids at 1.7 m (available thermocouple was at 1.5 m). In the freeboard above the solid inlet, a decrease of the gas temperature was observed due to heat losses through the reactor wall. A variation of the model parameter α in the range of 0.8–0.98 was carried out to adjust the fluidized bed temperature by changing the proportion of wake mixed within each discretization cell. From Figure 6, it can be seen that, with a value of α = 0.95 (red line), a temperature profile could be achieved, which corresponds to the measured temperature values. An almost vertical profile confirmed a homogeneous particle mixing in the fluidized bed. In the model, a heat transfer coefficient of k_{FluidB} = 12.9 Wm^{-2}K^{-1} was considered for the fluidized bed area. For the particle–gas heat transfer, a heat transfer coefficient value k_p = 160.7 Wm^{-2}K^{-1} and for the freeboard k_f = 3.4 Wm^{-2}K^{-1} were determined to describe the given temperature profile.

3.2. Validation with Experimental Data

After setting the model parameters, we carried out a verification based on temperature along the gasifier height, the syngas composition, *LHV*, reactions rates, and the *M* module, in order to compare the simulation results with experimental data over a temperature range from 600 °C to 850 °C.

3.2.1. Effect of Gasification Temperature on Fractions of Syngas Components

In Figure 7a–e, the simulated volume fractions of synthesis gas components (H_2, CO, CO_2, and CH_4) and non-condensable hydrocarbons (in the form of C_2H_4) are plotted over a gasification temperature between 600–850 °C. For each synthesis gas component, the curves of three different make-up flows (0.2 kg/h, 6.6 kg/h, and 15 kg/h) are shown to demonstrate its influence on the gas volume fraction. Additionally, experimental results derived from both 200 kW (Experiment 1) and 20 kW (Experiment 2) DFB systems are included to evaluate the simulated volume fractions.

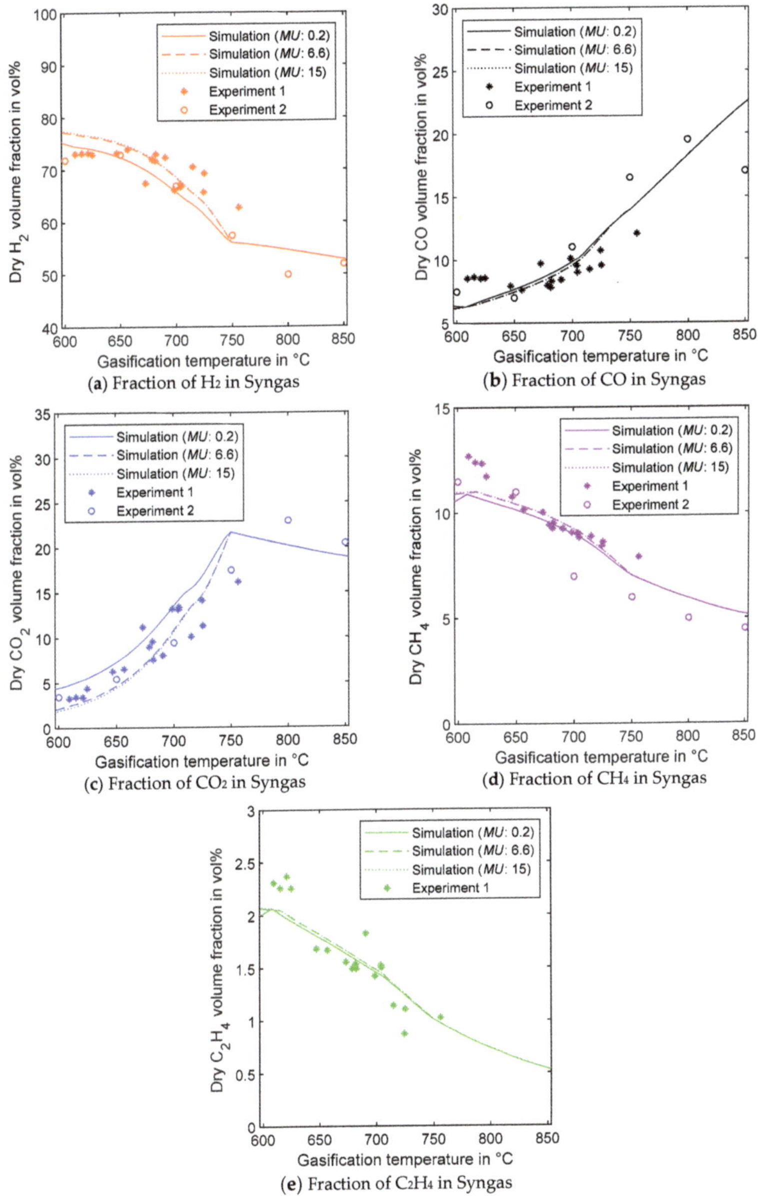

Figure 7. (**a–e**) Simulated syngas components: (**a**) H_2, (**b**) CO, (**c**) CO_2, (**d**) CH_4, and (**e**) C_2H_4 for gasification temperatures in the range of 600–850 °C (lines for limestone make-up of 0.2, 6.6, and 15 kg/h); comparison with experimental results from 200 kW (Experiment 1) and 20 kW (Experiment 2) DFB systems (data points).

For sorption enhanced gasification, one important characteristic is the strong dependency of the gas composition on the gasification temperature, which is affected by CO_2 capture through the carbonation reaction [19,22]. Beside gasification reactions, the pyrolysis step also has an important impact on the initial gas composition in the fluidized bed [34]. The results from a 20 kW system were additionally included in the present work, as it enables operation temperatures up to 850 °C due to its electrical heating system. For low gasification temperatures, there is a larger distance between the actual CO_2 concentration and the equilibrium curve for the carbonation/calcination regime [37,38]. This leads to a strong capture of CO_2 and, hence, low CO_2 concentrations in the syngas. With higher temperatures, the CO_2 capture rate decreases and, consequently, the CO_2 concentration in the syngas rises. Furthermore, this influences the water–gas shift reaction, resulting in decreased H_2 concentrations and increased CO concentrations. When the CO_2 concentration reaches the equilibrium concentration at around 750 °C, an inflection in the concentrations of CO_2 and H_2 can be observed. This demonstrates the strong coupling of the carbonation reaction with the water–gas shift reaction. Particularly for low gasification temperatures, there is also a distinctive influence of the make-up flow. Presumably, the reason for this effect is a reduced circulation mass flow of fresh CaO from the regenerator with a simultaneously higher CO_2 capture rate due to lower temperatures. For instance, at a gasification temperature of 600 °C, the delivered circulation mass flow is around ten times lower, compared to that when operating at 850 °C. This can lead to a higher content of carbonated particles if the bed inventory is hardly exchanged. In this operating range, an increase of the limestone make-up rate can increase the activity in the bed and, thus, the CO_2 capture rate.

3.2.2. Effect of Gasification Temperature on LHV

As seen in Section 3.2.1, the gas composition is strongly affected by gasification temperature, due to the temperature dependence of (i) products released from pyrolysis, (ii) Arrhenius approaches to describe the gasification reactions, and (iii) carbonation/calcination equilibrium. Based on the lower heating values (*LHV*) of pure syngas components, the *LHV* of the gas mixture was calculated and compared with experimental data. In Figure 8, simulation results are shown over a temperature range of 600–850 °C and for different make-up flow rates.

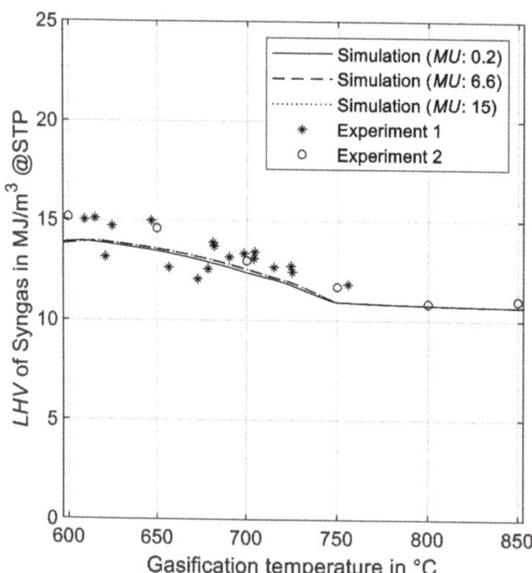

Figure 8. Simulated lower heating value (*LHV*) of syngas for gasification temperatures between 600 °C and 850 °C (lines for limestone make-up of 0.2, 6.6, and 15 kg/h); comparison with experimental results from a 200 kW (Experiment 1) and 20 kW (Experiment 2) DFB system.

The results show a good correlation with the experimental data and only a slight variation for different make-up flow rates is noticed. Furthermore, the effect of CO_2 capture and its limitation at around 750 °C, recognizable as an inflection, can be described with this model. For higher temperatures, the *LHV* remains the same at a value of 10.9 MJ/m^3 @STP. The highest values of *LHV*—around 14.5 MJ/m^3 @ STP—can be reached at temperatures lower than 650 °C.

3.2.3. Effect of Make-Up Flow on Reaction Rate

In Figure 9, reaction rates (in mol m^{-3} s^{-1}) are shown over a temperature range from 600 °C to 850 °C. Additionally, a variation of the make-up flows (0.2 kg/h, 6.6 kg/h, and 15 kg/h) was included to investigate the influence of sorbent deactivation on all considered reaction rates for the gasification process.

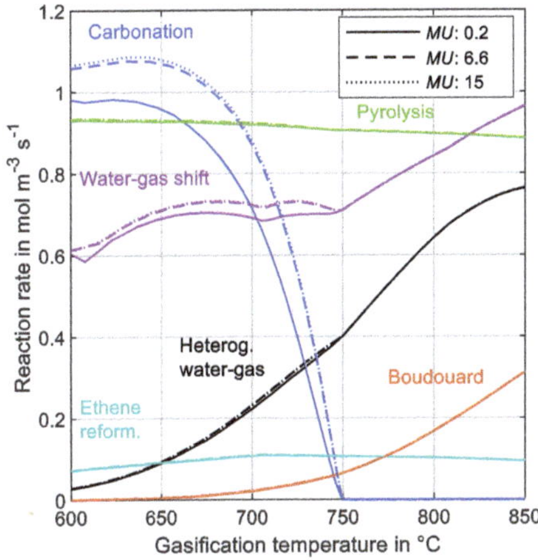

Figure 9. Reaction rates over temperature; influence of effective make-up of fresh limestone (lines for 0.2, 6.6, and 15 kg/h) to bed activity.

As the gasification temperature varies with changes of the circulation mass flow, the sorbent residence time differs and, according to Equation (35), the sorbent activity is also affected. It can be seen that deactivation mostly influenced the carbonation (reaction 4) and water–gas shift (reaction 5) reactions. For instance, for a constant temperature and a constant circulation mass flow, the carbonation reaction rate is higher with larger amounts of fresh limestone.

Considering the influence of the gasification temperature on the reaction rates, it can be seen that the reaction rates of the water–gas shift (reaction 5), the heterogeneous water–gas (reaction 2), and the Boudouard (reaction 3) reactions increased with higher temperatures. In contrast, for the pyrolysis (reaction 1) reaction, a minor decrease can be observed, even though an increase should be expected with higher temperatures. The reason for this behavior can be explained as follows: The amount of biomass in the fluidized bed system is limited by a constant fuel input flow. When the gasification temperature is increased, the reaction rate increases but, at the same time, a higher solid circulation is required for the higher temperature. Hence, more unreacted biomass is extracted from the gasifier through the loop seal, thus reducing the gas yield. Kinetic parameters of the Arrhenius approach influence the gradient of reactions for increasing temperatures. Comparing the Arrhenius parameters listed in Table 1, it can be seen that the influence of temperature on the heterogeneous water–gas shift reaction (reaction 2) was higher than that of the pyrolysis reaction (reaction 1).

3.2.4. Effect of Gasification Temperature and Make-Up of Fresh Limestone on M Module

The M module from Equation (37) relates the gas concentrations of hydrogen, carbon dioxide, and carbon monoxide and has been considered as an important parameter which dictates the application of syngas [50]. For validity of an ideal gas, it can be written with volume fractions.

$$M = \frac{y_{H_2} - y_{CO_2}}{y_{CO_2} + y_{CO}} \tag{37}$$

For instance, a M module of two is required for full stoichiometric conversion into dimethyl ether [50] and, for methane synthesis, a value of three can be derived from methanation reactions. Higher values are mostly interesting for hydrogen production. Figure 10 shows the influence of the gasification temperature on the M module. Additionally, three different simulation results, with a limestone make-up of 0.2 kg/h, 6.6 kg/h, and 15 kg/h, are compared with experimental data. It can be seen that a higher make-up flow enables higher M modules under the same gasification temperature.

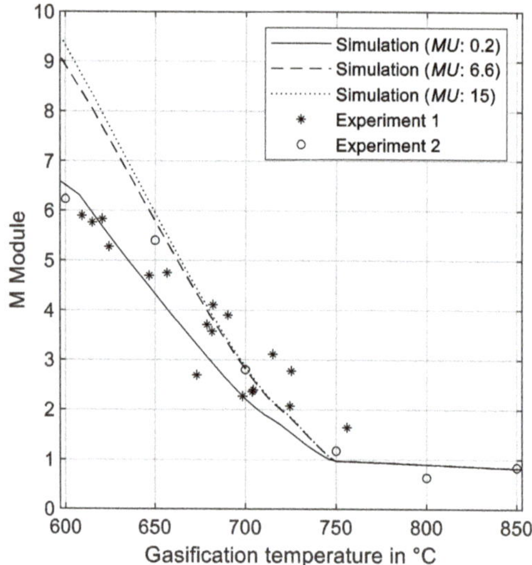

Figure 10. Influence of gasification temperature and make-up mass flow (lines for 0.2, 6.6, and 15 kg/h) on M module.

For a low gasification temperature, the experimental data can be reached with a make-up flow of 0.2 kg/h. At a gasification temperature above 650 °C, the simulation model with a make-up flow of 6.6 kg/h precisely describes the experimental data. This parametric study reveals that the make-up flow is an important factor when optimizing the gasification process for a certain application. When considered from an economic point of view, operation strategies with low make-up rates are preferable.

4. Sensitivity Analysis of Fuel Feeding Rate

For a realistic and flexible operation of the biomass gasification process, it is important that the process allows for safe operations under a wide load range. In this section, the effects of different fuel feeding rates on the bed height, the fluidization level (gas velocity ratio: superficial gas velocity u based on minimum fluidization velocity u_{mf}), and the power of the syngas, depending on the gasification temperature, are investigated.

4.1. Effect of Biomass Feeding Rate on Bed Height and Gas Velocity Ratio

The model was also used to study the influence of different biomass feeding rates on the hydrodynamics in the fluidized bed. In Figure 11, the gas velocity ratio (superficial gas velocity/minimum fluidization gas velocity) is shown over a height of 3 m of the gasifier focusing on the fluidized bed and lower part of the freeboard. In this figure, the biomass feeding rate is considered as curve parameter in the range of 22–40 kg/h, whereby the operating point considered in Section 3.1 (feeding rate: 29.7 kg/h) was additionally drawn as a red line. Since the fluidized bed expands with larger fluidization volume flows, the diagram also shows the height of the fluidized bed for the different biomass feeding rates as blue dots. This allows the gas velocity at the surface of the fluidized bed to be read directly from the diagram.

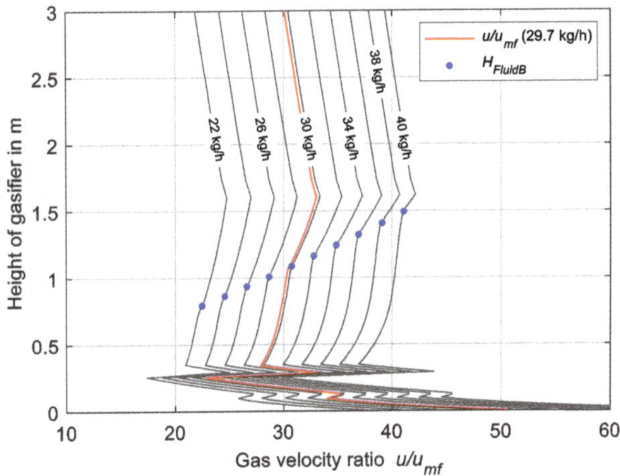

Figure 11. Simulated effect of biomass feeding rate (22–40 kg/h) on gas velocity ratio (superficial gas velocity/minimum fluidization gas velocity) over reactor height and position of bed height (H_{FluidB}).

It can be seen that with a higher biomass feeding rate, the gas velocity ratio increased at any position in the reactor. The reason is that when biomass particles pyrolyze in a fluidized bed, the released gas contributes to the reactor fluidization. In addition, a constant S/C ratio of 2.2 was selected for the simulations, in order to maintain a stable syngas quality [51] and to ensure comparability with the experimental data [11]. This changes the amount of steam supplied and, hence, the gas velocity ratio. In Figure 11, the red line corresponds to the same case which was considered for model verification in Section 3.1, with the reactor temperature in the axial direction. At the zero position in y-axis, the primary steam inlet is located. Due to the smallest cross-section in this area (compare Figure 4a), the highest gas velocity ratios were found here. On higher levels of the gasifier, the diameter of the reactor increases, which leads to a decrease of the gas velocity ratio due to the continuity equation. However, increasing velocities can be observed resulting from the gas release due to biomass pyrolysis ($h = 0.2$ m) and the secondary steam inlet ($h = 0.285$ m). At $h \geq 0.35$ m, the reactor has a cylindrical shape and from the constant cross section in combination with further biomass pyrolysis, the gas velocity ratio slightly increases. In the freeboard above the inlet of solids, the temperatures decrease due to heat losses, and hence, the gas velocity ratio decreases. This is indicated by an inflection in the gas velocity ratio curves at a height of 1.7 m. Looking at the fluidized bed height for operation with 22 kg/h and 40 kg/h biomass feeding rate, one can see that the height doubles. For the case with 40 kg/h, the bed height (blue dots) almost reaches the area of the inlet for the solid circulation at 1.7 m. In order to ensure stable operation with this reactor geometry, operating modes that lead to a further increase in bed height should be avoided.

4.2. Performance Diagram for Gasifier Operation

Based on the results derived from this work, a performance diagram of the bubbling fluidized bed gasifier was created in Figure 12. This relates the selected gasification temperature (based on downstream requirements of the syngas composition) and fuel feeding rate to the power of the syngas. Additionally, the gas velocity ratio from the superficial gas velocity at the fluidized bed surface and the gas velocity ratio for the lowest velocity in the fluidized bed are depicted.

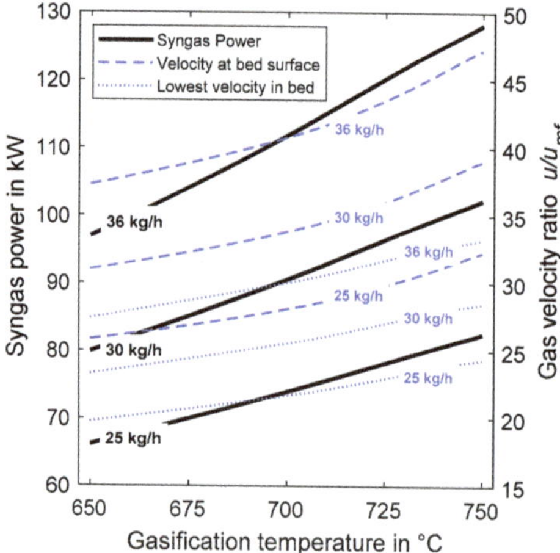

Figure 12. Power of syngas for gasification temperatures between 650 and 750 °C for water-free fuel input of 25 kg/h, 30 kg/h, and 36 kg/h (left side); Corresponding gas velocity ratio (related to u_{mf}) at the top of the fluidized bed and for the position where the lowest velocity in the fluidized bed occurs (right side).

By combining the gasification chemistry (which lead to the syngas power) and the hydrodynamic information (represented as the gas velocity ratio), it is possible to identify a realistic operational range of the gasifier.

With a gasification temperature of 650 °C and 25 kg/h fuel input, the lowest gas velocity ratio of 20 could be identified. This means that, in any position of the fluidized bed, the superficial gas velocity was 20 times higher than the minimum fluidization velocity. Hence, good mixing of the bed inventory can be assured. At the bed surface, the superficial gas velocity was 26 times higher than the minimum gas velocity and only a little particle extraction can be expected. Increasing the fuel feeding rate to 36 kg/h at 650 °C, the syngas power increased and reached almost 100 kW. At this point, the velocity at the surface of the fluidized bed increased by a factor of 37 (related to u_{mf}) and by a factor of 27 (related to u_{mf}) for the lowest velocity in the bed. By increasing the gasification temperature and the gasifier feeding rate up to 750 °C and 36 kg/h, respectively, good mixing in the fluidized bed is guaranteed. However, due to the high velocity at the bed surface (a factor of 47 related to u_{mf}), a high particle extraction has to be accounted for. To avoid this negative effect, the fuel input or the gasification temperature can be modified. In that case, the syngas composition is not very important for downstream applications, an identical syngas power (around 100 kW) can be reached for a gasification temperature of 750 °C and 30 kg/h fuel input or for an operation with 650 °C and with a fuel input of 36 kg/h. However, the operational point at 650 °C reduced particle extraction from the gasifier due to the non-linear behavior of the gas velocity ratio curves.

5. Conclusions

The one-dimensional sorption enhanced gasification model developed in this study was verified with experimental data obtained from a 200 kW facility at IFK, University of Stuttgart. The results showed that the model is able to successfully predict the performance of the pilot plant at different operation conditions. With this model, the influence of important process parameters, such as gasification temperature, steam-to-carbon ratio, solid inventory, and fuel mass flow, can be simulated. On the basis of gas composition (H_2, CO, CO_2, CH_4, and C_2H_4), *LHV*, and the *M* module, the model was validated over the whole SEG temperature range. As the activity of the limestone sorbent decreases after several carbonation/calcination cycles, an additional model was integrated, which adapts the carbonation reaction kinetics depending on the circulation rate and molar flow of fresh limestone. With this possibility, three different fresh limestone make-up flow rates (0.2 kg/h, 6.6 kg/h, and 15 kg/h) were simulated. A parametric study revealed a larger dependence on the limestone make-up, especially for gasification temperatures below 650 °C. This effect probably has to do with the lower circulation rate between gasifier and regenerator and the reduced transfer of fresh CaO into the gasifier with different CO_2 capture activities. Increasing the make-up flow rate also increases the bed activity for the same quantity of mass transferred into the gasifier. At higher temperatures, it can be assumed that this effect is reduced by limitations of the carbonation reaction equilibrium. Considering the reaction rates in the temperature range between 600 and 750 °C, a strong dependency of the limestone make-up on the carbonation reaction can be identified. Furthermore, the water–gas shift reaction is influenced due to the CO_2 capture. For the other reactions considered, only a minor influence from the limestone make-up was observed.

Variation of the fuel feeding rate (22 kg/h to 40 kg/h) with a constant *S/C* ratio (2.2 mol/mol) revealed an increase of the bed height by a factor of 2. From the gas velocity ratio (u/u_{mf}) along the reactor height, different fluidization states can be recognized. While a low fuel input led to low mixing in certain areas of the fluidized bed (u was only higher than u_{mf} by a factor of 20), high fuel input increased mixing in the whole fluidized bed. However, the entrainment of particles was also higher. Based on these evaluations, the syngas power and the gas velocities of the bubbling fluidized gasifier were described in a performance diagram which is dependent on the gasification temperature and the fuel input. Therefore, the developed model can be used as a fast and reliable engineering tool for reactor design or scale-up purposes.

Author Contributions: M.B. is responsible for conceptualization, simulation studies, validation, and original draft preparation. M.B. and M.S. designed the methodology of the software. Writing—review and editing, A.M.P.; supervision, G.S. All authors have read and agreed to the published version of the manuscript.

Funding: This research was funded by Ministry of the Environment Baden-Württemberg, Germany, grant number BWE13008.

Acknowledgments: The authors gratefully acknowledge financial supports from the project BioenergieFlex (BWE13008) funded by the Ministry of the Environment Baden-Württemberg, Germany.

Conflicts of Interest: The authors declare no conflict of interest.

Nomenclature

A	area (m^2)
Ar	Archimedes number (-)
$C_{D,b}$	drag coefficient of a bubble (-)
c_j^{mass}	concentration of component j (kg/m^3)
c_j	concentration of component j (mol/m^3)
c_p	specific heat capacity ($J\,kg^{-1}K^{-1}$)
d_{sv}	Sauter diameter (m)
d	diameter (m)
$\overline{E}$	mean Error (-)
F_0	flow of fresh $CaCO_3$ (mol/s)

F_R	flow of circulating CaO (mol/s)
F_C	flow of elemental carbon in fuel flow (mol/s)
g	acceleration of gravity (m/s^2)
H	height (m)
$\dot{H}$	enthalpy flow (J/s)
h	enthalpy (J/kg)
h	reactor co-ordinate (m)
I	maximum number of chemical reactions (-)
J	maximum number of components (-)
K_{db}	mass transfer coefficient (m/s)
k	heat transfer coefficient (W m^{-2}K^{-1})
k	empirical constant for sorbent deactivation (-)
k_{elut}	elutriation rate constant (here in g/s)
LHV	lower heating value (MJ/m^3)
M	mass (kg)
M	syngas module (mol/mol)
M	maximum number of cells in freeboard (-)
$\dot{M}$	mass flow (kg/h, g/s)
MU	make-up flow of fresh limestone (kg/h)
MW	molar weight (kg/kmol)
m	discretization cell in freeboard (-)
N	number of calcination-carbonation cycles (-)
N	maximum number of cells in fluidized bed (-)
n	discretization cell in fluidized bed (-)
n_{RZ}	parameter for descent rate of single particle in suspension phase
p_j	partial pressure of component j (-)
$\dot{Q}$	heat flux (J/s)
Re	Reynolds number (-)
R	reaction rate (mol/s, kg/s)
r	reaction rate (mol s^{-1}m^{-3}, kg s^{-1}m^{-3})
S/C	steam-to-carbon ratio (mol H$_2$O/mol carbon)
T	Temperature (K)
t	time (s)
u	superficial gas velocity (m/s)
u_{mf}	minimum fluidization gas velocity (m/s)
u_b^*	ascent velocity of bubble (m/s)
$\dot{V}$	volume flow (m^3/h)
v_p	particle falling velocity (m/s)
$WHSV$	weight hourly space velocity (1/h)
X_{ave}	average CO$_2$ carrying capacity (mol CaCO$_3$/mol Ca)
X_N	CO$_2$ carrying capacity after N cycles (mol CaCO$_3$/mol Ca)
X_r	empirical constant for sorbent deactivation (-)
x	mass fraction (-)
y	volume fraction (-)
α	fraction of mass exchange in calculation cell (-)
δ	precision of numerical calculation
ε	porosity (-)
ν	kinematic viscosity (m^2/s)
ν	stoichiometric coefficient (-)
ρ	density (kg/m^3)
ψ	sphericity of particles (-)
ω_j	mass fraction of pyrolysis product j (-)

BM	biomass
b	bubble phase
c	cross section in reactor
d	emulsion phase
dr	drift
eq	equilibrium
elut	elutriation
empty	condition in empty reactor tube
f	freeboard
g	gas
i	index of chemical reaction
in	inlet flow via boundary condition
j	index of gas component
k	index of solid component
mf	condition at minimum fluidization
p	particle
r	reactor
w	wake
wf	water-free

Appendix A

Table A1. Mass fractions (ω_j) of products from biomass pyrolysis in the temperature range of 600–800 °C; values from Fagbemi et al. [34] adapted for pyrolysis of wood pellets.

Component	600 °C	650 °C	700 °C	750 °C	800 °C	850 °C
Ash	0.0032	0.0032	0.0032	0.0032	0.0032	0.0032
Char	0.2473	0.2483	0.2451	0.2435	0.2412	0.2324
H_2O	0.1654	0.1508	0.1330	0.1312	0.1291	0.1243
CO_2	0.2475	0.2433	0.2807	0.2514	0.2293	0.1890
CO	0.1706	0.2017	0.1851	0.2246	0.2556	0.3145
CH_4	0.0523	0.0581	0.0632	0.0631	0.0629	0.0619
H_2	0.0251	0.0269	0.0284	0.0296	0.0308	0.0326
C_2H_4	0.0399	0.0457	0.0509	0.0482	0.0453	0.0407
$C_{10}H_8$	0.0488	0.0221	0.0104	0.0052	0.0027	0.0014

Table A2. Comparison of experimental and simulated data with method of mean error.

Simulated Values	Mean Error in %, *MU*: 0.2 kg/h	Mean Error in %, *MU*: 6.6 kg/h	Mean Error in %, *MU*: 15 kg/h
H_2—Composition vs. Temperature (°C)	5.5	4.9	5.0
CO—Composition vs. Temperature (°C)	15.7	15.3	15.3
CO_2—Composition vs. Temperature (°C)	33.9	22.8	25.1
CH_4—Composition vs. Temperature (°C)	7.2	6.6	6.5
C_2H_4—Composition vs. Temperature (°C)	14.1	14.3	14.2
LHV vs. Temperature (°C)	6.7	5.9	5.9
M module vs. Temperature (°C)	23.4	29.9	32.2

Partially higher mean errors can also be caused by large fluctuations in the measured values, as no filtering was performed.

References

1. Statistics, I.E.A. *CO₂ Emissions from Fuel Combustion: Highlights*; International Energy Agency: Paris, France, 2017.
2. Olivier, J.G.J.; Schure, K.M.; Peters, J.A.H.W. *Trends in Global CO₂ and Total Greenhouse Gas Emissions*; PBL Netherlands Environmental Assessment Agency: The Hague, The Netherlands, 2017.
3. Göransson, K.; Söderlind, U.; He, J.; Zhang, W. Review of syngas production via biomass DFBGs. *Renew. Sustain. Energy Rev.* **2011**, *15*, 482–492. [CrossRef]

4. Rauch, R.; Hofbauer, H. Wirbelschicht-Wasserdampf-Vergasung in der Anlage Güssing (A); Betriebserfahrungen aus zwei Jahren Demonstrationsbetrieb. In Proceedings of the 9th International Symposium "Energetische Nutzung nachwachsender Rohstoffe", Freiberg, Germany, 4–5 September 2003; FAU: Nuremberg, Germany, 2003.

5. Thunman, H.; Larsson, A.; Hedenskog, M. Commissioning of the GoBiGas 20 MW Biomethane Plant. Proceedings of Tcbiomass 2015, Chicago, IL, USA, 2–5 November 2015; GTI: Des Plaines, IL, USA, 2015.

6. Fiaschi, D.; Michelini, M. A two-phase one-dimensional biomass gasification kinetics model. *Biomass Bioenergy* **2001**, *21*, 121–132. [CrossRef]

7. Lü, P.; Kong, X.; Wu, C.; Yuan, Z.; Ma, L.; Chang, J. Modeling and simulation of biomass air-steam gasification in a fluidized bed. *Front. Chem. Eng. China* **2008**, *2*, 209–213. [CrossRef]

8. Puig-Arnavat, M.; Bruno, J.C.; Coronas, A. Review and analysis of biomass gasification models. *Renew. Sustain. Energy Rev.* **2010**, *14*, 2841–2851. [CrossRef]

9. Gordillo, E.D.; Belghit, A. A two phase model of high temperature steam-only gasification of biomass char in bubbling fluidized bed reactors using nuclear heat. *Int. J. Hydrog. Energy* **2011**, *36*, 374–381. [CrossRef]

10. Agu, C.E.; Pfeifer, C.; Eikeland, M.; Tokheim, L.-A.; Moldestad, B.M.E. Detailed one-dimensional model for steam-biomass gasification in a bubbling fluidized bed. *Energy Fuels* **2019**, *33*, 7385–7397. [CrossRef]

11. Hawthorne, C.; Poboss, N.; Dieter, H.; Gredinger, A.; Zieba, M.; Scheffknecht, G. Operation and results of a 200-kWth dual fluidized bed pilot plant gasifier with adsorption-enhanced reforming. *Biomass Convers. Biorefinery* **2012**, *2*, 217–227. [CrossRef]

12. Schweitzer, D.; Albrecht, F.G.; Schmid, M.; Beirow, M.; Spörl, R.; Dietrich, R.-U.; Seitz, A. Process simulation and techno-economic assessment of SER steam gasification for hydrogen production. *Int. J. Hydrog. Energy* **2018**, *43*, 569–579. [CrossRef]

13. Rauch, R.; Hrbek, J.; Hofbauer, H. Biomass gasification for synthesis gas production and applications of the syngas. *WIREs Energy Environ.* **2014**, *3*, 343–362. [CrossRef]

14. Inayat, A.; Ahmad, M.M.; Yusup, S.; Mutalib, M.I.A. Biomass Steam Gasification with In-Situ CO_2 Capture for Enriched Hydrogen Gas. Production: A Reaction Kinetics Modelling Approach. *Energies* **2010**, *3*, 1472–1484. [CrossRef]

15. Sreejith, C.C.; Muraleedharan, C.; Arun, P. Air–steam gasification of biomass in fluidized bed with CO_2 absorption: A kinetic model for performance prediction. *Fuel Process. Technol.* **2015**, *130*, 197–207. [CrossRef]

16. Hejazi, B.; Grace, J.R.; Bi, X.; Mahecha-Botero, A. Kinetic model of steam gasification of biomass in a bubbling fluidized bed reactor. *Energy Fuels* **2017**, *31*, 1702–1711. [CrossRef]

17. Hejazi, B.; Grace, J.R.; Mahecha-Botero, A. Kinetic modeling of lime-enhanced biomass steam gasification in a dual fluidized bed reactor. *Ind. Eng. Chem. Res.* **2019**, *58*, 12953–12963. [CrossRef]

18. Pitkäoja, A.; Ritvanen, J.; Hafner, S.; Hyppänen, T.; Scheffknecht, G. Simulation of a sorbent enhanced gasification pilot reactor and validation of reactor model. *Energy Convers. Manag.* **2020**, *204*, 112318. [CrossRef]

19. Poboß, N. Experimentelle Untersuchung der Sorptionsunterstützten Reformierung. Ph.D. Thesis, Universität Stuttgart, Stuttgart, Germany, 18 May 2016.

20. Wang, C.; Zhou, X.; Jia, L.; Tan, Y. Sintering of limestone in calcination/carbonation cycles. *Ind. Eng. Chem. Res.* **2014**, *53*, 16235–16244. [CrossRef]

21. Pfeifer, C.; Puchner, B.; Hofbauer, H. Comparison of dual fluidized bed steam gasification of biomass with and without selective transport of CO_2. *Chem. Eng. Sci.* **2009**, *64*, 5073–5083. [CrossRef]

22. Koppatz, S.; Pfeifer, C.; Rauch, R.; Hofbauer, H.; Marquard-Moellenstedt, T.; Specht, M. H_2 rich product gas by steam gasification of biomass with in situ CO_2 absorption in a dual fluidized bed system of 8 MW fuel input. *Fuel Process. Technol.* **2009**, *90*, 914–921. [CrossRef]

23. Schmid, M.; Beirow, M.; Schweitzer, D.; Waizmann, G.; Spörl, R.; Scheffknecht, G. Product gas composition for steam-oxygen fluidized bed gasification of dried sewage sludge, straw pellets and wood pellets and the influence of limestone as bed material. *Biomass Bioenergy* **2018**, *117*, 71–77. [CrossRef]

24. Soukup, G.; Pfeifer, C.; Kreuzeder, A.; Hofbauer, H. In situ CO_2 capture in a dual fluidized bed biomass steam gasifier—Bed material and fuel variation. *Chem. Eng. Technol.* **2009**, *32*, 348–354. [CrossRef]

25. Schweitzer, D.; Beirow, M.; Gredinger, A.; Armbrust, N.; Waizmann, G.; Dieter, H.; Scheffknecht, G. Pilot-scale demonstration of Oxy-SER steam gasification: Production of syngas with pre-combustion CO_2 capture. *Energy Procedia* **2016**, *86*, 56–68. [CrossRef]

26. Hilligardt, K. Zur Strömungsmechanik von Grobkornwirbelschichten. Ph.D. Thesis, Technische Universität Hamburg, Hamburg, Germany, 1986.

27. Tepper, H. Zur Vergasung von Rest-und Abfallholz in Wirbelschichtreaktoren für Dezentrale Energieversorgungsanlagen. Ph.D. Thesis, Universität Magdeburg, Magdeburg, Germany, 2005.

28. Richardson, J.F.; Zaki, W.N. Sedimentation and fluidisation: Part I. *Chem. Eng. Res. Des.* **1997**, *75*, S82–S100. [CrossRef]

29. Werther, J. Experimentelle Untersuchungen zur Hydrodynamik von Gas./Festoff-Wirbelschichten. Ph.D. Thesis, Technische Fakultät der Universität Erlangen-Nürnberg, Erlangen, Germany, 1972.

30. Davidson, J.F. Bubble formation at an orifice in an inviscid liquid. *Trans. Inst. Chem. Eng.* **1960**, *38*, 335–342.

31. Wen, C.Y.; Chen, L.H. Fluidized bed freeboard phenomena: Entrainment and elutriation. *AIChE J.* **1982**, *28*, 117–128. [CrossRef]

32. Dietz, S. Wärmeübergang in Blasenbildenden Wirbelschichten. Ph.D. Thesis, Technische Fakultät der Universität Erlangen-Nürnberg, Erlangen, Germany, 1994.

33. Zhang, D.; Koksal, M. Heat transfer in a pulsed bubbling fluidized bed. *Powder Technol.* **2006**, *168*, 21–31. [CrossRef]

34. Fagbemi, L.; Khezami, L.; Capart, R. Pyrolysis products from different biomasses: Application to the thermal cracking of tar. *Appl. Energy* **2001**, *69*, 293–306. [CrossRef]

35. Roberts, A.F.; Clough, G. Thermal decomposition of wood in an inert atmosphere. In Proceedings of the Ninth Symposium (International) on Combustion, New York, NY, USA, 27 August–1 September 1963.

36. Hemati, M.; Laguerie, C. Determination of the kinetics of the wood sawdust steam gasification of charcoal in a thermobalance. *Entropie* **1988**, *142*, 29–40.

37. Kramb, J.; Konttinen, J.; Gómez-Barea, A.; Moilanen, A.; Umeki, K. Modeling biomass char gasification kinetics for improving prediction of carbon conversion in a fluidized bed gasifier. *Fuel* **2014**, *132*, 107–115. [CrossRef]

38. Charitos, A.; Rodríguez, N.; Hawthorne, C.; Alonso, M.; Zieba, M.; Arias, B.; Kopanakis, G.; Scheffknecht, G.; Abanades, J.C. Experimental Validation of the calcium looping CO_2 capture process with two circulating fluidized bed carbonator reactors. *Ind. Eng. Chem. Res.* **2011**, *50*, 9685–9695. [CrossRef]

39. Stanmore, B.R.; Gilot, P. Review—calcination and carbonation of limestone during thermal cycling for CO_2 sequestration. *Fuel Process. Technol.* **2005**, *86*, 1707–1743. [CrossRef]

40. Di Blasi, C. Modeling wood gasification in a countercurrent fixed-bed reactor. *AIChE J.* **2004**, *50*, 2306–2319. [CrossRef]

41. Mostafavi, E.; Pauls, J.H.; Lim, C.J.; Mahinpey, N. Simulation of high-temperature steam-only gasification of woody biomass with dry-sorption CO_2 capture. *Can. J. Chem. Eng.* **2016**, *94*, 1648–1656. [CrossRef]

42. Dong, J.; Nzihou, A.; Chi, Y.; Weiss-Hortala, E.; Ni, M.; Lyczko, N.; Tang, Y.; Ducousso, M. Hydrogen-rich gas production from steam gasification of bio-char in the presence of CaO. *Waste Biomass Valorization* **2017**, *8*, 2735–2746. [CrossRef]

43. Alvarez, D.; Abanades, J.C. Pore-size and shape effects on the recarbonation performance of calcium oxide submitted to repeated calcination/recarbonation cycles. *Energy Fuels* **2005**, *19*, 270–278. [CrossRef]

44. Grasa, G.S.; Abanades, J.C. CO_2 capture capacity of CaO in long series of carbonation/calcination cycles. *Ind. Eng. Chem. Res.* **2006**, *45*, 8846–8851. [CrossRef]

45. Hawthorne, C.; Charitos, A.; Perez-Pulido, C.A.; Bing, Z.; Scheffknecht, G. Design of a dual fluidised bed system for the post-combustion removal of CO_2 using CaO. Part. I: CFB carbonator reactor model. In Proceedings of the 9th International Conference on Circulating Fluidized Beds, Hamburg, Germany, 13–16 May 2008; pp. 759–764.

46. Zhen-shan, L.; Ning-sheng, C.; Croiset, E. Process analysis of CO_2 capture from flue gas using carbonation/calcination cycles. *AIChE J.* **2008**, *54*, 1912–1925. [CrossRef]

47. Abanades, J.C. The maximum capture efficiency of CO_2 using a carbonation/calcination cycle of $CaO/CaCO_3$. *Chem. Eng. J.* **2002**, *90*, 303–306. [CrossRef]

48. Pauls, J.H.; Mahinpey, N.; Mostafavi, E. Simulation of air-steam gasification of woody biomass in a bubbling fluidized bed using Aspen Plus: A comprehensive model including pyrolysis, hydrodynamics and tar production. *Biomass Bioenergy* **2016**, *95*, 157–166. [CrossRef]

49. Nikoo, M.B.; Mahinpey, N. Simulation of biomass gasification in fluidized bed reactor using ASPEN PLUS. *Biomass Bioenergy* **2008**, *32*, 1245–1254. [CrossRef]

50. Dybkjær, I.; Aasberg-Petersen, K. Synthesis gas technology large-scale applications. *Can. J. Chem. Eng.* **2016**, *94*, 607–612. [CrossRef]
51. Poboss, N.; Zieba, M.; Scheffknecht, G. Experimental investigation of affecting parameters on the gasification of biomass fuels in a 20 kWth dual fluidized bed. In Proceedings of the International Conference on Polygeneration Strategies (ICPS10), Leipzig, Germany, 7–9 September 2010.

MDPI

St. Alban-Anlage 66

4052 Basel

Switzerland

Tel. +41 61 683 77 34

Fax +41 61 302 89 18

www.mdpi.com

Applied Sciences Editorial Office

E-mail: applsci@mdpi.com

www.mdpi.com/journal/applsci

www.ingramcontent.com/pod-product-compliance
Lightning Source LLC
LaVergne TN
LVHW071248170726
843515LV00006BA/1351